CHEMISTRY

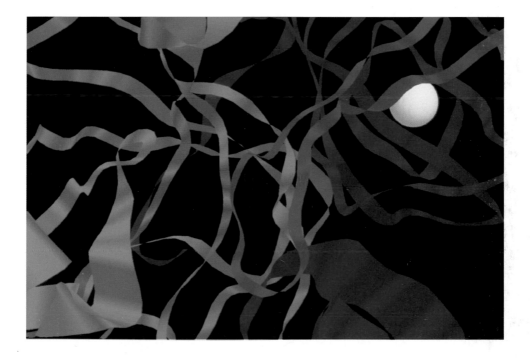

Andrew Hunt

Hodder & Stoughton

A MEMBER OF THE HODDER HEADLINE GROUP

Orders: please contact Bookpoint Ltd, 130 Milton Park, Abingdon, Oxon OX14 4SB.
Telephone: (44) 01235 827720, Fax: (44) 01235 400454. Lines are open from
9.00 – 6.00, Monday to Saturday, with a 24 hour message answering service.
Email address: orders@bookpoint.co.uk

British Library Cataloguing in Publication Data
A catalogue record for this title is available from The British Library

ISBN 0 340 79062 8

First published 2001
Impression number   10 9 8 7 6 5 4 3
Year   2006 2005 2004

Cover photo from Science Photo Library shows a computer graphic of the enzyme
trypsin. Ribbons represent chains of the amino acids that make up the protein structure of
the enzyme. The yellow sphere is a calcium ion.

Illustrated by Jeff Edwards and Tom Cross.
Typeset by Multiplex Techniques Ltd, Brook Industrial Park, Mill Brook Road,
St. Mary Cray, Kent BR5 3SR.
Printed in Dubai for Hodder & Stoughton Educational, a division of Hodder Headline Plc,
338 Euston Road, London NW1 3BH.

Contents

Acknowledgements

I would like to acknowledge the suggestions from teachers who commented on the initial plans and draft chapters including John Payne, Deirdre Cawthorne, Carole Fisher and Margaret Shears.

I am most grateful for the support of the team at Hodder and Stoughton including Ruth Hughes, Suzanne O'Farrell, Lynda King and Elisabeth Tribe.

The collaboration with New Media, publisher of the well-known *Multimedia Science School* and *Chemistry Set* CD-ROMs, has made it possible to produce a CD-ROM and web site to accompany this book. My thanks particularly to Dick Fletcher and David Tymm.

Finally I would like to acknowledge my debt to the many chemistry teachers and chemists I have worked with while contributing to the Nuffield Chemistry, SATIS 16–19 and Salters Chemistry projects. Writing and editing for these projects has clarified my understanding of chemical ideas and helped me to find ways to explain them clearly.

Andrew Hunt

Section one
Studying Chemistry

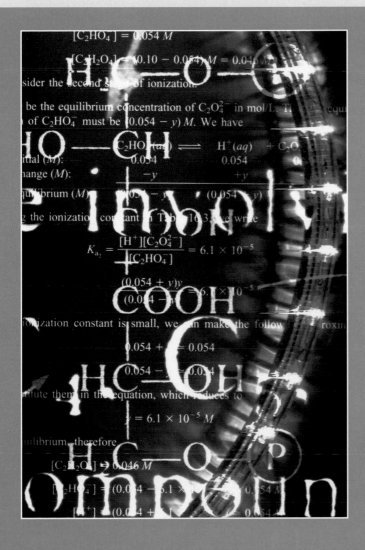

Contents

1.1 How to use this book

This is your guide to the second year of an advanced chemistry course. As you extend your travels in the world of chemistry much will be familiar but you will face new and more demanding challenges as the subject becomes more quantitative. You will develop your own mental map of the landscape of chemistry and make your own connections between ideas.

The book

Following a short introductory section, we have divided the book into three main sections followed by a reference section. You can find your way through the book guided by the coloured strip at the edge of each page. The first page of each section has a table of contents listing the short topics. This is followed by a double page which gives you an overview of the main ideas in the section.

There are many links in the text and diagrams to chemical reactions, tests and preparations which you will carry out in a laboratory.

The Test Yourself questions will encourage you to think about what you are studying. Many of the questions will help you to gain an overview of the subject because they ask you to draw on ideas from your AS course or from earlier chapters in this book. Check your answers at the end of the book to find out how well your understanding of chemistry is developing.

The Review pages at the end of each section will also help you to see what it is that you have to learn and understand.

The CD-ROM

The CD-ROM which accompanies this text will help you to make your study of chemistry active and rewarding. The main features on the CD-ROM are:
- a comprehensive database of the properties of all the elements and compounds which feature in A2 chemistry courses together with a data analyser to display the data,
- an interactive periodic table which helps you to examine periodic patterns and trends,
- three-dimensional models of molecules and crystals which you can rotate and view from different angles,
- short videos of a large number of inorganic and organic reactions to remind you of the observations you have made while carrying out practical work in the laboratory.

The web site

The web site (www.a2chemistry.co.uk) will keep you up-to-date with the requirements for your course and show you how to use the book, the CD-ROM and the Internet to study and learn each of the modules in your specification.

You can download slideshows from the web site. These are sequences of bookmarks for the CD-ROM illustrating important topics in your course. The web site also has links to pages in other web sites with useful information, software and illustrations to help you to improve your understanding of chemistry and extend your awareness of the frontiers of the subject.

The web site offers tutorial support. Consult the '*Ask Andrew*' section when faced with one of those tricky problems which many students find puzzling. Also try the 'Test Yourself' section for a quick check of your understanding of the key ideas.

The Icons

There are two icons which appear in this book:

 You will find this icon alongside many of the Test Yourself exercises. It tells you that to answer some of the questions you need to refer to the data tables at the end of the book, or to the data on the CD-ROM.

CD-ROM You will find this icon at the beginning of some sub-sections if the CD-ROM is useful for the whole topic.

1.2 Showing what you know and can do

Advanced chemistry specifications include assessment objectives which tell you and your teachers what the examiners will be testing in the exams they set. You have to develop knowledge and skills so that you can show the examiners that you have achieved the objectives. The Test Yourself questions in this section are based on the content of your AS course.

Knowledge with understanding

Facts, terminology, principles, concepts and practical techniques

During an advanced chemistry course you take a tour of a fascinating new world. A world with its own language and symbols; a world with new ideas; a world where people can see unexpected and colourful changes. Those who work in this world have to have a knowledge of the language, familiarity with the key ideas and first-hand experience of the changes and events seen in laboratories but not observed in everyday life.

A key objective for a student of the subject is to know enough of chemistry to be able to be effective in this world of the unexpected. Many of the questions in tests and examinations check that you have learnt this basic knowledge.

Test yourself

Facts

1 Do you know what happens when:
   a) chlorine reacts with water
   b) sodium reacts with ethanol
   c) hydrogen bromide reacts with ethene?

Terminology

2 Do you know how to:
   a) name alcohols
   b) describe a reaction in which there are changes of oxidation state
   c) define the standard enthalpy change of formation of a compound?

Principles

3 Do you know:
   a) the principles which determine the electron configuration of atoms
   b) the factors which influence the rates of reactions
   c) Le Chatelier's principle
   d) how Hess's law can be applied to calculate enthalpy changes which cannot be measured?

Concepts

4 Can you give examples to illustrate these theoretical ideas:
   a) the difference between a giant structure and a molecular structure
   b) the effect of hydrogen bonding on the properties of water
   c) isomerism in organic compounds?

Figure 1.2.1 ▲
The use of high-pressure liquid chromatography during a forensic analysis of a drug sample

Practical techniques

5 Do you know how to:

   a) measure the concentration of an acid in solution

   b) identify gases given off from chemical reactions

   c) separate a liquid product from a reaction mixture?

Chemistry in society

6 Can you give examples of:

   a) a practical application of catalysts

   b) the uses of chemicals made from oil

   c) the application of electrolysis to extract a metal from its ore

   d) the use of an organic chemical in medicine

   e) a new material made by polymerisation?

Making sense of data

7 Do you know:

   a) what information you need to determine the formula of a compound

   b) how to display and interpret the trend in first ionisation enthalpies across period 3

   c) how to draw a diagram to show the shape of a molecule?

The uses of chemistry in society

Part of the attraction of chemistry is the opportunity to experience the unusual and unexpected. Many people, however, study the subject because of its great practical importance in food science, medicine, pharmacy, material science, the chemical industry, biotechnology and in the study of the environment. A knowledge of chemistry is essentially background for anyone wishing to work in these fields.

Chemicals and the chemical industry have a bad reputation with many people. Yet an understanding of chemistry is vital if society is to deal with the issues of sustainable development and care of the environment.

Select, organise and present information clearly

There are agreed ways of gathering, handling and displaying information in chemistry. A successful chemist understands and can recall relevant information when trying to solve a problem or explain some observed changes.

Application of knowledge and understanding, analysis and evaluation

You have to do more than recall facts, applications and techniques.

Explaining phenomena

Explanations in chemistry often call for imagination to make sense of observations in terms of invisible atoms, molecules and ions. Chemists make use of diagrams and 3D models to make it easier to picture the meaning of abstract explanations.

Interpreting information

Chemists have to translate information from one form to another so that the information is displayed in a way which is easier to understand. Energy changes are described using energy level diagrams. Graphs help to see what is happening during experiments to investigate reactions. While chemical equations summarise the changes during reactions.

Carrying out calculations

Matching experimental results to quantitative predictions based on theory help to convince chemists that their ideas make sense. As part of your advanced chemistry course you have to be able to carry out a number of types of calculation to find formulas, to measure concentrations of solutions, to estimate the yields of reactions, to determine how fast and how far chemical reactions will go and to make sense of energy changes.

Applying ideas in unfamiliar situations

Applying ideas successfully creates confidence. Many advanced courses include topics about applied chemistry such as the synthesis of polymers, the manufacture of important chemicals, the medical value of pharmaceuticals and the nature of biochemical molecules.

Assessing the validity of information

Chemists have to be able to study experimental results, read reports and consider the conclusions based on results. They have to be able to decide whether the information is likely to be accurate and whether the statements based on the evidence make sense.

Experiment and investigation

You are expected to be able to plan for experimental activities, carry out practical techniques skilfully and safely, make precise observations and measurements and to interpret and evaluate results.

Showing understanding of links between different areas of chemistry

You are expected to be able to bring together knowledge, principles and concepts from different areas of chemistry and apply them to answer questions and solve problems. This means that you have to be able to use ideas about chemical structure, types of reaction, energy changes, rates and equilibrium are frequently used to make sense of the behaviour of inorganic and organic chemistry. You have to show that you can apply your knowledge and understanding to interpret data.

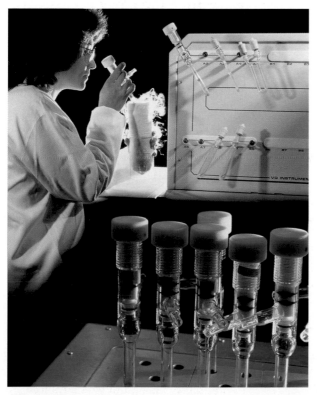

Figure 1.2.2 ▲
Preparing samples for analysis in a mass spectrometer as part of an investigation of the Earth's climatic history by studying samples collected from Antarctic ice cores

Test yourself

Making use of knowledge and understanding

8 Can you:
   a) explain why a catalyst speeds up a chemical reaction
   b) identify trends in a table showing the physical properties of elements
   c) calculate the concentration of a solution from the results of a titration
   d) explain the choice of conditions in the manufacture of chemicals such as ammonia or sulphuric acid
   e) use your knowledge of covalent bonding to explain the reactivity of the alkenes?

Definition

Synoptic assessment tests your ability to draw together and apply knowledge, understanding and skills from different topics in a course.

Studying Chemistry

Section one

Section two

Physical Chemistry

Contents

2.1 Physical chemistry in action

Physical chemists have developed theoretical ideas and models which can help to explain how fast reactions go, in which direction and how far. With the help of these theories, biochemists can account for changes in living cells, chemical engineers can design manufacturing plants to make high yields of pure products, while chemists can interpret their observations as they study the chemical elements and their compounds.

Rates of change

Several factors influence the rate of chemical change. These include the concentration of the reactants, their surface area, the temperature of the reaction mixture and the presence of a catalyst. Chemists have found that they can learn much more about reactions by studying these effects quantitatively. They can then set up models to simulate the data and make predictions about the impact of changing the conditions. Chemists apply these models to drug design and to the formulation of medicines to make sure that patients receive treatments which are effective and do not cause harmful side-effects. The models can also account for the damage arising from pollutants in the atmosphere and help chemists to suggest ways for reducing or preventing the resulting problems.

Figure 2.1.1 ▲
Chemists use data gathered at high altitudes during research into the hole in the Arctic ozone layer. Here a scientist in Sweden is wearing a suit for protection from radiation during an inspection of a radar dome. The radar tracks the balloons as they carry the measuring instruments to the upper atmosphere

The direction and extent of change

What makes things go? Why do they go one way and not another? How far do they go? Seeking answers to such questions for chemical reactions lies at the heart of all chemistry. Chemists have developed a number of quantitative theories for answering these questions.

◄ Figure 2.1.2
Understanding the factors which determine the rate and direction of chemical change are essential to the design of productive, safe and profitable chemical plants. This chemist is comparing the effects of two catalysts on the same reaction

The equilibrium law

All reactions are reversible, at least to some extent, and, given time, will reach a state of dynamic equilibrium. A law which describes equilibrium states is particularly helpful for the study of reversible reactions in solution, such as acid–base reactions. With the help of this law chemists can formulate mixtures for medical products or cosmetics which are 'pH balanced' meaning that they help to prevent the damaging extremes of acidity or alkalinity.

Electrochemistry

An electrochemical cell is a device for producing a voltage from a redox reaction. The development of new types of small, rechargeable cell have made possible many kinds of portable electronic devices such as calculators, mobile phones and laptop computers. The search is still on for a cell which will make electric vehicles really feasible. The tried-and-tested lead acid cells in car batteries work fine for starting the engine but they are too heavy to be the main source of power to drive vehicles except in special circumstances.

Figure 2.1.3 ▶
Assembly of a solid polymer fuel cell during testing. A fuel cell produces electricity directly from fuel and an oxidizer. This is much more efficient than generating electricity by burning the fuel and then using the energy to drive a turbine

Electrode systems can also be the basis of sensors to measure the concentrations of key chemicals in the blood and other body fluids. One example is the sensors which diabetics can now use to measure the concentration of glucose sugar in their blood.

Any redox reaction can, at least in principle, happen in a chemical cell. Chemists have found it valuable to measure the cell voltage of a wide range of cells, even cells of no practical value as sources of power. They have tabulated data as a list of electrode potentials which makes it possible to predict which redox reactions are possible.

Energy and change

The applications of the equilibrium law and the study of electrodes provide very useful ways of interpreting the behaviour of specific reactions. It turns out, however, that the underlying principles which determine the direction and extent of change involve concepts developed in the field of thermochemistry. Thermochemists account for the stability or instability of compounds and the expected direction of change in terms of enthalpy changes, entropy changes and free energy changes.

Figure 2.1.4 ▶
A technician using a large flame to manufacture a large glass tube. The silvered suit and metallic visor reflect radiation and protect the technician from the heat of the flame. What makes the gas burn so fast and give out so much energy?

2.2 Reaction kinetics – the effect of concentration changes

Reaction kinetics is the study of the rates of chemical reactions. Chemists have found that they can learn much about the mechanisms of reactions when they make measurements to find out how reaction rates vary with factors such as concentration and temperature. The quantitative study of rates also provide results which can help chemists to control reactions.

Measuring rates

Ideally chemists look for methods for measuring reaction rates which do not interfere with the reaction mixture, as shown in Figures 2.2.1–2.2.3. Sometimes, however, it is necessary to withdraw samples of the reaction mixture at regular intervals and analyse the concentration of a reactant or product by titration as illustrated in Figure 2.2.4.

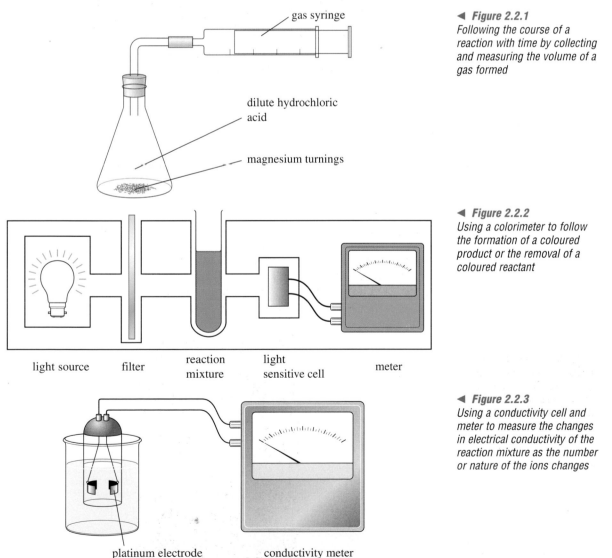

gas syringe

dilute hydrochloric acid

magnesium turnings

◀ *Figure 2.2.1*
Following the course of a reaction with time by collecting and measuring the volume of a gas formed

light source   filter   reaction mixture   light sensitive cell   meter

◀ *Figure 2.2.2*
Using a colorimeter to follow the formation of a coloured product or the removal of a coloured reactant

platinum electrode   conductivity meter

◀ *Figure 2.2.3*
Using a conductivity cell and meter to measure the changes in electrical conductivity of the reaction mixture as the number or nature of the ions changes

Figure 2.2.4 ▶
Following the course of the reaction catalysed by an acid by removing measured samples of the mixture at intervals. The reaction is stopped by running the sample into an alkali and then the concentration of one reactant or product is determined by titration. Further samples are taken at regular intervals

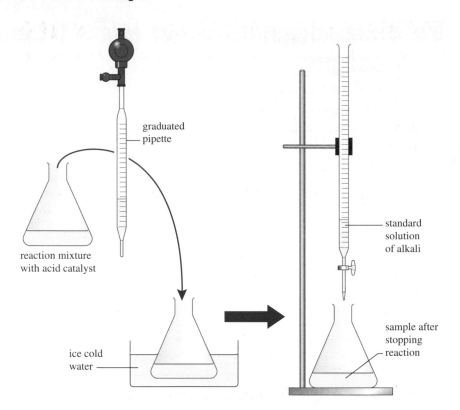

graduated pipette

standard solution of alkali

reaction mixture with acid catalyst

ice cold water

sample after stopping reaction

Chemists define the rate of reaction as the change in concentration of a product, or a reactant, divided by the time for the change. Usually the rate is not constant but varies as the reaction proceeds. So the usual method for analysing results of experiments is to plot a concentration–time graph. The gradient (or slope) of the graph at any point gives a measure of the rate of reaction at that time.

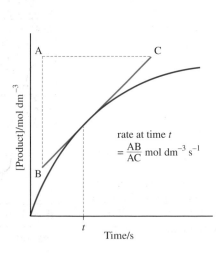

rate at time t
$= \dfrac{AB}{AC}$ mol dm^{-3} s^{-1}

Figure 2.2.5 ▲
Concentration–time graph for the formation of a product. The rate of formation of product at time t is the gradient (or slope) of the curve at this point

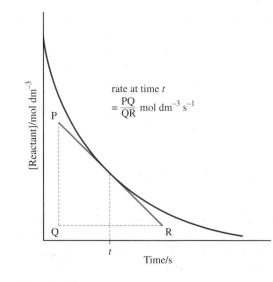

rate at time t
$= \dfrac{PQ}{QR}$ mol dm^{-3} s^{-1}

Figure 2.2.6 ▲
Concentration–time graph for the disappearance of a reactant. The rate of loss of reactant at time t is the gradient (or slope) of the curve at this point

Test yourself

1 Each of the following factors can change the rate of a reaction. Give an example of a reaction to illustrate each one.

   a) the concentration of reactants in solution

   b) the pressure of gaseous reactants

   c) the surface area of a solid

   d) the temperature

   e) the presence of a catalyst

2 How does collision theory account for the effects of altering each of the factors a) to e) in question 1?

3 Suggest a suitable method for measuring the rate of each of these reactions:

   a) $CaCO_3(s) + 2HCl(aq) \longrightarrow CaCl_2(aq) + CO_2(g)$

   b) $Br_2(aq) + HCO_2H(aq) \longrightarrow 2HBr(aq) + CO_2(g)$

   c) $CH_3CO_2CH_3(l) + H_2O(l) \longrightarrow CH_3CO_2H(aq) + CH_3OH(aq)$

   d) $C_4H_9Br(l) + H_2O(l) \longrightarrow C_4H_9OH(l) + H^+(aq) + Br^-(aq)$

4 In a study of the hydrolysis of an ester, the concentration of the ester fell from 0.55 mol dm^{-3} to 0.42 mol dm^{-3} in 15 seconds. What was the average rate of reaction in that period?

5 The gaseous oxide N_2O_5 decomposes to NO_2 gas and oxygen.

   a) Write a balanced equation for the reaction.

   b) If the rate of disappearance of N_2O_5 is 3.5×10^{-4} mol dm^{-3} s^{-1}, what is the rate of formation of NO_2?

Rate equations

Chemists have found that they can summarise the results of investigating the rate of reaction in the form of a rate equation. A rate equation shows how changes in the concentrations of reactants affect the rate of a reaction.

Take the example of a general reaction for which x mol A react with y mol B to form products:

$$xA + yB \longrightarrow \text{products}$$

The rate equation takes this form:

$$\text{rate} = k[A]^n[B]^m,$$

where [A] and [B] represent the concentrations of the reactants in moles per litre.

The powers n and m are the reaction orders. The reaction above is order n with respect to A and order m with respect to B. The overall order is (n + m).

The rate constant, k, is only constant for a particular temperature. The value of k varies with temperature (see pages 17–18). The units of the rate constant depend on the overall order of the reaction.

zero	mol dm^{-3} s^{-1}
first	s^{-1}
second	mol^{-1} dm^3 s^{-1}

Figure 2.2.7 ▲

Physical Chemistry

Section two

Definition

Writing the formula of a chemical in square brackets is the usual shorthand for **concentration in mol dm^{-3}**. For example: [X] represents the concentration of X in mol dm^{-3}.

Worked example

The decomposition of ethanal to methane and carbon monoxide is second order with respect to ethanal. When the concentration of ethanal in the gas phase is 0.20 mol dm^{-3}, the rate of reaction is 0.080 mol dm^{-3} s^{-1} at a certain temperature. What is the value of the rate constant at this temperature?

Notes on the method

Start by writing out the rate equation based on the information given. There is no need to write the equation for the reaction because the rate equation cannot be deduced from the balanced chemical equation.

Substitute values in the rate equation including the units. Then rearrange the equation to find the value of k. Check the units are as expected for a second order reaction.

Answer

The rate equation: rate = k[ethanal]^2

Substituting: 0.080 mol dm^{-3} s^{-1} = $k \times$ (0.20 mol dm^{-3})^2

Rearranging: $k = \dfrac{0.080 \text{ mol dm}^{-3} \text{ s}^{-1}}{(0.20 \text{ mol dm}^{-3})^2}$

Hence: k = 2.0 mol^{-1} dm^3 s^{-1}

Test yourself

6   The rate of decomposition of di(benzenecarbonyl) peroxide is first order with respect to the peroxide. Calculate the rate constant for the reaction at 107 °C if the rate of decomposition of the peroxide at this temperature is 7.4×10^{-6} mol dm^{-3} s^{-1} when the concentration of peroxide is 0.02 mol dm^{-3}.

7   The hydrolysis of the ester methyl ethanoate in alkali is first order with respect to the ester and first order with respect to hydroxide ions. The rate of reaction is 0.00069 mol dm^{-3} s^{-1}, at a given temperature, when the ester concentration is 0.05 mol dm^{-3} and the hydroxide ion concentration is 0.10 mol dm^{-3}. Write out the rate equation for the reaction and calculate the rate constant.

Definition

The **half-life** of a reaction is the time for the concentration of one of the reactants to fall by half.

First order reactions

A reaction is first order with respect to a reactant if the rate of reaction is proportional to the concentration of that reactant. The concentration term for this reactant is raised to the power one in the rate equation.

$$\text{Rate} = k[\text{X}]^1 = k[\text{X}]$$

This means that doubling the concentration of the chemical X leads to a doubling of the rate of reaction.

One of the easiest ways to spot a first order reaction is to plot a concentration–time graph and then study the time taken for the concentration to fall by half. At a constant temperature, the half-life of a first order reaction is the same wherever it is measured on a concentration–time graph.

Figure 2.2.8 ▶
Variation of concentration of a reactant plotted against time for a first order reaction. The gradient of this graph at any point is a measure of the rate of reaction. The half-life for a first order reaction is a constant, so it is the same wherever it is read off the curve. It is independent of the initial concentration

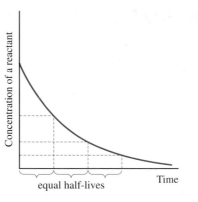

Figure 2.2.9 ▲
Variation of reaction rate with concentration for a first order reaction. The graph is a straight line through the origin showing that the rate is proportional to the concentration of the reactant

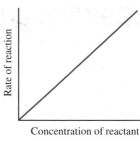

Physical Chemistry

Section Two

Test yourself

8 The data in the table refers to the decomposition of hydrogen peroxide, H_2O_2.

a) Plot a concentration–time graph for the decomposition reaction.

b) Read off several half-lives from the graph and show that this is a first order reaction.

c) Draw tangents to the curve in your graph at five different concentrations and hence find the gradient of the curve at each point.

d) Plot a graph of rate against concentration using your results from **c)** and hence find the value for the rate constant at the temperature of the experiment.

Time/s	$[H_2O_2]$/mol dm^{-3}
0	20.0×10^{-3}
12×10^3	16.0×10^{-3}
24×10^3	13.1×10^{-3}
36×10^3	10.6×10^{-3}
48×10^3	8.6×10^{-3}
60×10^3	6.9×10^{-3}
72×10^3	5.6×10^{-3}
96×10^3	3.7×10^{-3}
120×10^3	2.4×10^{-3}

Second order reactions

A reaction is second order with respect to a reactant if the rate of reaction is proportional to the concentration of that reactant squared. This means that the concentration term for this reactant is raised to the power two in the rate equation. At its simplest the rate equation for a second order reaction takes this form:

$$\text{Rate} = k[X]^2$$

This means that doubling the concentration of X increases the rate by a factor of four.

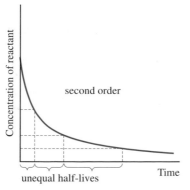

◀ **Figure 2.2.10**
Variation of concentration of a reactant plotted against time for a second order reaction. The half-life for a second order reaction is not a constant. The time for the concentration to fall from c to c/2 is half the time for the concentration to fall from c/2 to c/4. The half-life is inversely proportional to the starting concentration

Test yourself

9 The rate of reaction of 1-bromopropane with hydroxide ions is first order with respect to the halogenoalkane and first order with respect to hydroxide ions.

a) Write out the rate equation for the reaction.

b) What is the overall order of reaction?

10 The results in the table come from a study of the rate of reaction of iodine with a large excess of hex-1-ene dissolved in ethanoic acid.

a) Plot a concentration–time graph and show that the half-life is not a constant.

b) From your graph find the rate of reaction at a series of concentrations.

c) Use your results from **b)** to plot a graph which will confirm that the reaction is second order with respect to iodine.

Time/s	$[I_2]$/mol dm^{-3}
0	20.0×10^{-3}
1×10^3	15.6×10^{-3}
2×10^3	12.8×10^{-3}
3×10^3	10.9×10^{-3}
4×10^3	9.4×10^{-3}
5×10^3	8.3×10^{-3}
6×10^3	7.5×10^{-3}
7×10^3	6.8×10^{-3}
8×10^3	6.2×10^{-3}

Note

Any term raised to the power zero equals 1. So $[X]^0 = 1$.

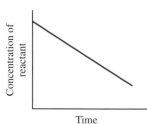

Figure 2.2.11 ▲
Variation of concentration of a reactant plotted against time for a zero order reaction. The gradient of this graph measures the rate of reaction. The gradient is a constant so the rate stays the same even though the concentration of the reactant is falling

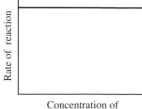

Figure 2.2.12 ▲
Variation of reaction rate with concentration for a zero order reaction

Zero order reactions

At first sight is seems odd that there can be zero order reactions. A reaction is zero order with respect to a reactant if the rate of reaction is unaffected by changes in the concentration of that reactant. Chemists have found a way to account for zero order reactions in terms of the mechanisms of these reactions (see page 19).

In a rate equation for a zero order reaction, the concentration term for the reactant is raised to the power zero.

$$\text{Rate} = k[X]^0 = k \text{ (a constant)}$$

The rate of reaction of iodine with propanone in the presence of acid is zero order with respect to iodine.

$$\text{Rate} = k[\text{propanone}]^1[\text{hydrogen ion}]^1[\text{iodine}]^0$$

If iodine reacts with a considerable excess of propanone in excess acid, then only the iodine concentration varies significantly while the propanone and hydrogen ion concentration remain effectively constant. Under these conditions the rate equation becomes:

$$\text{Rate} = k'[\text{iodine}]^0$$

Determining reaction orders

The most general method for determining reaction orders is the initial-rate method. The method is based on finding the rate immediately after the start of a reaction. This is the one point when all the concentrations are known.

The investigator makes up a series of mixtures in which all the initial concentrations are the same except one. A suitable method is used to measure the change of concentration with time for each mixture (see page 9). The results are used to plot concentration–time graphs. The initial rate for each mixture is then found by drawing tangents to the curve at the start and calculating their gradients.

Worked example

The initial rate method was used to study the reaction;

$$BrO_3^-(aq) + 5Br^-(aq) + 6H^+(aq) \longrightarrow 3Br_2(aq) + 3H_2O(l)$$

The initial rate was calculated from four graphs plotted to show how the concentration of $BrO_3^-(aq)$ varied with time for different initial concentrations of reactants.

Experiment	Initial concentration of BrO_3^-/mol dm^{-3}	Initial concentration of Br$^-$/mol dm^{-3}	Initial concentration of H$^+$/mol dm^{-3}	Initial rate of reaction/ mol dm^{-3} s^{-1}
1	0.10	0.10	0.10	1.2×10^{-3}
2	0.20	0.10	0.10	2.4×10^{-3}
3	0.10	0.30	0.10	3.6×10^{-3}
4	0.20	0.10	0.20	9.6×10^{-3}

What is
a) the rate equation for the reaction,
b) the value of the rate constant?

Notes on the method

Recall that the rate equation cannot be worked out from the balanced equation for the reaction.

First study the experiments in which the concentration of BrO_3^- varies but the concentration of the other two reactants stays the same. How does doubling the concentration of BrO_3^- affect the rate?

Then study in turn the experiments in which the concentrations of first Br$^-$ and then H$^+$ vary while the concentrations of the other two rectants stay the same. How does doubling or tripling the concentration of a reactant affect the rate?

Substitute values for any one experiment in the rate equation to find the value of the rate constant, k. Take care with the units.

Answer

From experiment 1 and 2: doubling $[BrO_3^-]_{initial}$ increases the rate by a factor of 2. So rate $\propto [BrO_3^-]^1$.

From experiment 1 and 3: tripling $[Br^-]_{initial}$ triples the rate. So rate $\propto [Br^-]^1$.

From experiment 2 and 4: doubling $[H^+]_{initial}$ increases the rate by a factor of 4 (2^2). So rate $\propto [H^+]^2$.

The reaction is first order with respect to BrO_3^- and Br$^-$ but second order with respect to H$^+$.

The rate equation is: rate = $k [BrO_3^-][Br^-] [H^+]^2$

Rearranging this equation, and substituting values from experiment 4:

$$k = \frac{rate}{[BrO_3^-][Br^-] [H^+]^2} = \frac{9.6 \times 10^{-3} \text{ mol dm}^{-3} \text{ s}^{-1}}{0.2 \text{ mol dm}^{-3} \times 0.1 \text{ mol dm}^{-3} \times (0.2 \text{ mol dm}^{-3})^2}$$

$$k = 12.0 \text{ mol}^{-3} \text{ dm}^9 \text{ s}^{-1}$$

Test yourself

11 Ammonia gas decomposes to nitrogen and hydrogen in the presence of a hot platinum wire. Experiments show that the reaction continues at a constant rate until all the ammonia has disappeared.

a) Sketch a concentration–time graph for the reaction.

b) Write both the balanced chemical equation and the rate equation for this reaction.

12 Hydrogen gas reacts with nitrogen monoxide gas to form steam and nitrogen. Doubling the concentration of hydrogen doubles the rate of reaction. Tripling the concentration of NO gas increases the rate by a factor of nine.

a) Write the balanced equation for the reaction.

b) Write the rate equation for the reaction.

13 This data refers to the reaction of the halogenoalkane 1-bromobutane (here represented as RBr) with hydroxide ions.

Experiment	[RBr]/mol dm^{-3}	[OH$^-$]/mol dm^{-3}	Rate of reaction/ mol dm^{-3} s^{-1}
1	0.020	0.020	1.36
2	0.010	0.020	0.68
3	0.010	0.005	0.17

a) Deduce the rate equation for the reaction.

b) Calculate the value of the rate constant.

14 This data refers to the reaction of the halogenoalkane 2-bromo-2-methylpropane (here represented as R'Br) with hydroxide ions.

Experiment	[R'Br]/mol dm^{-3}	[OH$^-$]/mol dm^{-3}	Rate of reaction/ mol dm^{-3} s^{-1}
1	0.020	0.020	40.40
2	0.010	0.020	20.19
3	0.010	0.005	20.20

a) Deduce the rate equation for the reaction.

b) Calculate the value of the rate constant.

2.3 Reaction kinetics – the effect of temperature changes

Raising the temperature often has a dramatic effect on the rate of a reaction especially a reaction which involves the breaking of strong covalent bonds. This explains why the practical procedures for most organic reactions involve heating the reaction mixtures. With the help of collision theory it is possible to make quantitative predictions about the effect of temperature changes on rates.

Transition state theory

The constant k in a rate equation is only a constant at a specified temperature. Generally, the value of the rate constant increases as the temperature rises and this means that the rate of reaction increases.

Collision theory accounts for the effect of temperature on reaction rates by supposing that chemical changes pass through a transition state. The transition state is at a higher energy than the reactants so there is an energy barrier or activation energy. Reactant molecules must collide with enough energy to overcome the activation energy barrier. This means that the only collisions which lead to a reaction are those with enough energy to break existing bonds and allow the atoms to rearrange to form new bonds in the product molecules.

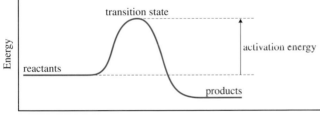

◄ *Figure 2.3.1*
Reaction profile showing the activation energy for a reaction

Activation energies account for the fact that reactions go much more slowly than would be expected if every collision in a mixture of chemicals led to a reaction. Only a very small proportion of collisions bring about chemical change because molecules can only react if they collide with enough energy to overcome the energy barrier. At room temperature, for many reactions, only a small proportion of molecules have enough energy to react.

The Maxwell–Boltzman curve describes the distribution of the kinetic energies of molecules. As Figure 2.3.2 shows, the proportion of molecules with energies greater than the activation energy is small at around 300 K.

The shaded areas in Figure 2.3.2 show the proportions of molecules having at least the activation energy for a reaction at two temperatures. This area is bigger at a higher temperature. So at a higher temperature there are more molecules with enough energy to react when they collide and the reaction goes faster.

> **Definitions**
>
> A **reaction profile** is a plot which shows how the total energy of the atoms, molecules or ions changes during the progress of a change from reactants to products.
>
> A **transition state** is the state of the reacting atoms, molecules or ions when they are at the top of the activation energy barrier for a reaction step.

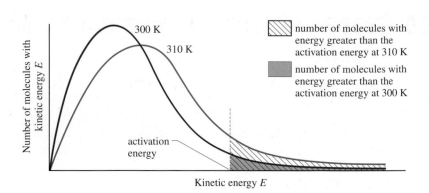

Figure 2.3.2 ▶
The Maxwell–Boltzman distribution of molecular kinetic energies in a gas at two temperatures. The modal speed gets higher as the temperature rises. The area under the curve gives the total number of molecules. This does not change as the temperature rises so the peak height falls as the curve widens

The effect of temperature changes on rate constants

The Swedish physical chemist Svante Arrhenius (1859–1927) found that he obtained a straight line if he plotted the logarithm of the rate constant against $1/T$ (the inverse of the absolute temperature).

The Arrhenius equation can take this form:

$$\ln k = \text{constant} - \frac{E_a}{RT}$$

where k is the rate constant, R is the gas constant, T the absolute temperature and E_a is the activation for the reaction.

Test yourself

1   Show that the Arrhenius equation signifies that:

   a) the higher the temperature the greater the value of k and hence the faster the reaction

   b) a reaction with a relatively high activation energy has a relatively small rate constant.

2   The rate constant for the decomposition of hydrogen peroxide is 4.93×10^{-4} s^{-1} at 295 K. It increases to 1.40×10^{-3} s^{-1} at 305 K. Estimate the activation energy for the reaction.

3   The activation energy for the decomposition of ammonia into nitrogen and hydrogen is 335 kJ mol^{-1} in the absence of a catalyst but 162 kJ mol^{-1} in the presence of a tungsten catalyst. Explain the significance of these values.

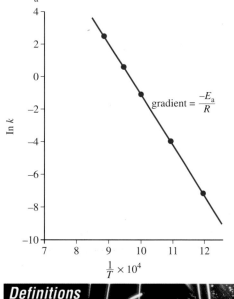

◀ **Figure 2.3.3**
Plot of ln k against 1/T for a reaction. The activation energy can be calculated from the gradient. The general equation for a straight line is y = mx + c where m is the gradient and c the intercept on the y axis. Here c = the constant in the Arrhenius equation and the gradient $m = -E_a/RT$

gradient $= \dfrac{-E_a}{R}$

Definitions

The **absolute temperature** is the temperature on the Kelvin scale. The absolute zero of temperature is at 0 K which is approximately −273 °C.

The **gas constant** is the constant R in the ideal gas equation $PV = nRT$. The value of the constant depends on the units used for pressure and volume. If all quantities are in SI units, then $R = 8.314$ J K^{-1} mol^{-1}.

Logarithms in chemistry are of two kinds – logarithms to base 10 (log) and natural logarithms to base e (ln). Chemists use logarithms to base 10 to handle values which range over several orders of magnitude (see page 35). Natural logarithms appear in relationships in chemical kinetics and thermochemistry. They follow similar mathematical rules as logarithms to base 10.

A useful rough guide, based on the Arrhenius equation, is that the value of the rate constant will double for each 10 degree rise in temperature for a reaction with an activation energy of about 50 kJ mol^{-1}.

2.4 Rate equations and reaction mechanisms

Rate equations were one of the first pieces of evidence which set chemists thinking about the mechanisms of reactions. They wanted to understand why a rate equation cannot be predicted from the balanced equation for the reaction. They were puzzled that similar reactions turned out to have different rate equations.

Multi-step reactions

The key to understanding reaction mechanisms was the realisation that most reactions do not take place in one step, as suggested by the balanced equation, but in a series of steps.

It seems puzzling that the decomposition of ammonia gas in the presence of a hot platinum catalyst is a zero order reaction. How can it be that the concentration of the only reactant does not affect the rate? A possible explanation is illustrated in Figure 2.4.1.

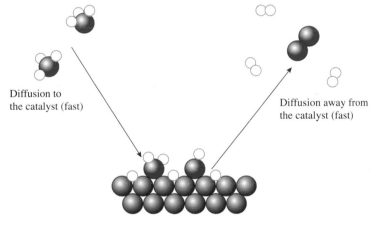

Diffusion to the catalyst (fast)

Diffusion away from the catalyst (fast)

Bonds breaking and new bonds forming. Rate determined by the surface area of the catalyst which is a constant (rate-determining)

◀ **Figure 2.4.1**
Three steps in the decomposition of ammonia gas in the presence of a platinum catalyst

Ammonia rapidly diffuses to the surface of the metal and is absorbed onto the surface. This happens quickly. Bonds break and atoms rearrange to make new molecules on the surface of the metal. This is the slowest process. Once formed, the nitrogen and hydrogen rapidly break away from the metal into the gas phase.

So there is a rate-determining step which can only happen on the surface of the platinum. The rate of reaction is determined by the surface area of the platinum which is a constant. This means that the rate of reaction is a constant as long as there is enough ammonia to be absorbed all over the metal surface. The rate is independent of the ammonia concentration.

Hydrolysis of halogenoalkanes

Another puzzle is that there are different rate equations for the reactions between hydroxide ions and two isomers with the formula C_4H_9Br (see questions 13 and 14 on page 16).

Hydrolysis of a primary halogenoalkane, such as 1-bromobutane, is overall second order. The rate equation has the form: rate = $k[C_4H_9Br][OH^-]$.

To account for this chemists have suggested a mechanism showing the C—Br bond breaking at the same time as the nucleophile, OH^-, forms a C—OH bond.

> **Definitions**
>
> The **mechanism of a reaction** describes how the reaction takes place, showing step by step the bonds which break and the new bonds which form.
>
> The **rate-determining step** in a multi-step reaction is the one with the highest activation energy.

2.4 Rate equations and reaction mechanisms

Figure 2.4.2 ▶
A one-step mechanism for the hydrolysis of 1-bromobutane (see also page 157)

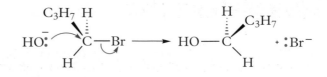

Hydrolysis of tertiary halogenoalkanes such as 2-bromo-2-methylpropane, however, is overall first order. The rate equation has the form:
rate = $k[C_4H_9Br]$.

Figure 2.4.3 ▶
A two-step mechanism for the hydrolysis of 2-bromo-2-methylpropane (see also page 158)

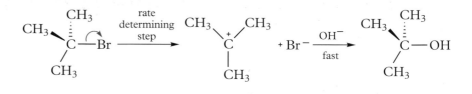

The suggested mechanism shows the C—Br bond breaking first to form an ionic intermediate. Then the nucleophile, OH⁻, rapidly forms a new bond with carbon.

What these examples show is that it is generally the molecules or ions involved (directly or indirectly) in the rate-determining step that appear in the rate equation for the reaction.

Test yourself

1 In the proposed two-step mechanism for the reaction of nitrogen dioxide gas with carbon monoxide gas, the first step is slow while the second step is fast:

$$2NO_2(g) \longrightarrow NO_3(g) + NO(g)$$
$$NO_3(g) + CO(g) \longrightarrow NO_2(g) + CO_2(g)$$

a) What is the overall equation for the reaction?

b) Suggest a rate equation which is consistent with the proposed mechanism.

c) What, according to your suggested rate equation, is the order of reaction with respect to carbon monoxide?

2 Hydrogen peroxide oxidises iodide ions to iodine in the presence of hydrogen ions. The other product is water. The reaction is first order with respect to hydrogen peroxide, first order with respect to iodide ions but zero order with respect to hydrogen ions.

a) Write a balanced equation for the reaction.

b) Write a rate equation for the reaction.

c) What is the overall order of the reaction?

d) A proposed mechanism for the reaction involves three steps:

$$H_2O_2 + I^- \longrightarrow H_2O + IO^-$$
$$H^+ + IO^- \longrightarrow HIO$$
$$HIO + H^+ + I^- \longrightarrow I_2 + H_2O$$

Which step is likely to be the rate-determining step and why?

2.5 The equilibrium law

All chemical reactions tend towards a state of dynamic equilibrium. Chemists have discovered a law which allows them to predict the concentrations of chemicals expected in equilibrium mixtures. This law is one of several approaches which chemists use to answer the questions: 'How far?' and 'In which direction?'. An understanding of equilibrium ideas help to explain changes in the natural environment, the biochemistry of living things and the conditions used in the chemical industry to manufacture new products.

Reversibility and equilibrium

Reversible reactions tend towards a state of balance. They reach equilibrium when neither the forward change nor the backward change is complete but both changes are still going on at equal rates. They cancel each other out and there is no overall change. This is dynamic equilibrium.

Under given conditions the same equilibrium state can be reached either by starting with the chemicals on one side of the equation for a reaction or by starting with the chemicals on the other side. Figures 2.5.1 and 2.5.2 illustrate this for the reversible reaction between hydrogen and iodine.

$$H_2(g) + I_2(g) \rightleftharpoons 2HI(g)$$

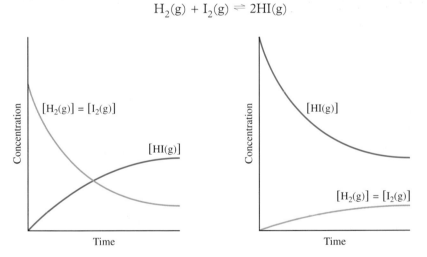

Figure 2.5.1 ▲
Reaching an equilibrium state by the reaction of equal amounts of hydrogen gas and iodine gas

Figure 2.5.2 ▲
Reaching the same equilibrium state by the decomposition of hydrogen iodide, HI(g) under the same conditions as for Figure 2.5.1

Test yourself

1 Draw and label a diagram (showing the atoms or molecules) to describe the dynamic equilibrium between a solid and a saturated solution of the solid.

2 Describe what is happening to the molecules in the gas mixtures described by Figures 2.5.1 and 2.5.2 from time zero to the time at which each mixture reaches equilibrium.

Note

Equilibrium constants are constant for a particular temperature.

Equilibrium constants

The equilibrium law is a quantitative law for predicting the amounts of reactants and products when a reversible reaction reaches a state of dynamic equilibrium.

In general, for a reversible reaction at equilibrium:

$$aA + bB \rightleftharpoons cC + dD$$

$$K_c = \frac{[C]^c[D]^d}{[A]^a[B]^b}$$

This is the form for the equilibrium constant, K_c, when the concentrations of the reactants and products are measured in moles per litre. [A], [B] and so on, are the equilibrium concentrations, sometimes written as $[A]_{eqm}$ and $[B]_{eqm}$ to make this clear.

The concentrations of the chemicals on the right-hand side of the equation appear on the top line of the expression. The concentrations of reactants on the left appear on the bottom line. Each concentration term is raised to the power of the number in front of its formula in the equation.

Worked example

Calculate the value of K_c for the reaction which forms an ester from ethanoic acid and ethanol (see page 178):

$$CH_3CO_2H(l) + C_2H_5OH(l) \rightleftharpoons CH_3CO_2C_2H_5(l) + H_2O(l)$$

given that when 0.50 mol of $CH_3CO_2H(l)$ is dissolved in 0.5 dm^3 of an organic solvent with 0.09 mol $C_2H_5OH(l)$ and allowed to come to equilibrium at 373 K, the amount of $CH_3CO_2C_2H_5(l)$ at equilibrium is 0.086 mol.

Notes on the method

Write down the equation. Underneath write first the initial amounts, then write the amounts at equilibrium. Use the equation to calculate the amounts not given.

Calculate the equilibrium concentrations given the volume of the solution. Substitute the values and units in the expression for K_c.

Answer

Equation $\quad CH_3CO_2H(l) + C_2H_5OH(l) \rightleftharpoons CH_3CO_2C_2H_5(l) + H_2O(l)$

Initial amounts/mol	0.50	0.09	0	0
Equilibrium amounts/mol	(0.50 – 0.086) = 0.414	(0.09 – 0.086) = 0.004	0.086	0.086
Equilibrium concentrations /mol dm^{-3}	0.414 ÷ 0.5 = 0.828	0.004 ÷ 0.5 = 0.008	0.086 ÷ 0.5 = 0.172	0.086 ÷ 0.5 = 0.172

$$K_c = \frac{[CH_3CO_2C_2H_5(l)][H_2O(l)]}{[CH_3CO_2H(l)][C_2H_5OH(l)]} = \frac{0.172 \text{ mol dm}^{-3} \times 0.172 \text{ mol dm}^{-3}}{0.828 \text{ mol dm}^{-3} \times 0.008 \text{ mol dm}^{-3}}$$

Hence $K_c = 4.47$

In this example the units cancel, so K_c has no units.

Test yourself

3 Show that these sets of equilibrium concentrations for mixtures of hydrogen, iodine and hydrogen iodide at 731 K are consistent with the equilibrium law by calculating K_c for each set of values for the reaction: $H_2(g) + I_2(g) \rightleftharpoons 2HI(g)$. Include the units of K_c, if any, in your answers.

Equilibrium concentrations/mol dm^{-3}		
[$H_2(g)$]	[$I_2(g)$]	[$HI(g)$]
0.0086	0.0086	0.0586
0.0114	0.0012	0.0252
0.0092	0.0022	0.0302

4 Calculate K_c for the reaction: $PCl_5(g) \rightleftharpoons PCl_3(g) + Cl_2(g)$ given that on mixing 1.68 mol $PCl_5(g)$ with 0.36 mol $PCl_3(g)$ in a 2.0 dm^3 container and allowing the mixture to reach equilibrium, the amount of PCl_5 in the equilibrium mixture was 1.44 mol.

5 $K_c = 170$ mol^{-1} dm^3, at 298 K for the equilibrium system: $2NO_2(g) \rightleftharpoons N_2O_4(g)$. If a 5 dm^3 flask contains 1.0×10^{-3} mol of NO_2 and 7.5×10^{-4} mol N_2O_4, is the system at equilibrium? Is there any tendency for the concentration of NO_2 to change and, if so, will it increase or decrease?

Equilibrium constants and balanced equations

An equilibrium constant always applies to a particular chemical equation and can be deduced directly from the equation.

There are two common ways of writing the reaction of sulfur dioxide with oxygen. As a result there are two forms for the equilibrium constant which have different values. So long as the matching equation and equilibrium constant are used in any calculation the predictions based on the equilibrium law are the same. For:

$$2SO_2(g) + O_2(g) \rightleftharpoons 2SO_3(g) \qquad \text{equation 1}$$

$$K_c = \frac{[SO_3(g)]^2}{[SO_2(g)]^2[O_2(g)]}$$

But for:

$$SO_2(g) + \tfrac{1}{2}O_2(g) \rightleftharpoons SO_3(g) \qquad \text{equation 2}$$

$$K_c = \frac{[SO_3(g)]}{[SO_2(g)][O_2(g)]^{\frac{1}{2}}}$$

So, it is important to write the balanced equation and the equilibrium constant together.

Reversing the equation also changes the form of the equilibrium constant because the concentration terms for the chemicals on the right-hand side of the equation always appear on the top of the expression for K_c.
So for:

$$2SO_3(g) \rightleftharpoons 2SO_2(g) + O_2(g) \qquad \text{equation 3}$$

$$K_c = \frac{[SO_2(g)]^2[O_2(g)]}{[SO_3(g)]^2}$$

Test yourself

6 Consider the equilibrium between sulfur dioxide, oxygen and sulfur trioxide.

a) Show that the units for the equilibrium constant, K_c, for equation 1 are mol^{-1} dm^3.

b) $K_c = 1.6 \times 10^6$ mol^{-1} dm^3 for equation 1 at a particular temperature. What is the value of K_c for equation 2?

c) What is the value of K_c for equation 3?

7 Write the expression for K_c for these equations and state the units of the equilibrium constant for each example.

a) $CO_2(g) + H_2(g) \rightleftharpoons CO(g) + H_2O(g)$

b) $N_2(g) + 3H_2(g) \rightleftharpoons 2NH_3(g)$

Equilibrium constants and the direction of change

If the value of the equilibrium constant is large, then the position of equilibrium is over to the right-hand side of the equation. Broadly speaking, if K_c is about 100 or larger at a given temperature then the products predominate.

Conversely if the value of the equilibrium constant is small then the position of equilibrium is over to the left-hand side of the equation. If K_c is about 0.01 or smaller at a given temperature then the reactants predominate.

If the value of K_c is close to 1, then there are significant quantities of both reactants and products present at equilibrium.

It is very important to keep in mind that equilibrium constants say nothing about the time it takes for a reaction mixture to reach equilibrium. The system may reach equilibrium rapidly or slowly. The value of K_c says nothing about the rate of change.

The value of K_c for the reaction of hydrogen with chlorine to make hydrogen chloride, for example, is about 10^{31} at room temperature but in the absence of a catalyst or ultraviolet light there is no reaction.

Definition

A **phase** is one of the three states of matter – solid, liquid or gas. Chemical systems often have more than one phase. Each phase is distinct but need not be pure:

■ a solid in equilibrium with its saturated solution is a two-phase system,

■ in the reactor for ammonia manufacture the mixture of nitrogen, hydrogen and ammonia gases is one phase with the iron catalyst being a separate solid phase.

Test yourself

8 What can you conclude about the direction and extent of change in each of these examples?

a) $Zn(s) + Cu^{2+}(aq) \rightleftharpoons Zn^{2+}(aq) + Cu(s)$ $K_c = 1 \times 10^{37}$ at 298 K

b) $2HBr(g) \rightleftharpoons H_2(g) + Br_2(g)$ $K_c = 1 \times 10^{-10}$ at 298 K

c) $N_2(g) + 3H_2(g) \rightleftharpoons 2NH_3(g)$ $K_c = 2.2 \ mol^{-2} \ dm^6$ at 623 K

9 In general, if the equilibrium constant for a forward reaction is large, what will be the size of the equilibrium constant for the reverse of this reaction?

Heterogeneous equilibria

In some equilibrium systems the substances involved are not all in the same phase. An example is the equilibrium state involving two solids and a gas formed on heating calcium carbonate in a closed container.

$$CaCO_3(s) \rightleftharpoons CaO(s) + CO_2(g)$$

The concentrations of solids do not appear in the expression for the equilibrium constant. Pure solids have, in effect, a constant 'concentration'.

$$K_c = [CO_2(g)]$$

The same applies to heterogeneous systems which have a separate pure liquid phase as one of the reactants or products.

Figure 2.5.3 ▶
Stalactites growing in the Florida Caverns, Marianna, Florida. The stalactites grow very slowly because this heterogeneous system is almost at equilibrium

Another example is the equilibrium state between solid calcium carbonate and a dilute solution containing dissolved carbon dioxide and calcium hydrogencarbonate.

$$CaCO_3(s) + CO_2(aq) + H_2O(l) \rightleftharpoons Ca^{2+}(aq) + 2HCO_3^-(aq)$$

This example illustrates another general rule. The K_c expression for dilute solutions does not include a concentration term for water. There is so much water present that its concentration is effectively constant.

So the expression for the equilibrium constant becomes:

$$K_c = \frac{[Ca^{2+}(aq)]\ [HCO_3^-(aq)]^2}{[CO_2(aq)]}$$

Test yourself **D**

10 Write the expression for K_c for these equilibria:

a) $2Fe(s) + 3H_2O(g) \rightleftharpoons Fe_2O_3(s) + 3H_2(g)$

b) $H_2(g) + S(l) \rightleftharpoons H_2S(g)$

c) $Ag^+(aq) + Fe^{2+}(aq) \rightleftharpoons Fe^{3+}(aq) + Ag(s)$

11 Calculate the concentration of water in water (in mol dm^{-3}) to show that it is reasonable to regard the concentration of water as a constant when writing the expression for K_c for equilibria in dilute aqueous solution.

The effect of changing concentrations on systems at equilibrium

The equilibrium law makes it possible to explain the effect of changing the concentration of one of the chemicals in an equilibrium mixture.

An example is the equilibrium in solution involving chromate(VI) and dichromate(VI) ions in water:

$$2CrO_4^{2-}(aq) + 2H^+(aq) \rightleftharpoons Cr_2O_7^{2-}(aq) + H_2O(l)$$
$$\text{yellow} \qquad\qquad\qquad \text{orange}$$

At equilibrium: $K_c = \dfrac{[Cr_2O_7^{2-}(aq)]}{[CrO_4^{2-}(aq)]^2\ [H^+(aq)]^2}$

where these are equilibrium concentrations.

In dilute solution $[H_2O(l)]$ is constant so it does not appear in the equilibrium law expression.

Adding a few drops of concentrated acid increases the concentration of $H^+(aq)$ on the left-hand side of the equation. This briefly upsets the equilibrium. For an instant after adding acid:

$$\frac{[Cr_2O_7^{2-}(aq)]}{[CrO_4^{2-}(aq)]^2\ [H^+(aq)]^2} < K_c$$

The system restores equilibrium as chromate(VI) ions react with hydrogen ions to produce more of the products. There is very soon a new equilibrium. Once again:

$$\frac{[Cr_2O_7^{2-}(aq)]}{[CrO_4^{2-}(aq)]^2\ [H^+(aq)]^2} = K_c$$

but now with new values for the various concentrations.

Physical Chemistry

Section two

Note

Changing concentrations does not alter the value of the equilibrium constant so long as the temperature stays constant.

Figure 2.5.4 ▲
On the left, a yellow solution of chromate(VI) ions in water. On the right, a solution of chromate(VI) ions in water after adding a few drops of strong acid – the solution has turned orange as more dichromate(VI) ions form

Chemists say that adding acid makes the 'position of equilibrium shift to the right'. The effect is visible because the yellow colour of the chromate(VI) ions turns to the orange colour of dichromate(VI) ions. This is as Le Chatelier's principle predicts. The advantage of using K_c is that it makes quantitative predictions possible.

Test yourself

12 Describe and explain the effect of adding alkali to a solution of dichromate(VI) ions.

13 a) Use the equilibrium law to predict and explain the effect of adding more pure ethanol to an equilibrium mixture of ethanoic acid, ethanol, ethyl ethanoate and water:

$$CH_3CO_2H(l) + C_2H_5OH(l) \rightleftharpoons CH_3CO_2C_2H_5(l) + H_2O(l)$$

b) Show that your prediction is consistent with Le Chatelier's principle.

The equilibrium law

2.6 Gaseous equilibria

Many important industrial processes involve reversible reactions between gases. Applying the equilibrium law to these reactions helps to determine the optimal conditions for manufacturing chemicals. When it comes to gas reactions it is often easier to measure pressures rather than concentrations and to use a modified form of the equilibrium law.

Gas mixtures and partial pressures

In any mixture of gases the total pressure of the mixture can be 'shared out' between the gases. It is possible to calculate a partial pressure for each gas in the mixture. In a mixture of gases A, B and C, the sum of the three partial pressures equals the total pressure.

$$p_A + p_B + p_B = p_{total}$$

Partial pressures are a useful alternative to concentrations when studying mixtures of gases and gas reactions.

In gas mixtures it is the amounts in moles which matter and not the chemical nature of the molecules. This means that the total pressure is shared between the gases simply according to their mole fractions in the mixture.

In a mixture of n_A moles of A with n_B moles of B and n_C moles of C, the mole fractions (symbol X) are given by the following:

$$X_A = \frac{n_A}{n_A + n_B + n_C} \qquad X_B = \frac{n_B}{n_A + n_B + n_C} \qquad X_C = \frac{n_C}{n_A + n_B + n_C}$$

So the mole fraction of A is the fraction of the total number of moles of all compounds which are moles of A.

The sum of all the mole fractions is 1, so $X_A + X_B + X_C = 1$.

On this basis the partial pressures of three gases A, B and C in a gas mixture with total pressure p are:

$$p_A = X_A p, \quad p_B = X_B p \text{ and } p_C = X_C p.$$

The partial pressure for each gas is the pressure it would exert if it was the only gas in the container under the same conditions. The ideal gas equation: $pV = nRT$ can be applied to each gas in the mixture. Rearranging the equation gives:

$$p_A = \frac{n_A RT}{V} \qquad p_B = \frac{n_B RT}{V} \qquad \text{and} \qquad p_C = \frac{n_C RT}{V}$$

Since R is a constant, these equations show that at constant temperature:

$$p_A \propto \frac{n_A}{V} \qquad p_B \propto \frac{n_B}{V} \qquad \text{and} \qquad p_C \propto \frac{n_C}{V}$$

So the partial pressures are proportional to the concentrations of the gases since n/V is the concentration in moles per unit volume. This is the justification of working in partial pressures when applying the equilibrium law to gas reactions.

Figure 2.6.1 ▲
The partial pressures of the gases in the air add up to the total pressure of the atmosphere

Note

Do not use square brackets when writing K_p expressions. In the context of the equilibrium law square brackets signify concentrations in moles per litre.

Note

Changing the total pressure or the composition of the gas mixture has no affect on the value of K_p so long as the temperature stays constant.

Definitions

Pressure is defined as force per unit area. The SI unit of pressure is the pascal (Pa) which is a pressure of one newton per square metre (1 N m^{-2}). The pascal is a very small unit so pressures are often quoted in kilopascals, kPa.

When studying gases the standard pressure is **atmospheric pressure** which is 101.3×10^3 Nm^{-2} = 101.3 kPa.

In accounts of chemical processes, multiples of atmospheric pressure give an indication of the extent to which gases are compressed.

Standard pressure for definitions in thermodynamics is now 1 bar which is 100 000 Nm^{-2} = 100 kPa

The **partial pressure** of a gas is a measure of its concentration in a mixture of gases.

K_p

K_p is the symbol for the equilibrium constant for an equilibrium involving gases with the concentrations measured by partial pressures. The rules for writing equilibrium expressions are the same for K_p as for K_c with partial pressures replacing concentrations as shown in Figure 2.6.2.

Equilibrium	K_p	Units of K_p (using the SI unit of pressure, Pa)
$H_2(g) + I_2(g) \rightleftharpoons 2HI(g)$	$K_p = \dfrac{(p_{HI})^2}{(p_{H_2})\,(p_{I_2})}$	no units
$N_2(g) + 3H_2(g) \rightleftharpoons 2NH_3(g)$	$K_p = \dfrac{(p_{NH_3})^2}{(p_{N_2})\,(p_{H_2})^3}$	Pa^{-2}
$N_2O_4(g) \rightleftharpoons 2NO_2(g)$	$K_p = \dfrac{(p_{NO_2})^2}{(p_{N_2O_4})}$	Pa
$HCl(g) + LiH(s) \rightleftharpoons H_2(g) + LiCl(s)$	$K_p = \dfrac{(p_{H_2})}{(p_{HCl})}$	no units

Figure 2.6.2 ▲
Examples of equilibrium expressions for K_p. Note that when writing an expression for K_p for a heterogeneous reaction the same rules apply as for K_c. The expression does not include terms for any separate pure solid phases

Worked example

An experimental study of the equilibrium mixture of $N_2(g)$, $H_2(g)$ and $NH_3(g)$ found that one equilibrium mixture contained 2.15 mol of $N_2(g)$, 6.75 mol of $H_2(g)$ and 1.41 mol of $NH_3(g)$ at a total pressure 1000 kPa. Calculate the value for K_p under the conditions that the measurements were taken.

Notes on the method

First work out the mole fractions of the gases.

Multiply the total pressure by the mole fractions to get the partial pressures.

Check that the sum of the partial pressures equals the total pressure.

Finally substitute in the expression for K_p and give the units.

Answer

Total number of moles = 2.15 mol + 6.75 mol + 1.41 mol

= 10.31 mol

Mole fraction of $N_2(g)$ = $\dfrac{2.15 \text{ mol}}{10.31 \text{ mol}}$ = 0.208

Mole fraction of $H_2(g)$ = $\dfrac{6.75 \text{ mol}}{10.31 \text{ mol}}$ = 0.655

Mole fraction of $NH_3(g)$ = $\dfrac{1.41 \text{ mol}}{10.31 \text{ mol}}$ = 0.137

Partial pressure of $N_2(g)$ = 0.208 × 1000 kPa = 208 kPa

Partial pressure of $H_2(g)$ = 0.655 × 1000 kPa = 655 kPa

Partial pressure of $NH_3(g)$ = 0.137 × 1000 kPa = 137 kPa

For the equilibrium:

$$N_2(g) + 3H_2(g) \rightleftharpoons 2NH_3(g)$$

$$K_p = \frac{(p_{NH_3})^2}{(p_{N_2})\,(p_{H_2})^3} = \frac{(137 \text{ KPa})^2}{(208 \text{ KPa})(655 \text{ KPa})^2} = 2.1 \times 10^{-4} \text{ kPa}^{-2}$$

Test yourself

1 Write the expression for K_p for these equilibria and give the units:

a) $2SO_2(g) + O_2(g) \rightleftharpoons 2SO_3(g)$

b) $4NH_3(g) + 3O_2(g) \rightleftharpoons 2N_2(g) + 6H_2O(g)$

c) $CaCO_3(s) \rightleftharpoons CaO(s) + CO_2(g)$

2 Calculate K_p for this equilibrium mixture: **CD-ROM**

$N_2O_4(g) \rightleftharpoons 2NO_2(g)$

at 330 K and 120 kPa pressure. Under these conditions a sample of the gas mixture consists of 8.1 mol $N_2O_4(g)$ and 3.8 mol $NO_2(g)$.

3 Calculate K_p for this reversible reaction at 1000 K and 180 kPa pressure:

$C_2H_6(g) \rightleftharpoons C_2H_4(g) + H_2(g)$

given that under these conditions, starting with just 5 mol ethane yields an equilibrium mixture containing 1.8 mol ethene.

4 Use the expression for K_p to predict the effect of the following changes on a equilibrium mixture of hydrogen, carbon monoxide and methanol:

$2H_2(g) + CO(g) \rightleftharpoons CH_3OH(g)$

a) adding more hydrogen to the gas mixture at constant total pressure

b) compressing the mixture to increase the total pressure

c) adding an inert gas such as argon while keeping the total pressure constant.

Physical Chemistry

Section Two

2.7 The effects of temperature changes and catalysts on equilibria

Temperature changes cause a shift in the position of equilibrium for a reaction. This happens because the value of the equilibrium constant changes. Often a balance has to be struck between raising the temperature of a process high enough for the rate to be faster whilst not decreasing the yield of the reaction. Catalysts help by speeding up reactions but they have no effect on the position of equilibrium.

Figure 2.7.1 ▶
In manufacturing industry the people controlling processes have to find the conditions which provide a good yield of products at equilibrium but at a rate which is fast enough to be practical and economical

Temperature changes

Le Chatelier's principle predicts that raising the temperature makes the equilibrium shift in the direction which is endothermic. During the manufacture of sulfuric acid, for example, raising the temperature lowers the percentage of sulfur trioxide at equilibrium.

$$2SO_2(g) + O_2(g) \rightleftharpoons 2SO_3(g) \qquad \Delta H^\ominus = -98 \text{ kJ mol}^{-1}$$

Note

If ΔH for the forward reaction is negative, then ΔH for the reverse reaction will have the same magnitude but the opposite sign. So if the forward reaction is exothermic, then the reverse reaction is endothermic.

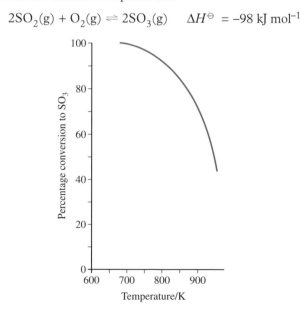

Figure 2.7.2 ▶
Effect of raising the temperature on the equilibrium between SO_2, O_2 and SO_3

The equilibrium shifts to the left as the temperature rises because this is the direction in which the reaction is endothermic.

The shift happens because the equilibrium constant varies with temperature.

$$K_p = \frac{(p_{SO_3})^2}{(p_{SO_2})^2 (p_{O_2})}$$

At 500 K, $K_p = 2.5 \times 10^{10}$ atm^{-1}, but at 700 K, $K_p = 3.0 \times 10^4$ atm^{-1}. The value of the equilibrium constant falls as the temperature rises. With a smaller value of K_p the proportion of $SO_3(g)$ falls while the proportions of $SO_2(g)$ and $O_2(g)$ rise at equilibrium.

The effect of catalysts

In industry, the reaction of sulfur dioxide with oxygen is a step in the manufacture of sulfuric acid from sulfur. The reaction takes place on the surface of a vanadium(V) oxide catalyst. The temperature effect on equilibrium means that raising the temperature lowers the percentage conversion to SO_3 but the temperature must be high enough to make the reaction go fast enough. The catalyst is not active below 380 °C and works best at a higher temperature.

The catalyst increases the rate of reaction but it has no effect on the position of equilibrium.

A similar compromise between rate and high yield has to be made in the manufacture of ammonia from nitrogen and hydrogen. The reaction is very slow at room temperature. Raising the temperature increases the rate of reaction but the reversible reaction is exothermic so the higher the temperature the lower the yield of ammonia at equilibrium. A catalyst makes it possible for the reaction to go faster without the temperature being so high that the yield is too low. As always the catalyst only affects the rate of reaction and not the position of equilibrium.

The process typically operates at pressures between 70 atmospheres and 200 atmospheres with temperatures in the range 400 °C to 600 °C.

> **Note**
>
> The value of an equilibrium constant for an exothermic reaction becomes smaller as the temperature rises.
>
> The value of the equilibrium constant for an endothermic reaction becomes larger as the temperature rises.

Physical Chemistry

Section two

Test yourself

1 For the reaction between hydrogen and iodine to form hydrogen iodide, the value of K_p is 794 at 298 K but 54 at 700 K. What can you deduce from this information?

2 For the decomposition of calcium carbonate to calcium oxide and carbon dioxide, the value of K_p is 1.6×10^{-21} kPa at 298 K but 1.0 kPa at 900 K. What can you deduce from this information?

3 Draw up a summary table to show the effects of changing the concentration, changing the temperature or adding a catalyst on:

   a) the position of equilibrium

   b) the rate of reaction.

2.8 Acids, bases and the pH scale

Acids and bases are very common not only in laboratories but also in living things, in the home and in the natural environment. Acid–base reactions are reversible and governed by the equilibrium law. This means that chemists are able to predict reliably and quantitatively how acids and bases behave. This is important for the supply of safe drinking water, the care of patients in hospital, the formulation of shampoos and cosmetics as well as the processing of food and many other aspects of life.

Figure 2.8.1 ▶
Limestone near Malham Cove in Yorkshire etched by rainwater made acidic by dissolved carbon dioxide

Definition

A **proton** is the nucleus of a hydrogen atom, so a hydrogen ion, H⁺, is just a proton.

Chemists have several theories to explain the behaviour of acids (see section 3.4) but the preferred theory for discussing acid–base equilibria is the Brønsted–Lowry theory. This theory describes acids as proton donors and bases as proton acceptors.

Acids

According to the Brønsted–Lowry theory, hydrogen chloride molecules give hydrogen ions (protons) to water molecules when they dissolve in water producing hydrated hydrogen ions called oxonium ions. The water acts as a base.

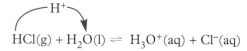

$$HCl(g) + H_2O(l) \rightleftharpoons H_3O^+(aq) + Cl^-(aq)$$

All acid–base reactions are reversible. In this example a proton from the oxonium ion can transfer back to the chloride ion to give hydrogen chloride and water.

Hydrogen chloride is a strong acid. What this means is that it readily gives up its protons to water molecules and the equilibrium in solution lies well over to the right. Hydrogen chloride is effectively completely ionised in solution. Other examples of strong acids are sulfuric acid and nitric acid.

Note

Chemists often represent hydrogen ions in aqueous solution as $H_3O^+(aq)$, the oxonium ion; but they also use a shorter symbol for hydrated protons, $H^+(aq)$.

Definitions

A **monoprotic acid** is an acid that can give away (donate) one proton per molecule. Examples of monoprotic acids are: hydrogen chloride, HCl; nitric acid, HNO_3 and ethanoic acid, CH_3CO_2H.

A **diprotic acid** is an acid which can give away two protons per molecule. Examples of diprotic acids are: sulfuric acid, H_2SO_4; and ethanedioic acid, $HO_2C–CO_2H$.

Test yourself

1 **a)** What type of bond links the water molecule to a proton in an oxonium ion?

b) Draw a dot-and-cross diagram to show the bonding in an oxonium ion.

c) Predict the shape of an oxonium ion.

2 Write a balanced, ionic equation for the reaction of 1 mol ethanedioic acid with 2 mol NaOH showing the displayed formulae (see page 143) for the acid and for the ethanedioate ion formed.

3 **a)** Identify the products of the reaction when concentrated sulfuric acid reacts with sodium chloride.

b) Show that this is a proton transfer reaction

c) Account for the fact that this reaction can give a good yield of hydrogen chloride gas.

Bases

A base is a molecule or ion which can accept a hydrogen ion (proton) from an acid. A base has a lone-pair of electrons which can form a dative covalent bond with a proton.

$$:\overset{\cdot\cdot}{\underset{\cdot\cdot}{O}}:^{2-} \overset{\frown}{} H^+ \longrightarrow :\overset{\cdot\cdot}{\underset{\cdot\cdot}{O}} - H^-$$

An ionic oxide, such as calcium oxide, reacts completely with water to form calcium hydroxide. The calcium ions do not change. But the oxide ions, which are powerful proton acceptors, all take protons from water molecules. An oxide ion is a strong base. Common bases include the oxide and hydroxide ions, ammonia, amines, as well as the carbonate and hydrogencarbonate ions.

There is a lone pair on the nitrogen atom of ammonia which allows it to act as a base:

$$NH_3(g) + HNO_3(g) \longrightarrow NH_4^+NO_3^-(s)$$
$$\text{base} \qquad \text{acid}$$

◀ *Figure 2.8.2*
Oxide ions have lone pairs of electrons which can form dative covalent bonds with hydrogen ions

Note

Many compounds of the group 1 and group 2 metals form alkaline solutions. This is because metals such as sodium, potassium, magnesium and calcium (unlike other metals) form oxides, hydroxides and carbonates which are soluble (to a greater or lesser extent) in water. Note that it is the oxide, hydroxide or carbonate ions in these compounds which are bases and not the metal ions.

Note

In biochemistry the term *base* often refers to one of the five nitrogenous bases which make up nucleotides and nucleic acids DNA and RNA. These compounds (adenine, guanine, cytosine, uracil and thymine) are bases in the chemical sense because they have lone pairs on nitrogen atoms which can accept hydrogen ions.

Test yourself

4 Show that the reactions between these pairs of compounds are acid–base reactions and identify as precisely as possible the molecules or ions which are the acid and the base in each example.

a) MgO + HCl

b) $H_2SO_4 + NH_3$

c) $NH_4NO_3 + NaOH$

d) $HCl + Na_2CO_3$

5 **a)** What type of bond links the ammonia molecule to a proton in an ammonium ion?

b) Draw a dot-and-cross diagram to show the bonding in an ammonium ion.

c) Predict the shape of an ammonium ion.

6 Identify and name the conjugate bases of these acids: HNO_3, CH_3CO_2H, H_2SO_4, HCO_3^-.

7 Identify the conjugate acid of each of these bases: O^{2-}, OH^-, NH_3, CO_3^{2-}, HCO_3^-, SO_4^{2-}.

Conjugate acid–base pairs

Any acid–base reaction involves competition for protons. This is illustrated by a solution of an ammonium salt, such as ammonium chloride, in water.

$$NH_4^+(aq) \; + \; H_2O(l) \; \rightleftharpoons \; NH_3(aq) \; + \; H_3O^+(aq)$$

$$\text{acid 1} \qquad \text{base 2} \qquad \text{base 1} \qquad \text{acid 2}$$

In this example there is competition for protons between ammonia molecules and water molecules. On the left-hand side of the equation the protons are held by lone pairs on the ammonia molecules. On the right-hand side they are held by lone pairs on water molecules. The position of equilibrium shows which of the two bases has the stronger hold on the protons.

Chemists use the term conjugate acid–base pair to describe a pair of molecules or ions which can be converted from one to the other by the gain or loss of a proton. An acid turns into its conjugate base when it loses a proton. A base turns into its conjugate acid when it gains a proton.

The equilibrium in a solution of an ammonium salt above involves two conjugate acid–base pairs:

■ NH_4^+ and NH_3
■ H_3O^+ and H_2O.

The pH scale

The concentration of oxonium ions in aqueous solutions commonly ranges from 2 mol dm^{-3} to 10^{-14} mol dm^{-3}. The concentration of aqueous hydrogen ions in dilute hydrochloric acid is about 100 000 000 000 000 times greater than the concentration of hydrogen ions in dilute sodium hydroxide solution. Given such a wide range of concentrations, chemists find it convenient to use a logarithmic scale to measure the concentration of aqueous hydrogen ions in acidic or alkaline solutions. This is the pH scale.

The definition of pH is:

Figure 2.8.3 ▼
The pH scale

$$pH = -\log_{10}[H_3O^+(aq)], \text{ which is often written as } pH = -\log[H^+(aq)]$$

pH	0	1	2	3	4	5	6	7	8	9	10	11	12	13	14
$[H_3O^+(aq)]$/mol dm^{-3}	10^0	10^{-1}	10^{-2}	10^{-3}	10^{-4}	10^{-5}	10^{-6}	10^{-7}	10^{-8}	10^{-9}	10^{-10}	10^{-11}	10^{-12}	10^{-13}	10^{-14}

increasingly acidic ← neutral → increasingly alkaline

Strictly speaking the definition of pH is given by:

$$pH = -\log_{10}\frac{[H^+(aq)]}{1 \text{ mol dm}^{-3}}$$

because mathematically it is only possible to take logarithms of numbers and not of quantities with units. Dividing by a standard concentration of 1 mol dm^{-3} does not change the value but cancels the units.

Worked example

What is the pH of 0.020 mol dm^{-3} hydrochloric acid?

Notes on the method

Hydrochloric acid is a strong acid so it is fully ionised. Note that 1 mol HCl gives 1 mol $H^+(aq)$.

Use the log button on your calculator. Do not forget the minus sign in the definition of pH.

Answer

$[H^+(aq)] = 0.020$ mol dm^{-3}

$pH = -\log(0.020) = 1.70$

Worked example

The pH of human blood is 7.40. What is the aqueous hydrogen ion concentration in blood?

Notes on the method

$pH = -\log [H^+(aq)]$

From the definition of logarithms this rearranges to $[H^+(aq)] = 10^{-pH}$

Use the inverse log button (10^x) on your calculator. Do not forget the minus sign in the definition of pH.

Answer

$pH = 7.4$

$[H^+(aq)] = 10^{-7.4} = 4.0 \times 10^{-8}$ mol dm^{-3}

The ionic product of water, K_w

There are hydrogen and hydroxide ions even in pure water because of a transfer of hydrogen ions between water molecules. This only happens to a very slight extent.

$$H_2O(l) + H_2O(l) \rightleftharpoons H_3O^+(aq) + OH^-(aq)$$

Which can be written more simply as:

$$H_2O(l) \rightleftharpoons H^+(aq) + OH^-(aq)$$

The equilibrium constant $K_c = \dfrac{[H^+(aq)][OH^-(aq)]}{[H_2O(l)]}$

There is such a large excess of water that $[H_2O(l)] = $ a constant, so the relationship simplifies to:

$$K_w = [H^+(aq)][OH^-(aq)]$$

where K_w is the ionic product of water.

The pH of pure water at 298 K is 7. So the hydrogen ion concentration at equilibrium: $[H^+(aq)] = 1 \times 10^{-7}$ mol dm^{-3}.

Also in pure water $[H_3O^+(aq)] = [OH^-(aq)]$
So the $[OH^-(aq)] = 1 \times 10^{-7}$ mol dm^{-3}.
Hence $K_w = 1 \times 10^{-14}$ mol^2 dm^{-6}
K_w is a constant in all aqueous solutions at 298 K. This makes it possible to calculate the pH of alkalis.

Worked example

What is the pH of a 0.05 mol dm^{-3} solution of sodium hydroxide?

Notes on the method

Sodium hydroxide is fully ionised in solution. So in this solution

$[OH^-(aq)] = 0.05$ mol dm^{-3}

$pH = -\log [H^+(aq)]$

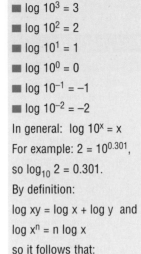

Definition

Chemists use **logarithms to base 10** to handle values which range over several orders of magnitude. Logarithms to base 10 are defined such that:

- $\log 10^3 = 3$
- $\log 10^2 = 2$
- $\log 10^1 = 1$
- $\log 10^0 = 0$
- $\log 10^{-1} = -1$
- $\log 10^{-2} = -2$

In general: $\log 10^x = x$

For example: $2 = 10^{0.301}$,

so $\log_{10} 2 = 0.301$.

By definition:

$\log xy = \log x + \log y$ and

$\log x^n = n \log x$

so it follows that:

$\log 1/x = \log x^{-1} = -\log x$.

Test yourself

8 What is the pH of solutions of hydrochloric acid with these concentrations?

   a) 0.1 mol dm^{-3}

   b) 0.01 mol dm^{-3}

   c) 0.001 mol dm^{-3}

9 Calculate the pH of a 0.08 mol dm^{-3} solution of nitric acid.

10 What is the concentration of oxonium ion in these solutions?

   a) orange juice with a pH of 3.3

   b) coffee with a pH of 5.4

   c) saliva with a pH of 6.7

   d) a suspension of an antacid in water with a pH of 10.5.

Definition

$pK_w = -\log K_w$

11 The value of K_w varies with temperature. At 273 K its value is 1.1×10^{-15} mol^2 dm^{-6} while at 303 K it is 1.5×10^{-14} mol^2 dm^{-6}.

   a) Is the ionisation of water an exothermic or an endothermic process?

   b) What happens to the oxonium ion concentration in pure water, and hence the pH, as the temperature rises?

12 Calculate the pH of these solutions:

   a) 1.0 mol dm^{-3} NaOH

   b) 0.02 mol dm^{-3} KOH

   c) 0.001 mol dm^{-3} Ba(OH)$_2$

Answer

For this solution:

$K_w = [H^+(aq)] \times 0.05 \text{ mol dm}^{-3} = 10^{-14} \text{ mol}^2 \text{ dm}^{-6}$

So $[H^+(aq)] = \dfrac{10^{-14} \text{ mol}^2 \text{ dm}^{-6}}{0.05 \text{ mol dm}^{-3}} = 2 \times 10^{-13} \text{ mol dm}^{-3}$

Hence pH $= -\log(2 \times 10^{-13}) = 12.7$

Working in logarithms

The logarithmic form of equilibrium constants is particularly useful for pH calculations. Taking logarithms produces a convenient small scale of values.

$$K_w = [H^+(aq)][OH^-(aq)] = 10^{-14} \text{ at 298 K}$$

Taking logarithms, and applying the rule that log xy = log x + log y, gives:

$$\log K_w = \log[H^+(aq)] + \log[OH^-(aq)] = \log 10^{-14} = -14$$

Multiplying through by -1, reverses the signs:

$$-\log K_w = -\log[H^+(aq)] - \log[OH^-(aq)] = 14$$

Hence: $pK_w = pH + pOH = 14$, where pOH is defined as $-\log[OH^-(aq)]$ by analogy with pH.

So: pH $= 14 -$ pOH which makes it easy to calculate the pH of alkaline solutions

Worked example

What is the pH of a 0.05 mol dm^{-3} solution of sodium hydroxide?

Notes on the method

Sodium hydroxide, NaOH, is a strong base so it is fully ionised.

Find the values of logarithms with the log button of a calculator.

Answer

$[OH^-] = 0.05 \text{ mol dm}^{-3}$

$pOH = -\log 0.05 = 1.3$

$pH = 14 - pOH = 14 - 1.3 = 12.7$

Figure 2.8.5 ▲
Bloody cranes-bill in limestone. These plants only flourish when growing in soil where the minerals, such as limestone, keep the soil water alkaline

Figure 2.8.4 ▲
These plants only flourish when growing in peaty soils where the breakdown of organic remains keeps the soil water acidic

2.9 Weak acids and bases

Most organic acids and bases only ionise to a slight extent in aqueous solution. Carboxylic acids (see page 176) such as ethanoic acid in vinegar, citric acid in fruit juices and lactic acid in sour milk are all weak acids. Ammonia and amines (see page 192) are weak bases.

Weak acids

Weak acids, such as ethanoic acid, are only slightly ionised when they dissolve in water. In a 0.1 mol dm^{-3} solution of ethanoic acid, for example, only about one in a hundred molecules react with water to form oxonium ions.

$$CH_3CO_2H(aq) + H_2O(l) \rightleftharpoons CH_3CO_2^-(aq) + H_3O^+(aq)$$

Note the important distinction between strength and concentration. Strength is the extent of ionisation. Concentration is the amount of acid in a litre (in mol dm^{-3}). It takes as much sodium hydroxide to neutralise 25 cm^3 of 0.1 mol dm^{-3} of a weak acid such as ethanoic acid (pH = 2.9) as it does to neutralise 25 cm^3 of 0.1 mol dm^{-3} of a strong acid such as hydrochloric acid (pH = 1).

Weak bases

Weak bases are only slightly ionised when they dissolve in water. In a 0.1 mol dm^{-3} solution of ammonia, for example, ninety nine in every hundred molecules do not react but remain as dissolved molecules. Only one molecule in a hundred reacts to form ammonium ions.

$$NH_3(aq) + H_2O(l) \rightleftharpoons NH_4^+(aq) + OH^-(aq)$$

As with weak acids, it is important to distinguish between strength and concentration.

Acid dissociation constants

Chemists use the equilibrium constant for the reversible ionisation of a weak acid as a measure of its strength. The equilibrium constant shows the extent to which acids dissociate into ions in solution.

For a weak acid represented by the general formula HA:

$$HA(aq) + H_2O(l) \rightleftharpoons H_3O^+(aq) + A^-(aq)$$

According to the equilibrium law, the equilibrium constant,

$$K_c = \frac{[H_3O^+(aq)][A^-(aq)]}{[HA(aq)][H_2O(l)]}$$

In dilute solution the concentration of water is effectively constant, so the expression can be written in this form:

$$K_a = \frac{[H_3O^+(aq)][A^-(aq)],}{[HA(aq)]}$$ where K_a is the acid dissociation constant.

Given the value for K_a it is possible to calculate the pH of a solution of a weak acid.

▲ **Figure 2.9.1**
Citric acid crystals seen through a microscope with polarized light. Citric acid is a weak acid

Test yourself

1 Explain why measuring the pH of a solution of an acid does not provide enough evidence to show whether or not the acid is strong or weak.

Physical Chemistry

Section Two

Worked example

Calculate the hydrogen ion concentration and the pH of a 0.01 mol dm^{-3} solution of propanoic acid. K_a for the acid is 1.3×10^{-5} mol dm^{-3}.

Notes on the method

Two approximations simplify the calculation.

1 The first assumption is that, at equilibrium: $[H_3O^+(aq)] = [A^-(aq)]$. In this example A^- is the propanoate ion $CH_3CH_2CO_2^-$. This assumption seems obvious from the equation for the ionisation of a weak acid but it ignores the hydrogen ions from the ionisation of water. Water produces far fewer hydrogen ions than most weak acids so its ionisation can be ignored. This assumption is acceptable so long as the pH of the acid is below 6.

2 The second assumption is that so little of the propanoic acid ionises in water that, at equilibrium, $[HA(aq)] \approx 0.01$ mol dm^{-3}. Here HA represents propanoic acid. This is a riskier assumption which has to be checked because in very dilute solutions the degree of ionisation may become quite large relative to the amount of acid in the solution. Chemists generally accept that this assumption is acceptable so long as less than 5% of the acid ionises.

Answer

$$CH_3CH_2CO_2H(aq) + H_2O(l) \rightleftharpoons H_3O^+(aq) + CH_3CH_2CO_2^-(aq)$$

$$K_a = \frac{[H_3O^+(aq)][\,CH_3CH_2CO_2^-(aq)]}{[CH_3CH_2CO_2H(aq)]} = \frac{[H_3O^+(aq)]^2}{0.01 \text{ mol dm}^{-3}}$$

$= 1.3 \times 10^{-5}$ mol dm^{-3}

Therefore $[H_3O^+(aq)]^2 = 1.3 \times 10^{-7}$ mol^2 dm^{-6}

So $[H_3O^+(aq)] = 3.6 \times 10^{-4}$ mol dm^{-3}

pH $= -\log [H_3O^+(aq)]$

$= -\log (3.6 \times 10^{-4})$

$= 3.4$

Check the second assumption: in this case less than 0.0004 mol dm^{-3} of the 0.0100 mol dm^{-3} of acid (4%) has ionised. In this instance the degree of ionisation is small enough to justify the assumption that $[HA(aq)] \approx$ the concentration of unionised acid.

One method which can, in principle, be used to measure K_a for a weak acid is to measure the pH of a solution when the concentration of the acid is accurately known. This is not a good method for determining the size of K_a because the pH values of dilute solutions are very susceptible to contamination – for example in dissolved carbon dioxide from the air.

Worked example

Calculate the K_a of lactic acid given that pH = 2.43 for a 0.10 mol dm^{-3} solution of the acid.

Notes on the method

The same two approximations simplify the calculation.

1   Assume that $[H_3O^+(aq)] = [A^-(aq)]$ where $A^-(aq)$ here represents the aqueous lactate ion. Since the pH is well below 6 this is certainly justified.

2   Also assume that so little of the lactic acid ionises in water that at equilibrium $[HA(aq)] \approx 0.01$ mol dm^{-3}. Here HA represents lactic acid. This is a riskier assumption which again can be checked during the calculation.

Answer

pH = 2.43

$[H_3O^+(aq)] = 10^{-2.43} = 3.72 \times 10^{-3}$ mol dm^{-3}

$[H_3O^+(aq)] = [A^-(aq)] = 3.72 \times 10^{-3}$ mol dm^{-3}

In this example less than 5% of the acid is ionised (less than 0.004 out of 0.100 mol in each litre.)

So $[HA(aq)] \approx 0.01$ mol dm^{-3}

Substituting in the expression for K_a:

$$K_a = \frac{[H_3O^+(aq)][A^-(aq)]}{[HA(aq)]} = \frac{(3.72 \times 10^{-3} \text{ mol dm}^{-3})^2}{0.01 \text{ mol dm}^{-3}}$$

$$K_a = 1.38 \times 10^{-3} \text{ mol dm}^{-3}$$

Test yourself

2   Calculate the pH of a 0.01 mol dm^{-3} solution of hydrogen cyanide given that $K_a = 4.9 \times 10^{-10}$ mol dm^{-3}.

3   Calculate the pH of a 0.05 mol dm^{-3} solution of ethanoic acid given that $K_a = 1.7 \times 10^{-5}$ mol dm^{-3}.

4   Calculate K_a for methanoic acid given that pH = 2.55 for a 0.050 mol dm^{-3} solution.

5   Calculate K_a for butanoic acid, $C_3H_7CO_2H$, given that pH = 3.42 for a 0.01 mol dm^{-3} solution.

Working in logarithms

Chemists find it convenient to define a quantity $pK_a = -\log K_a$ when working with weak acids. Books of data tabulate pK_a values. The relationship between acid strength and pH can be expressed more simply because both are logarithmic quantities.

$$K_a = \frac{[H_3O^+(aq)][A^-(aq)]}{[HA(aq)]}$$

Note

Do not try to remember this logarithmic form of the equilibrium law. The relationship is easy to use, but only apply it if you can derive it quickly from first principles as shown here. Do not forget that this form of the law has two built-in assumptions so it only applies when these assumptions are acceptable.

The two common assumptions when using this expression in calculations are that:

■ $[H_3O^+(aq)] = [A^-(aq)]$
■ $[HA(aq)] = c_A$, where c_A = the concentration of the un-ionised acid.

Substituting in the expression for K_a gives:

$$K_a = \frac{[H_3O^+(aq)]^2}{c_A}$$

Hence: $K_a \times c_A = [H_3O^+(aq)]^2$

Taking logarithms: $\log (K_a \times c_A) = \log [H_3O^+(aq)]^2$

Applying the rules that $\log xy = \log x + \log y$ and the $\log x^n = n\log x$, gives:

$$\log K_a + \log c_A = 2\log [H_3O^+(aq)]$$

which on multiplying by −1 becomes:

$$-\log K_a - \log c_A = -2\log [H_3O^+(aq)]$$

Hence $pK_a - \log c_A = 2 \times pH$
This shows that for a solution of a weak acid which is less than 5% ionised:

$$pH = \tfrac{1}{2} (pK_a - \log c_A)$$

Test yourself

6   What is the value of pK_a for methanoic acid given that $K_a = 1.6 \times 10^{-4}$?

7   What is the value of K_a for benzoic acid given that $pK_a = 4.2$?

8   Arrange these acids in order of acid strength putting the strongest first: ethanoic acid ($pK_a = 4.8$), chlorethanoic acid ($pK_a = 2.9$), dichlorethanoic acid ($pK_a = 1.3$) and trichloroethanoic acid ($pK_a = 0.8$).

9   Show that the logarithmic relationship $pK_a = 2pH + \log c_A$, gives the same answers from the data as the methods used in the worked examples on pages 38 and 39.

2.10 Acid–base titrations

The equilibrium law helps to explain what happens during acid–base titrations and it provides a rationale for the selection of the right indicator for a titration.

pH changes during titrations

The pH changes during a titration as a solution of an alkali runs from a burette and mixes with an acid in a flask. Plotting pH against volume of alkali added gives a shape determined by the nature of the acid and the base. Usually there is a marked change in pH near the equivalence point and it is this which makes it possible to detect the end-point of the titration with an indicator.

Titration of a strong acid with a strong base

Strong acids and bases are fully ionised in solution. Figure 2.10.2 shows the shape of the pH curve for a titration of a strong acid, such as hydrochloric acid, with a strong base, such as sodium hydroxide.

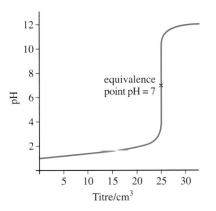

▲ **Figure 2.10.1**
Apparatus for measuring the changes in pH during an acid–base titration

CD-ROM

Figure 2.10.2 ▲
The pH change on adding 0.1 mol dm^{-3} of $NaOH(aq)$ from a burette to 25 cm^3 of a 0.1 mol dm^{-3} of $HCl(aq)$. Note the sharp change of pH around the equivalence point.

Test yourself

1 Show that pH = 1 for a solution of 0.1 mol dm^{-3} of $HCl(aq)$.

2 Why does pH = 7 at the equivalence point of a titration of a strong acid with a strong base?

3 Calculate the pH of 25 cm^3 of a solution of sodium chloride after adding:

   a) 0.05 cm^3 (1 drop) of 0.1 mol dm^{-3} of $HCl(aq)$,

   b) 0.05 cm^3 (1 drop) of 0.1 mol dm^{-3} of $NaOH(aq)$

   (In both instances assume that the volume change on adding 1 drop is insignificant.)

4 Calculate the pH of the solution produced by adding 5 cm^3 of 0.1 mol dm^{-3} of $NaOH(aq)$ to 25 cm^3 of a solution of sodium chloride.

5 Show that your answers to questions 1, 2, 3 and 4 are consistent with Figures 2.10.2, 2.10.3 and 2.10.4.

Definitions

The **end point** during a titration is the point at which a colour change shows that enough of the solution in the burette has been added to react with the amount of the chemical in the flask. In a well planned titration the colour change observed at the end-point corresponds exactly with the equivalence point.

The **equivalence point** is the point during any titration when the amount in moles of one reactant added from a burette is just enough to react exactly with all of the measured amount of chemical in the flask as shown by the balanced equation.

Physical Chemistry

Section two

41

Titration of a weak acid with a strong base

If the acid in the titration flask is weak, then the equilibrium law applies and the pH curve up to the equivalence point has to be calculated with the help of the expression for K_a.

Consider, for example, the reaction of ethanoic acid with sodium hydroxide during a titration. At the start the flask contains the pure acid.

$$CH_3CO_2H(aq) + H_2O(l) \rightleftharpoons H_3O^+(aq) + CH_3CO_2^-(aq)$$

(The methods in section 2.9 show how to use K_a to calculate the pH of a pure solution of a weak acid.) As strong alkali runs in from the burette, some of the ethanoic acid reacts to produce sodium ethanoate. Once some alkali has been added $[H_3O^+(aq)] \neq [CH_3CO_2^-(aq)]$ and the method of calculating the pH changes to account for this.

Figure 2.10.3 ▶
The pH change on adding 0.1 mol dm^{-3} of NaOH*(aq) from a burette to 25 cm^3 of a 0.1 mol dm^{-3} of* CH$_3$CO$_2$H*(aq). Note the sharp change of pH around the equivalence point.*

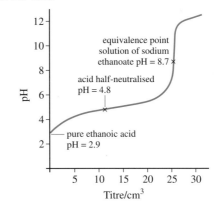

Worked example

What is the pH of a mixture formed during a titration after adding 20.0 cm^3 of 0.10 mol dm^{-3} NaOH(aq) to 25.0 cm^3 of a 0.10 mol dm^{-3} of CH$_3$CO$_2$H(aq) if pK_a for the acid = 4.8?

Notes on the method

The pH of the mixture can be estimated quite accurately using the equilibrium law by assuming that the concentration of:

■ ethanoic acid molecules at equilibrium is determined by the amount of acid which has yet to be neutralised

■ ethanoate ions is determined by the amount of acid converted to sodium ethanoate.

Answer

$$K_a = \frac{[H_3O^+(aq)][CH_3CH_2CO_2^-(aq)]}{[CH_3CH_2CO_2H(aq)]}$$

Rearranging gives: $[H_3O^+(aq)] = K_a \dfrac{[CH_3CH_2CO_2H(aq)]}{[CH_3CH_2CO_2^-(aq)]}$

The total volume of the solution = 45.0 cm^3.

5.0 cm^3 of the 0.1 mol dm^{-3} ethanoic acid remains not neutralised this is now diluted to a total volume of 45 cm^3 solution.

So the concentration of ethanoic acid molecules $= \dfrac{5.0 \text{ cm}^3}{45.0 \text{ cm}^3} \times 0.10 \text{ mol dm}^{-3}$

Also the concentration of ethanoate ions $= \dfrac{20.0 \text{ cm}^3}{45.0 \text{ cm}^3} \times 0.10 \text{ mol dm}^{-3}$

So the ratio: $\dfrac{[CH_3CH_2CO_2H(aq)]}{[CH_3CH_2CO_2^-(aq)]} = \dfrac{5.0}{20.0}$

Substituting: $[H_3O^+(aq)] = K_a \times \dfrac{5.0}{20.0}$

Taking logarithms, and multiplying through by minus 1 gives:

$$pH = pK_a - \log\left(\dfrac{5.0}{20.0}\right) = 5.35$$

Note that half way to the equivalence point in Figure 2.10.3, the added alkali converts half of the weak acid to its salt. In this example, at this point: $[CH_3CH_2CO_2H(aq)] = [CH_3CH_2CO_2^-(aq)]$.

So: $[H_3O^+(aq)] = K_a \dfrac{[CH_3CH_2CO_2H(aq)]}{[CH_3CH_2CO_2^-(aq)]} = K_a$

Hence $pH = pK_a$ half way to the equivalence point.

At the equivalence point itself the solution contains sodium ethanoate. As Figure 2.10.3 shows, the solution at this point is not neutral. A solution of a salt of a weak acid and a strong base is alkaline. Sodium ions have no effect on the pH of a solution, but ethanoate ions are basic. The ethanoate ion is the conjugate base of a weak acid.

Beyond the equivalence point the curve is determined by the excess of strong base and so the shape is the same as for Figure 2.10.2.

Titration of a strong acid with a weak base

At first the flask contains a strong acid and the titration curve follows the same line as in Figure 2.10.2. In a titration of hydrochloric acid with ammonia solution, the salt formed at the equivalence point is ammonium chloride. Since ammonia is a weak base the ammonium ion is an acid. So a solution of ammonium chloride is acidic and the pH is below 7 at the equivalence point.

After the equivalence point the curve rises less far than in Figure 2.10.2 because the excess alkali is a weak base and not fully ionised.

Test yourself

6 Calculate the pH of a 0.1 mol dm^{-3} solution of $CH_3CO_2H(aq)$.

7 Calculate the pH of a mixture formed during a titration after adding 10.0 cm^3 of 0.10 mol dm^{-3} NaOH(aq) to 25.0 cm^3 of a 0.10 mol dm^{-3} of $CH_3CO_2H(aq)$.

8 Write an ionic equation to explain why a solution of sodium ethanoate is alkaline.

9 Why is the equivalence point reached at 25 cm^3 in both of the titrations illustrated in Figures 2.10.2 and 2.10.3.

◄ *Figure 2.10.4*
The pH change on adding 0.1 mol dm^{-3} of NH_3(aq) from a burette to 25 cm^3 of a 0.1 mol dm^{-3} of HCl(aq). Note the sharp change of pH around the equivalence point

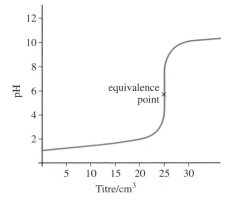

Titre/cm^3

Physical Chemistry

Section two

10 Write an ionic equation to explain why a solution of ammonium chloride is acidic.

Titration of a weak acid with a weak base

In practice it is not usual to titrate a weak acid with a weak base. As shown in Figure 2.10.5 the change of pH around the equivalence point is gradual and not very marked. This means that it is hard to fix the end-point precisely. If the dissociation constants for the weak acid and for the weak base are approximately equal (as is the case for ethanoic acid and ammonia) then the salt formed at the equivalence point is neutral and pH = 7 at this point.

Figure 2.10.5 ▶
The pH change on adding 0.1 mol dm^{-3} of NH_3(aq) from a burette to 25 cm^3 of a 0.1 mol dm^{-3} of CH_3CO_2H(aq). Before the end-point the curve is essentially the same as in Figure 2.10.3 while after the end point it is as in Figure 2.10.4. Note the resulting small change of pH around the equivalence point

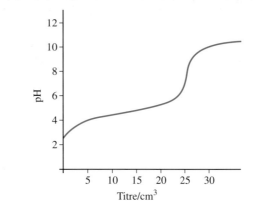

11 Write a balanced equation for the neutralisation of ethanoic acid by ammonia solution.

Working with logarithms

The worked example on pages 42–43 shows the advantage of working with a logarithmic form of the equilibrium law when calculating the pH of a mixture of a weak acid and one of its salts.

In general, for a weak acid HA: $HA(aq) + H_2O(l) \rightleftharpoons H_3O^+(aq) + A^-(aq)$

$$K_a = \frac{[H_3O^+(aq)][A^-(aq)]}{[HA(aq)]}$$

This rearranges to give: $[H_3O^+(aq)] = \dfrac{K_a[HA(aq)]}{[A^-(aq)]}$

Taking logs and substituting pH for $-\log[H_3O^+(aq)]$ and pK_a for $-\log K_a$, gives:

$$pH = pK_a + \log\left(\frac{[A^-(aq)]}{[HA(aq)]}\right) \text{ because } -\log\left(\frac{[HA(aq)]}{[A^-(aq)]}\right) = +\log\left(\frac{[A^-(aq)]}{[HA(aq)]}\right)$$

In a mixture of a weak acid and its salt, the weak acid is only slightly ionised while the salt is fully ionised, so it is often accurate enough to assume that all the anions come from the salt present and all the unionised molecules from the acid.

Hence: $pH = pK_a + \log\left(\dfrac{[\text{salt}]}{[\text{acid}]}\right)$

This form of the equilibrium law helps to make sense of the properties of acid–base indicators (section 2.11) and to account for the behaviour of buffer solutions (section 2.12).

2.11 Indicators

Acid–base indicators change colour when the pH changes. They signal the end-point of a titration. No one indicator is right for all titrations and the equilibrium law can help chemists to choose the indicator which will give accurate results.

CD-ROM

The indicator chosen for a titration must change colour completely in the pH range of the near vertical part of the pH curve (see Figures 2.10.2–2.10.4). This is essential if the visible end-point is to correspond to the equivalence point when exactly equal amounts of acid and base are mixed.

Figure 2.11.4 gives some data for four common indicators. Note that each indicator changes colour over a range of pH values which differs from one indicator to the next.

Indicators are themselves weak acids or bases which change colour when they lose or gain hydrogen ions. When added to a solution, an indicator gains or loses protons depending on the pH of the solution. It is conventional to represent a weak acid indicator as HIn where In is a shorthand for the rest of the molecule. In water:

$$HIn(aq) \; + \; H_2O(l) \rightleftharpoons H_3O^+(aq) \; + \; In^-(aq)$$

un-ionised
indicator   indicator after
                                              losing a proton

colour 1   colour 2

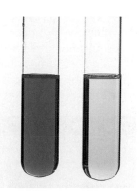

Figure 2.11.1 ▲
The colours of methyl orange indicator at pH 3 and at pH 5

Note that an analyst only adds a drop or two of indicator during a titration. This means that there is so little indicator that it cannot affect the pH of the mixture. The pH is determined by the titration (as shown in Figures 2.10.2–2.10.4). The position of the equilibrium for the ionisation of the indicator shifts one way or the other as dictated by the pH of the solution in the titration flask.

indicator	pK_a	colour change HIn/In$^-$	pH range over which colour change occurs
methyl orange	3.6	red/yellow	3.2 – 4.2
methyl red	5.0	yellow/red	4.2 – 6.3
bromothymol blue	7.1	yellow/blue	6.0 – 7.6
phenolphthalein	9.4	colourless/red	8.2 – 10

Figure 2.11.4 ▲

Figure 2.11.2 ▲
The colours of phenolphthalein indicator at pH 7 and pH 11

The pH range over which an indicator changes colour is determined by its strength as the acid. Typically the range is given roughly by p$K_a \pm 1$.

The logarithmic form of the equilibrium law derived on page 44 shows why this is so. For an indicator it takes this form:

Hence: $pH = pK_a + \log\left(\dfrac{[In^-]}{[HIn]}\right)$

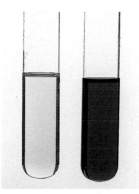

Figure 2.11.3 ▲
The colours of bromothymol blue indicator at pH 5 and pH 8

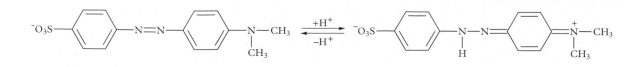

When pH = pK_a, [HIn] = [In⁻] and the two different colours of the indicator are present in equal amounts. The indicator is mid-way through its colour change.

Add a few drops of acid and the pH falls. The characteristic acid colour of the indicator is distinct when [HIn] = 10 × [In⁻].

At this point: pH = pK_a + log 0.1 = pK_a − 1, since log 0.1 = −1.

Add a few drops of alkali and the pH rises. The characteristic alkaline colour of the indicator is distinct when [In⁻] = 10 × [HIn].

At this point: pH = pK_a + log 10 = pK_a + 1, since log 10 = +1.

▲ Figure 2.11.5
The structures of methyl orange in acid and alkaline solutions. In acid solution the added hydrogen ion (proton) localises two electrons to form a covalent bond. In alkaline solution the removal of the hydrogen ion allows the two electrons to join the other delocalised electrons (see page 161). The change in the number of delocalised electrons causes a shift in the peak of the wavelengths of light absorbed, so the colour changes (see pages 230–231) and the molecule acts as an indicator

Test yourself

1 Explain, qualitatively with the help of Le Chatelier's principle, why methyl orange is red when the pH = 3 but is yellow when pH = 5.

2 **a)** Why is methyl orange an unsuitable indicator for the titration illustrated by Figure 2.10.3?

   b) Why is phenolphthalein an unsuitable indicator for the titration illustrated by Figure 2.10.4?

   c) Identify an indicator which can be used to detect the equivalence point of the titrations illustrated by Figures 2.10.2, 2.10.3 and 2.10.4.

   d) Explain why it is not possible to use an indicator to give a sharp and accurate end-point for the titration illustrated by Figure 2.11.5.

3 Suggest an explanation for the fact that the indicators shown in Figure 2.11.4 do not all change colour over a pH range of 2 units.

Titrations with two end-points

A titration of sodium carbonate with hydrochloric acid

Figure 2.11.6 shows the change in pH during a titration of sodium carbonate with dilute hydrochloric acid being added from the burette. Note that there are two end-points.

Test yourself

4 **a)** Write ionic equations for the two reactions taking place during the titration illustrated by Figure 2.11.6.

   b) Why are the equivalence points at 25.0 and 50.0 cm³?

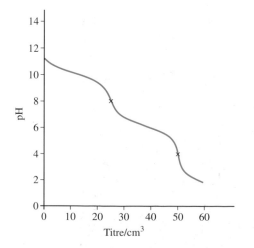

◄ Figure 2.11.6
The pH curve during a titration of 25.0 cm³ of 0.1 mol dm⁻³ Na₂CO₃(aq) with 0.1 mol dm⁻³ HCl.
pK_a = 6.4 for carbonic acid and pK_a = 10.3 for the hydrogencarbonate ion

The solution starts alkaline because the carbonate ion is a base. Phenolphthalein is added at the start and it turns red. On adding acid from the burette the phenolphthalein becomes colourless at the first end point. This corresponds to the conversion of all $CO_3^{2-}(aq)$ ions to $HCO_3^-(aq)$ ions.

Next, methyl orange indicator is added so that the solution becomes yellow. More acid is run in from the burette and the indicator turns red at the second end point. The second end-point corresponds to the complete neutralisation of the $HCO_3^-(aq)$ ions.

A titration of ethanedioic acid with sodium hydroxide

Ethanedioic acid is a diprotic acid which can form two salts with sodium hydroxide. Figure 2.11.7 shows the change in pH during a titration of ethanedioic acid solution with aqueous sodium hydroxide being added from the burette. Note that there are two end-points.

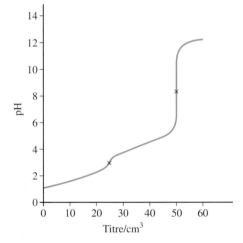

◀ **Figure 2.11.8**
Pipettes for volumetric analysis

Test yourself

5 **a)** Write ionic equations for the two reactions taking place during the titration illustrated by Figure 2.11.7.

   b) Why are the equivalence points at 25.0 and 50.0 cm^3?

6 What, in theory, is the pH of the mixture in the titration flask when the volume of acid added in Figure 2.11.7 is:

   a) 12.5 cm^3

   b) 37.5 cm^3?

◀ **Figure 2.11.7**
The pH curve during a titration of 25.0 cm^3 0.1 mol dm^{-3} $HO_2C-CO_2H(aq)$ with 0.1 mol dm^{-3} NaOH. pK_a = 1.2 for $HO_2C-CO_2H(aq)$ and pK_a = 4.2 for the $HO_2C-CO_2Na(aq)$

Physical Chemistry

Section two

2.12 Buffer solutions

Buffer solutions are mixtures of molecules and ions in solution which help to keep the pH more or less constant. Buffer solutions help to stabilise the pH of blood, medicines, shampoos, swimming pools and of many other solutions in living things, domestic products and in the environment.

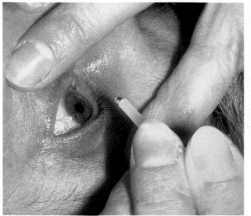

Figure 2.12.1 ▲
Eye drops contain a buffer solution to make sure that they do not irritate the sensitive surface of the eye

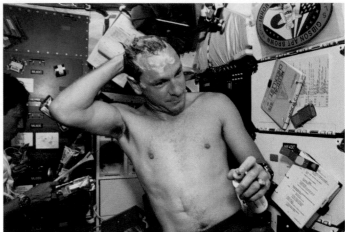

Figure 2.12.2 ▲
The pilot of the space shuttle Endeavour *washes his hair in zero gravity. Many shampoos contain a buffer solution. They are marketed as pH balanced shampoos*

A buffer solution cannot prevent pH changes but it evens out the large swings in pH which can happen without a buffer.

Buffers are important in living organisms. The pH of blood, for example, is closely controlled by buffers within the narrow range 7.38 to 7.42. Chemists use buffers when they want to investigate chemical reactions at a fixed pH.

Buffers are equilibrium systems which illustrate the practical importance of the equilibrium law. A typical buffer mixture consists of a solution of a weak acid and one of its salts. For example a mixture of ethanoic acid and sodium ethanoate. There must be plenty of both the acid and its salt.

Figure 2.12.3 ▶
The action of a buffer solution. Note that Le Chatelier's principle provides a qualitative explanation of the buffering action. Adding a little strong acid temporarily increases the concentration of $H^+(aq)$ so the equilibrium shifts to the left to counteract the change. Adding a little strong alkali temporarily decreases the concentration of $H^+(aq)$ so the equilibrium shifts to the right to counteract the change

$CH_3CO_2H(aq)$ acid molecules are a reservoir of H^+ ions $+ H_2O(l) \rightleftharpoons$ $CH_3CO_2^-(aq)$ base ions – with the capacity to accept H^+ ions $+ H_3O^+(aq)$ stays roughly constant so the pH hardly changes

plenty of weak acid to supply more H^+ ions if alkali is added

plenty of the ions from the salt able to combine with H^+ ions if acid is added

By choosing the right weak acid, it is possible to prepare buffers at any pH value throughout the pH scale. If the concentrations of the weak acid and its salt are the same, then the pH of the buffer is equal to pK_a for the acid.

The pH of a buffer mixture can be calculated with the logarithmic form of the equilibrium law derived on page 44.

$$pH = pK_a + \log\left(\frac{[\text{salt}]}{[\text{acid}]}\right)$$

Diluting a buffer solution with water does not change the ratio of the concentrations of the salt and acid, so the pH does not change (unless the dilution is so great that the assumptions made when deriving the equation no longer apply).

Worked example

What is the pH of a buffer solution containing 0.40 mol dm^{-3} methanoic acid and 1.00 mol dm^{-3} sodium methanoate?

Notes on the answer

Look up the value of pK_a in a book of data. The pK_a of methanoic acid is 3.8.

Answer

$$pH = 3.8 + \log\left(\frac{1.00}{0.40}\right)$$

$$pH = 4.20$$

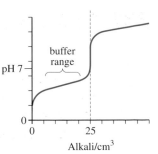

Figure 2.12.4 ▲
pH changes during the titration of ethanoic acid with sodium hydroxide. In the buffering range the pH changes little on adding substantial volumes of strong alkali. Over this range the flask contains significant amounts of both the acid and the salt formed when the acid is neutralised

Note

Make sure that you can derive the logarithmic form of the equilibrium law as shown on page 44. Also check the assumptions made deriving this form of the law so that you know when it applies.

Test yourself

1 Calculate the pH of these buffer mixtures given the pK_a values for these weak acids

   a) A solution containing equal amounts in moles of $H_2PO_4^-$(aq) and HPO_4^{2-}(aq). pK_a for the dihydrogenphosphate(v) ion is 7.2.

   b) A solution containing 12.2 g benzenecarboxylic acid ($C_6H_5CO_2H$) and 7.2 g of sodium benzenecarboxylate in 250 cm^3 solution. pK_a for benzenecarboxylic acid is 4.2.

   c) A solution containing 12.2 g benzenecarboxylic acid ($C_6H_5CO_2H$) and 7.2 g of sodium benzenecarboxylate in 1000 cm^3 solution.

2 What must be the ratio of the concentrations of the ethanoate ions and ethanoic acid molecules in a buffer solution with pH 5.4 if pK_a = 4.8 for ethanoic acid?

Note

The theory of buffer solutions and the theory of acid–base indicators is essentially the same. The difference is that a large amount of buffer mixture is added to dictate the pH of a solution whereas an analyst only adds a few drops of indicator during a titration – too little to affect the pH.

Physical Chemistry

Section two

2.13 Neutralisation reactions

Chemists use the term neutralisation to describe any reaction in which an acid reacts with a base to form a salt, even when the pH does not equal 7 on mixing equivalent amounts of the acid and the alkali.

Test yourself

1 Predict whether the salt formed on mixing equivalent amounts of these acids and alkalis gives a solution with pH = 7, pH above 7 or pH below 7:

   a) nitric acid and potassium hydroxide

   b) chloric(I) acid and sodium hydroxide

   c) hydrobromic acid and ammonia

   d) propanoic acid and sodium hydroxide.

Mixing equal amounts (in moles) of hydrochloric acid with sodium hydroxide produces a neutral solution of sodium chloride. Strong acids, such as hydrochloric acid, and strong bases, such as sodium hydroxide, are fully ionised in solution. So is the salt formed from the reaction of hydrochloric acid and sodium hydroxide, sodium chloride. Writing ionic equations for these examples shows that neutralisation is essentially a reaction between aqueous hydrogen ions and hydroxide ions. This is supported by the values for enthalpies of neutralisation.

$$H_3O^+(aq) + OH^-(aq) \longrightarrow 2H_2O(l)$$

The surprise is that 'neutralisation reactions' do not always produce neutral solutions. 'Neutralising' a weak acid, such as ethanoic acid, with an equal amount, in moles, of a strong base, such as sodium hydroxide, produces a solution of sodium ethanoate which is alkaline.

'Neutralising' a weak base, such as ammonia, with an equal amount of the strong acid hydrochloric acid produces a solution of ammonium chloride which is acidic.

Where a salt has either a 'parent acid' or a 'parent base' which is weak it dissolves to give a solution which is not neutral. The 'strong parent' in the partnership 'wins':

■ weak acid/strong base – the salt is alkaline in solution
■ strong acid/weak base – the salt is acidic in solution.

Enthalpy change of neutralisation

Strong acids and bases

The enthalpy change of neutralisation is the enthalpy change for the reaction when an acid neutralises an alkali.

$$HCl(aq) + NaOH(aq) \longrightarrow NaCl(aq) + H_2O(l)$$

$$\Delta H^{\ominus}_{neutralisation} = -57.5 \text{ kJ mol}^{-1}$$

The standard enthalpy of neutralisation for dilute solutions of strong acid with strong base is always close to 57.5 kJ mol^{-1}. The reason is that these acids and alkalis are fully ionised. So in every instance the reaction is the same:

$$H^+(aq) + OH^-(aq) \longrightarrow H_2O(l) \quad \Delta H^{\ominus} = -57.5 \text{ kJ mol}^{-1}$$

Enthalpies of neutralisation can be measured approximately by mixing solutions of acids and alkalis in a calorimeter.

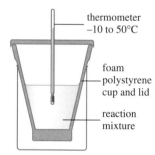

thermometer
–10 to 50°C

foam
polystyrene
cup and lid

reaction
mixture

Figure 2.13.1 ▲
Apparatus for measuring the enthalpy change for the neutralisation of an acid by a base

Worked example

50 cm^3 of 1.0 mol dm^{-3} dilute nitric acid was mixed with 50 cm^3 of 1.0 mol dm^{-3} dilute potassium hydroxide solution in an expanded polystyrene cup. The temperature rise was 6.7 °C. Calculate the enthalpy of neutralisation for the reaction.

Notes on the method

Note that the total volume of solution on mixing is 100 cm^3.

Assume that the density and specific heat capacity of the solutions is the same as pure water.

The density of water is 1 g cm^{-3} so the mass of 50 cm^3 is 50 g.

The specific heat capacity of water is 4.18 J g^{-1} K^{-1}.

A change of 6.7 °C is the same as the temperature change of 6.7 K on the Kelvin scale.

The energy from the exothermic reaction is trapped in the system by the expanded polystyrene so it heats up the mixture.

Answer

The energy change $= 4.18$ J g^{-1} K$^{-1} \times 100$ g $\times 6.7$ K $= 2800$ J

Amount of acid neutralised $= \dfrac{50}{1000}$ dm$^3 \times 1.0$ mol dm$^{-3} = 0.05$ mol

$$HNO_3(aq) + KOH(aq) \longrightarrow KNO_3(aq) + H_2O(l)$$

$$\Delta H_{neutralisation} = -\frac{2800 \text{ J}}{0.05 \text{ mol}} = -56\,000 \text{ J mol}^{-1} = -56 \text{ kJ mol}^{-1}$$

Weak acids and bases

The standard enthalpy changes for neutralisation reactions involving weak acids and weak bases are less negative than those for neutralisation reactions between strong acids and bases. The standard enthalpy changes for the neutralisation of ethanoic acid by sodium hydroxide, for example, is −56.1 kJ mol^{-1}.

   This is partly because the weak acids and bases are not fully ionised at the start so that the neutralisations cannot be described simply as reactions between aqueous hydrogen ions and aqueous hydroxide ions. Also because the solutions are not neutral at the equivalence point.

2 Account for the discrepancy between the value calculated in the worked example from experimental results and the expected value of about −57.5 kJ mol^{-1}.

3 Suggest an explanation for the difference in the values of $\Delta H^{\ominus}_{neutralisation}$ for HCl/NaOH and CH_3CO_2H/NaOH.

4 Here are three pairs of acids and bases which can react to form salts: HBr/NaOH, HCl/NH$_3$, CH_3CO_2H/NH$_3$.

   Here are three values for the standard enthalpy change of neutralisation: −50.4 kJ mol^{-1}, −53.4 kJ mol^{-1}, −57.6 kJ mol^{-1}.

   Write the equations for the three neutralisation reactions and match them with the corresponding value of $\Delta H^{\ominus}_{neutralisation}$.

Physical Chemistry

Section Two

2.14 Electrode potentials

Redox reactions, like all other reactions, tend towards a state of dynamic equilibrium. Redox reactions involve electron transfer. Chemists have found that it is possible to make electrochemical cells based on redox changes. Some cells of this kind are of great practical importance. Measurements of the voltages of cells provides a way of answering the questions 'How far?' and 'In which direction?' for redox reactions.

Test yourself

1 Write two ionic half-equations and the overall balanced equation for each of these redox reactions. In each example, state which molecule or ion is oxidised and which is reduced.

 a) magnesium metal with copper(II) sulfate solution

 b) aqueous chlorine with a solution of potassium bromide

 c) a solution of silver nitrate with copper metal.

Redox reactions

Redox reactions involve the transfer of electrons from the reducing agent to the oxidising agent. Writing half-equations for redox reactions help to show electron transfer. (For a further discussion of redox reactions see section 3.5 on pages 102–108.)

Chemists write ionic half-equations to describe either the gain or the loss of electrons during a redox process. Two half-equations combine to give an overall balanced equation for a redox reaction.

Iron(III) ions, for example, can oxidise iodide ions to iodine. This can be shown as two half-equations:

electron gain (reduction) $Fe^{3+}(aq) + e^- \longrightarrow Fe^{2+}(aq)$

electron loss (oxidation) $2I^-(aq) \longrightarrow I_2(s) + 2e^-$

The number of electrons gained must equal the number lost. So the first half-equation must be doubled to arrive at the overall balanced equation:

$$2Fe^{3+}(aq) + 2e^- \longrightarrow 2Fe^{2+}(aq)$$

$$2I^-(aq) \longrightarrow I_2(s) + 2e^-$$
$$\overline{2Fe^{3+}(aq) + 2I^-(aq) \longrightarrow 2Fe^{2+}(aq) + I_2(s)}$$

Electrochemical cells

Instead of just mixing two reagents it is possible to carry out a redox reaction in a cell so that electron transfer takes place along a wire connecting two electrodes. This harnesses the energy from the redox reaction to produce an electric potential difference (voltage).

One of the first practical cells was based on the reaction of zinc metal with aqueous copper(II) ions. Zinc is oxidised to zinc ions as copper(II) ions are reduced to copper metal.

$$Zn(s) \longrightarrow Zn^{2+}(aq) + 2e^-$$

$$Cu^{2+}(aq) + 2e^- \longrightarrow Cu(s)$$

In an electrochemical cell the two half-reactions happen in separate half-cells. The electrons flow from one cell to the other through a wire connecting the electrodes. A salt bridge connecting the two solutions completes the electric circuit.

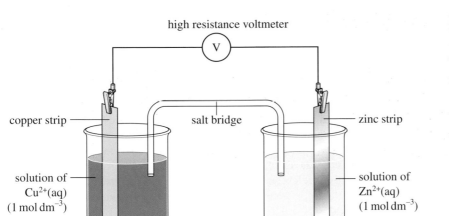

high resistance voltmeter

V

copper strip — salt bridge — zinc strip

solution of
$Cu^{2+}(aq)$
$(1 \, mol \, dm^{-3})$

solution of
$Zn^{2+}(aq)$
$(1 \, mol \, dm^{-3})$

◀ **Figure 2.14.1**
Diagram of an electrochemical cell based on the reaction of zinc metal with aqueous copper(II) ions. Note that electrons tend to flow from the negative zinc electrode to the positive copper electrode in the external circuit

The salt bridge is the ionic connection between the solutions of the two half-cells which make up an electrochemical cell. A salt bridge makes an electrical connection between the two halves of the cell by allowing ions to flow while preventing the two solutions from mixing.

At its simplest, a salt bridge consists of a strip of filter paper soaked in potassium nitrate solution and folded over each of the two beakers. Potassium salts and nitrates are soluble so the salt bridge does not react to produce precipitates with the ions in the half-cells. In more permanent cells a salt bridge may consist of a porous solid such as sintered glass.

Chemists measure the tendency for the current to flow in the external circuit between the two electrodes of the cell. They use a high resistance voltmeter to measure the cell e.m.f. when no current is flowing.

In Figure 2.14.1 electrons tend to flow out of the zinc electrode (negative) round the circuit and into the copper electrode (positive). The e.m.f. of the cell is 1.10 V under standard conditions (298 K and concentrations of $1.0 \, mol \, dm^{-3}$).

When measuring the e.m.f.s of cells the standard conditions are:

- temperature 298 K
- solutions at a concentration of $1.0 \, mol \, dm^{-3}$ and
- any gases at a pressure of 10^5 Pa (100 kPa = 1 bar).

There is a convenient shorthand for describing cells. The standard e.m.f. of the cell, $E^{\ominus}_{cell}$ is written alongside the cell diagram. The agreed convention is that the sign of E is the charge on the right-hand electrode.

$$Zn(s) \,|\, Zn^{2+}(aq) \,\|\, Cu^{2+}(aq) \,|\, Cu(s) \qquad E^{\ominus}_{cell} = +1.10 \, V$$

If the cell e.m.f. is positive the reaction tends to go according to the cell diagram reading from left to right. As a current flows in a circuit connecting the two electrodes, zinc atoms turn into zinc ions and dissolve, while copper ions turn into copper atoms and deposit on the copper electrode.

$$Cu^{2+}(aq) + 2e^- \longrightarrow Cu(s)$$

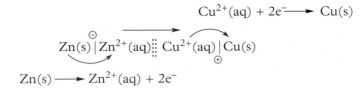

$$Zn(s) \,|\, Zn^{2+}(aq) \,\|\, Cu^{2+}(aq) \,|\, Cu(s)$$

$$Zn(s) \longrightarrow Zn^{2+}(aq) + 2e^-$$

◀ **Figure 2.14.2**
The direction of change in an electrochemical cell

Test yourself

2 Consider a cell based on this redox reaction which tends to go in the direction shown. The potential difference between the electrodes is 0.46 V.

$$Cu(s) + 2Ag^+(aq) \longrightarrow Cu^{2+}(aq) + 2Ag(s)$$

a) Write the half-equations for the electrode processes when the cell supplies a current.

b) Write out the conventional cell diagram for the cell including the value for $E^{\ominus}_{cell}$.

Measuring electrode potentials

The study of many cells has shown that a half-electrode, such as the $Cu^{2+}(aq)|Cu(s)$ electrode, makes the same contribution to the cell e.m.f. in *any* cell, so long as the measurements are taken under the same conditions. Unfortunately there is no way to measure the e.m.f. of an isolated electrode because the leads from the meter have to be in contact with the solution of ions as well as the metal electrode. Dipping a connecting electrode into the solution immediately creates a second electrode system.

Chemists have solved this problem by selecting a standard electrode system as a reference electrode against which they compare all other electrode systems. The chosen reference electrode is the standard hydrogen electrode. By convention the electrode potential of the standard hydrogen electrode is zero, $E^{\ominus}{}_{\frac{1}{2}H_2|H^+} = 0.00\,V$.

The standard electrode potential for any half-cell is measured relative to a standard hydrogen electrode under standard conditions. A standard hydrogen electrode sets up an equilibrium between hydrogen ions in solution (1 mol dm^{-3}) and hydrogen gas (at 1 bar pressure) all at 298 K on the surface of a platinum electrode coated with platinum black.

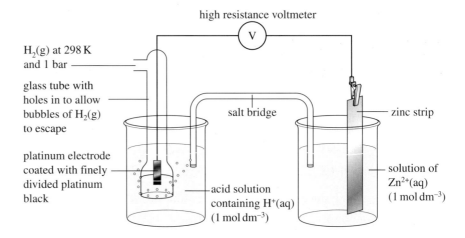

high resistance voltmeter

H$_2$(g) at 298 K and 1 bar

glass tube with holes in to allow bubbles of H$_2$(g) to escape

platinum electrode coated with finely divided platinum black

salt bridge

zinc strip

acid solution containing H$^+$(aq) (1 mol dm^{-3})

solution of Zn^{2+}(aq) (1 mol dm^{-3})

By convention, when a standard hydrogen electrode is the left-hand electrode in an electrochemical cell, the cell e.m.f. is the electrode potential of the right-hand electrode.

This is the conventional diagram to represent the cell which defines the standard electrode potential of the $Cu^{2+}(aq)|Cu(s)$ electrode.

$$Pt[H_2(g)]\,|\,2H^+(aq)\,\vdots\,Cu^{2+}(aq)\,|\,Cu(s) \qquad E^{\ominus} = +0.34\,V$$

The electrode and its electrode potential are often represented in this way:

$$Cu^{2+}(aq) + 2e^- \rightleftharpoons Cu(s) \qquad E^{\ominus} = +0.34\,V$$

$\qquad\qquad$ oxidised form $\qquad\qquad$ reduced form

A hydrogen electrode is difficult to use so it is much easier to use a secondary standard such as a silver/silver chloride electrode or a calomel electrode as a **reference electrode**. These electrodes are available commercially and are reliable to use. They have been calibrated against a standard hydrogen electrode. Calomel is an old-fashioned name for mercury(I) chloride. This is the cell reaction and electrode potential relative to a hydrogen electrode for a calomel electrode:

$$Hg_2Cl_2(s) + 2e^- \rightleftharpoons 2Hg(l) + 2Cl^-(aq) \qquad E^{\ominus} = +0.27\,V$$

It is possible to measure electrode potentials for electrode systems in which both the oxidised and reduced forms are ions in solution.

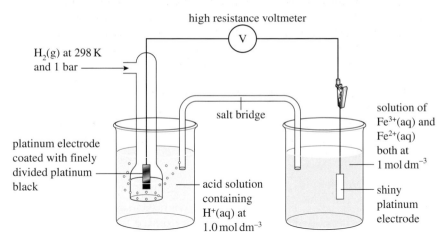

◄ **Figure 2.14.4**
Diagram of a cell for measuring the electrode potential for the redox reaction:
$Fe^{3+}(aq) + e^- \rightleftharpoons Fe^{2+}(aq)$.
Note that in this cell the right hand electrode consists of shiny platinum

The conventional cell diagram for the cell in Figure 2.14.4 is:

$$\text{Pt}[\text{H}_2(\text{g})] \,|\, 2\text{H}^+(\text{aq}) \,\vdots\, \text{Fe}^{3+}(\text{aq}), \text{Fe}^{2+}(\text{aq}) \,|\, \text{Pt} \qquad E^{\ominus} = +0.77\,\text{V}$$

Test yourself

3 What are the half-equations and standard electrode potentials of the right-hand electrode of each of these cells?

   a) $\text{Pt}[\text{H}_2(\text{g})] \,|\, 2\text{H}^+(\text{aq}) \,\vdots\, \text{Sn}^{2+}(\text{aq}) \,|\, \text{Sn}(\text{s})$ $E^{\ominus} = -0.14\,\text{V}$

   b) $\text{Pt}[\text{H}_2(\text{g})] \,|\, 2\text{H}^+(\text{aq}) \,\vdots\, \text{Br}_2(\text{aq}), 2\text{Br}^-(\text{aq}) \,|\, \text{Pt}(\text{s})$ $E^{\ominus} = +1.07\,\text{V}$

   c) $\text{Pt} \,|\, [2\text{Hg}(\text{l}) + 2\text{Cl}^-(\text{aq})], \text{Hg}_2\text{Cl}_2(\text{s}) \,\vdots\, \text{Cr}^{3+}(\text{aq}) \,|\, \text{Cr}(\text{s})$ $E^{\ominus} = -1.01\,\text{V}$

4 The standard electrode potential for the $\text{Cu}^{2+}(\text{aq})|\text{Cu}(\text{s})$ is $+0.34$ V. For this cell:

   $$\text{Cu}(\text{s}) \,|\, \text{Cu}^{2+}(\text{aq}) \,\vdots\, \text{Pb}^{2+}(\text{aq}) \,|\, \text{Pb}(\text{s}) \qquad E^{\ominus}_{\text{cell}} = -0.47\,\text{V}$$

   What is the standard electrode potential for the $\text{Pb}^{2+}(\text{aq}) \,|\, \text{Pb}(\text{s})$ electrode?

Chemists use standard electrode potentials:

■ to calculate the e.m.f.s of electrochemical cells, and
■ as the basis of the electrochemical series for predicting the direction of redox reactions.

Cell e.m.f.s and the direction of change

The tables on pages 249–250 in the reference section list redox half-reactions in order of their standard electrode potentials.

Some chemists list the half-cell with the most negative electrode potential at the top of the list (as in table on page 249). The most negative electrode has the greatest tendency to give up electrons, so it is powerfully reducing. Others choose to list the electrode systems in the opposite order with the most powerful oxidising agent at the top of the list and the most powerful reducing at the bottom of the list (as in table on page 250).

With a table of electrode potentials it is possible to work out $E^{\ominus}_{\text{cell}}$ values for any redox reaction. This provides a way of predicting the expected direction of chemical change. As explained on page 53, a reaction goes in the direction shown by the conventional cell diagram if the sign of the right-hand electrode is positive.

$$E^{\ominus}_{\text{cell}} = E^{\ominus}_{\text{(right-hand electrode)}} - E^{\ominus}_{\text{(left-hand electrode)}}$$

Worked example

Draw the conventional cell diagram for a cell based on these two half-equations. Work out the e.m.f of the cell and write the overall reaction for the reaction which tends to go (the spontaneous reaction).

$$Fe^{3+}(aq) + e^- \rightleftharpoons Fe^{2+}(aq) \qquad E^{\ominus} = +0.77\,V$$

$$Cu^{2+}(aq) + 2e^- \rightleftharpoons Cu(s) \qquad E^{\ominus} = +0.34\,V$$

Notes on the method

Write the cell diagram with the more positive electrode on the right.

Use this equation:

$$E^{\ominus}_{cell} = E^{\ominus}_{(right\text{-}hand\ electrode)} - E^{\ominus}_{(left\text{-}hand\ electrode)}$$

Recall that any cell takes the form:

reduced form – oxidised form – salt bridge – oxidised form – reduced form.

Answer

The $Fe^{3+}(aq)$, $Fe^{2+}(aq)$ electrode is the more positive. Both the oxidised and reduced forms are in solution so a shiny platinum electrode is needed.

$$Fe^{3+}(aq), Fe^{2+}(aq)\,|\,Pt(s)$$

The left-hand electrode is a metal/metal ion system, so the solid metal can also be the conducting electrode.

$$Cu(s)\,|\,Cu^{2+}(aq)\,\vdots\,Fe^{3+}(aq), Fe^{2+}\,|\,Pt(s)$$

$$E^{\ominus}_{cell} = (+\,0.77\,V) - (+\,0.34\,V) = +\,0.43\,V$$

The right-hand electrode is positive so it tends to take in electrons from the negative left-hand electrode.

$$Cu(s) + 2Fe^{3+}(aq) \longrightarrow Cu^{2+}(aq) + 2Fe^{2+}(aq)^+$$
$$\underset{2e-}{\curvearrowright}$$

Test yourself   D

5   Draw the conventional cell diagram for a cell based on each of these pairs of half-equations. For each example look up the standard electrode potentials, work out the e.m.f. of the cell and write the overall reaction for the reaction which tends to happen (the spontaneous reaction).

a) $V^{3+}(aq) + e^- \rightleftharpoons V^{2+}(aq)$ $Zn^{2+}(aq) + 2e^- \rightleftharpoons Zn(s)$

b) $Br_2(aq) + 2e^- \rightleftharpoons 2Br^-(aq)$ $I_2(aq) + 2e^- \rightleftharpoons 2I^-(aq)$

c) $Cl_2(aq) + 2e^- \rightleftharpoons 2Cl^-(aq)$ $PbO_2(s) + 4H^+(aq) + 2e^- \rightleftharpoons Pb^{2+}(aq) + 2H_2O(l)$

6   With the help of the data on pages 249–250, arrange the following sets of metals in order of increasing strength as reducing agents:

a) Ca, K, Li, Mg, Na

b) Cu, Fe, Pb, Sn, Zn.

7   With the help of the data on pages 249–250, arrange the following sets of molecules or ions in order of increasing strength as oxidising agents in acid solution:

a) $Cr_2O_7^{2-}$, Fe^{3+}, H_2O_2, MnO_4^-

b) Br_2, Cl_2, ClO^-, H_2O_2, O_2.

The electrochemical series

A series of electrode systems set out in order of their electrode potentials (see pages 249 and 250) is a useful guide to the behaviour of oxidising and reducing agents. It is an electrochemical series.

The metal ion | metal electrodes with highly negative electrode potentials involve half-reactions for group 1 metal ions and metals. Lithium is the most reactive of the metals when it reacts as a reducing agent forming metal ions.

The metal | metal ion electrodes with positive electrode potentials involve half reactions for d-block metal ions and metals such as copper and silver. These metals are relatively unreactive as reducing agents and they do not react with dilute hydrochloric acids to form hydrogen gas.

This broadly corresponds to the reactivity series for metals and the order of reaction shown by metal/metal ion displacement reactions.

The electrode potentials for the half-reactions involving halogen molecules and halide ions are positive. The $F_2(aq) | 2F^-(aq)$ system is the most positive showing that fluorine is the most reactive of these halogens when it acts as an oxidising agent forming fluoride ions. Iodine is the least reactive. This corresponds to the order of reactivity of the halogens as shown by their displacement reactions.

An electrochemical series based on electrode potentials can be used to predict the direction of change in redox reactions without writing out the cell diagram.

Worked example

Show that manganate(VII) ions can oxidise iron(II) to iron(III) in acid conditions?

Notes on the method

First identify the relevant half-equations in the table of standard electrode potentials on page 249. Write them down one above the other. The more positive half-reaction tends to go from left to right taking in electrons, while the more negative half-reaction goes from right to left giving out electrons.

Answer

Stronger reducing agent

$$Fe^{3+}(aq) + e^- \rightleftharpoons Fe^{2+}(aq) \qquad E^{\ominus} = +0.77 \text{ V} \quad \text{less positive (more negative) electrode}$$

$$MnO_4^-(aq) + 8H^+(aq) + 5e^- \rightleftharpoons Mn^{2+}(aq) + 4H_2O(l) \quad E^{\ominus} = +1.51 \text{ V} \quad \text{more positive electrode}$$

Stronger oxidising agent

The standard electrode potentials show that the reaction does tend to go. It is spontaneous. The balanced equation for the reaction which tends to go is:

$$MnO_4^-(aq) + 8H^+(aq) + 5Fe^{2+}(aq) \longrightarrow Mn^{2+}(aq) + 4H_2O(l) + 5Fe^{2+}(aq)$$

Note

In whatever order electrode potentials are tabulated, it is always true that:

- the half-cell with the most positive electrode potential has the greatest tendency to gain electrons so it is the most powerfully oxidising.

- the half-cell with the most negative electrode potential has the greatest tendency to give up electrons so it is the most powerfully reducing.

Note

It is possible to calculate electrode potentials from other experimental data for electrode systems which cannot be set up in cells as in Figure 2.14.3.

◀ *Figure 2.14.5*
Using half-reactions and $E^{\ominus}$ *values to predict the direction of change under acid conditions*

Electrode potentials can predict whether or not a disproportionation reaction is likely to occur (see page 134). Copper(I) ions disproportionate in aqueous solution but iron(II) ions do not tend to disproportionate.

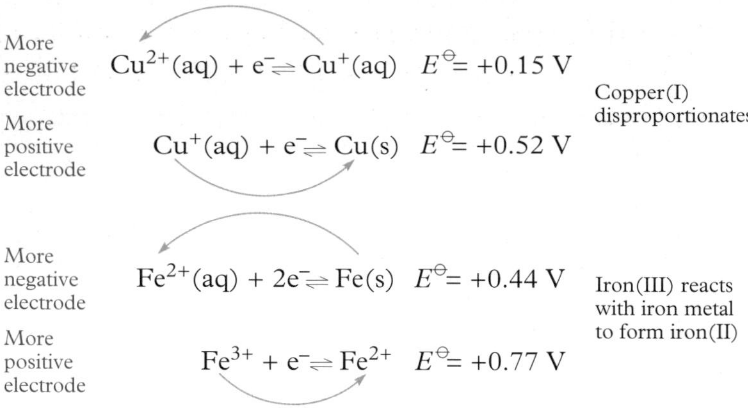
More negative electrode
$Cu^{2+}(aq) + e^- \rightleftharpoons Cu^+(aq)$ $E^\ominus = +0.15$ V
More positive electrode
$Cu^+(aq) + e^- \rightleftharpoons Cu(s)$ $E^\ominus = +0.52$ V
Copper(I) disproportionates
More negative electrode
$Fe^{2+}(aq) + 2e^- \rightleftharpoons Fe(s)$ $E^\ominus = +0.44$ V
More positive electrode
$Fe^{3+} + e^- \rightleftharpoons Fe^{2+}$ $E^\ominus = +0.77$ V
Iron(III) reacts with iron metal to form iron(II)

Figure 2.14.6 ▶ *Using standard electrode potential values to decide whether or not ions are likely to disproportionate*

Standard electrode potentials predict the direction of change but say nothing about the *rate* of change. A reaction which is feasible may not in fact happen because it is so slow. Also the predictions only apply under standard conditions. Changing the concentrations can alter the direction of change.

Test yourself

D

8 Why is not possible to measure the electrode potential for the $Na^+(aq)|Na(s)$ using the method illustrated in Figure 2.14.3?

9 Predict which of these pairs of reagents will react. Write the overall balanced equation for the reactions which you predict will tend to go. Point out any examples in which the reaction is likely to be very slow even though it tends to happen.

a) $Zn(s) + Ag^+(aq)$

b) $Fe(s) + Ca^{2+}(aq)$

c) $Cr(s) + H^+(aq)$

d) $Ca(s) + H^+(aq)$

e) $Ag(s) + H^+(aq)$

f) $I_2(aq) + Cl^-(aq)$

g) $Cl_2(aq) + I^-(aq)$

10 Does hydrogen peroxide tend to disproportionate under standard conditions in the presence of acid according to the electrode potentials for these two half-reactions?

$$2H^+(aq) + O_2(g) + 2e^- \rightleftharpoons H_2O_2(aq)$$

$$H_2O_2(aq) + 2H^+(aq) + 2e^- \rightleftharpoons 2H_2O(aq)$$

11 Do $Au^+(aq)$ ions tend to disproportionate under standard conditions?

2.15 Cells and batteries

Some electrochemical cells are designed for practical use. Examples include the cells in torch batteries and the rechargeable cells used in motor vehicles, laptop computers and mobile phones.

Ordinary cells such as an alkaline cell can only be used once. Once all the chemicals in the cell have reacted the cell e.m.f. falls to zero and the cell is 'flat' and can only be thrown away. Other cells can be recharged and so are used to store electricity. They are storage cells.

Lead–acid cells

The commonest storage cells are lead–acid cells used in car batteries and back-up lighting systems. These electrochemical cell are rechargeable because the chemical changes at the electrodes are reversible. The working of lead–acid cells involves lead in three oxidation states as Pb(o), Pb(II) and Pb(IV).

Figure 2.15.1 ▲
An assortment of electrochemical cells. These cells cannot be recharged. Once used they cannot be used again

◄ **Figure 2.15.2**
An electric car in Hawaii, USA with solar panels on its roof which help to recharge its battery

In a fully charged lead–acid cell, the negative electrode consists of lead metal. The positive electrode consists of lead coated with lead(IV) oxide. The electrodes dip into a solution of sulfuric acid. These are the electrode processes as current flows:
at the negative electrode:

$$Pb(s) + SO_4^{2-}(aq) \longrightarrow PbSO_4(s) + 2e^- \quad E^{\ominus} = -0.36 \text{ V}$$

at the positive electrode:

$$PbO_2(s) + SO_4^{2-}(aq) + 4H^+(aq) + 2e^- \longrightarrow PbSO_4(s) + 2H_2O(l) \quad E^{\ominus} = +1.69 \text{ V}$$

The equations help to explain why the cell is rechargeable. The changes which happen as a current flows from the cell both produce an insoluble solid, lead sulfate. This traps the lead(II) ions beside the electrode instead of dissolving in the electrolyte. So when the current is reversed to charge up the cell, the two reactions can be reversed turning lead(II) back to lead metal at one electrode and back to PbO_2 at the other.

Physical Chemistry

Section Two

59

Figure 2.15.3 ▲
*A lithium battery in a camera.
A rechargeable lithium battery
has an e.m.f. of about 4 V*

Test yourself

1 **a)** In a lead–acid storage cell, identify the chemical or chemicals used to make:

(i) the negative electrode

(ii) the positive electrode

(iii) the electrolyte.

b) Write the balanced equation for the overall reaction in the cell when it is supplying current.

c) What is the e.m.f. of the cell under standard conditions?

d) How are lead–acid cells combined to make a 12 V car battery?

e) Write the half-equations for the processes at the two electrodes when the cell is being recharged.

Nickel–cadmium cells

Nickel–cadmium cells are now widely used as rechargeable 'ni–cad' cells. As in lead–acid cells, when a current flows the products formed at the electrodes are insoluble solids which stay put instead of dissolving in the electrolyte. This means that the electrode processes can be reversed as the cell is recharged. The electrode processes as a ni–cad cell supplies a current are:

anode (oxidation): $Cd(s) + 2OH^-(aq) \longrightarrow Cd(OH)_2(s) + 2e^-$

cathode (reduction): $NiO(OH)(s) + H_2O(l) + e^- \longrightarrow Ni(OH)_2(s) + OH^-(aq)$.

Unlike lead–acid cells, 'ni–cad' cells must be regularly discharged fully and then recharged if they are to retain their full capacity.

Lithium cells

Modern mobile phones and laptop computers use lithium batteries. The advantages of electrodes based on lithium are that the metal has a low density, therefore cells based on lithium electrodes can be relatively light. Also lithium is very reactive, this means that the electrode potential of a lithium electrode is higher so that each cell has a large e.m.f.

The difficulty to overcome is that lithium is so reactive that it readily combines with oxygen in the air. A layer of non-conducting oxide quickly forms on the surface of the metal. The metal also reacts rapidly with water. Research workers have solved the technical problems of working with lithium by developing electrodes with lithium ions inserted into the crystal lattice of other materials. Also the electrode is a polymeric material rather than an aqueous solution.

Figure 2.15.4 ▶
*Schematic diagram of a lithium
battery showing the battery
discharging. The electrode
processes are reversible so
that the battery can be
recharged*

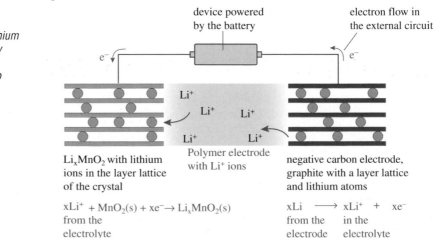

device powered
by the battery

electron flow in
the external circuit

Li_xMnO_2 with lithium
ions in the layer lattice
of the crystal

Polymer electrode
with Li^+ ions

negative carbon electrode,
graphite with a layer lattice
and lithium atoms

$xLi^+ + MnO_2(s) + xe^- \rightarrow Li_xMnO_2(s)$
from the
electrolyte

$xLi \longrightarrow xLi^+ + xe^-$
from the   in the
electrode   electrolyte

2.16 Corrosion

Corrosion is a redox process in which oxygen, water and acids attack metals. The most familiar and economically serious example of corrosion is the rusting of iron. Rusting is an electrochemical reaction.

◀ *Figure 2.16.1*
The effect of corrosion on a steel chain. Corrosion causes large-scale damage. The costs of preventing corrosion and of repair or replacement are high

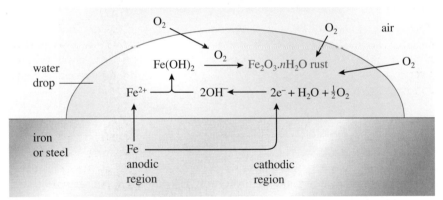

◀ *Figure 2.16.2*
An electrochemical cell on the surface of iron in contact with water and the air. Regions rich in oxygen are cathodic. In these regions oxygen is reduced to hydroxide ions. Other parts of the metal surface are anodic. In these regions iron is oxidised to iron(II) ions. Iron(II) ions and hydroxide ions diffuse together and form a precipitate of iron(II) hydroxide which is then further oxidised to rust, hydrated iron(III) oxide

Iron atoms turn to iron(II) ions at the anodic regions on the surface of the iron. At the cathodic regions oxygen is reduced to hydroxide ions in the presence of hydrogen ions.

Definitions

The **cathode** is always the electrode at which reduction takes place. Note that electrons flow into the cathode both when a current is drawn from a chemical cell and during electrolysis.

In an electrochemical cell the reduction process at the cathode takes in electrons making it the *positive* terminal of the cell.

The **anode** is always the electrode at which oxidation takes place. Note that electrons flow out of the anode to the external circuit both when a current is drawn from a chemical cell and during electrolysis.

In electrochemical cells the oxidation process at the anode forces electrons onto the electrode which becomes the *negative* terminal of the cell.

Preventing corrosion

Cathodic protection is an electrochemical method of preventing corrosion. It is used with pipelines, oil rigs and the hulls of ships. Steel corrodes where it is oxidised. This happens wherever the metal is an anode. In anodic regions the iron atoms give up electrons turning into ions. Attaching a more reactive metal, such as zinc or magnesium, to iron creates an electrochemical cell in which the iron is the cathode. The more reactive metal is the anode and it is this metal which corrodes.

Figure 2.16.3 ▶
Cathodic protection of a steel pipeline. The more reactive magnesium becomes the 'sacrificial anode' and corrodes as it protects the iron

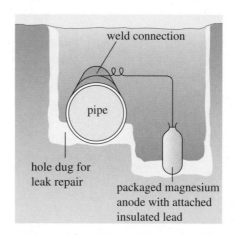

Figure 2.16.4 ▶
Cathode attached to oil rig to prevent corrosion

Galvanising is a form of cathodic protection. A layer of zinc on the surface of the iron or steel prevents corrosion. The zinc corrodes protecting the iron.

Chromium and aluminium are protected from corrosion by a thin layer of oxide on the metal surface. The main use of chromium is to make alloys with iron to produce stainless steels.

Test yourself   D

1 Show that the corrosion of iron is expected under standard conditions (as shown in Figure 2.16.2) with the help of the standard electrode potentials on pages 249–250.

2 Why does the presence of dissolved salt in the water speed up corrosion?

3 With the help of standard electrode potentials, explain why zinc and magnesium can give cathodic protection to steel but not tin.

2.17 Enthalpy changes, bonding and stability

Exploring enthalpy changes gives chemists yet another way of seeking answers to the questions: 'How far?' and 'In which direction?' Experience suggests that it is generally true that the changes or reactions which tend to happen are the ones which are exothermic. Some compounds are stable while others decompose to their elements or to other compounds. An analysis of bond breaking and bond forming in terms of enthalpy changes can help to explain why compounds differ in their stability.

Enthalpy changes and ionic bonding

Atoms into ions

Simplified accounts of ionic bonding show electron transfer from a metal atom to a non-metal atom forming charged ions which attract each other.

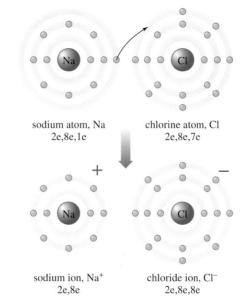

sodium atom, Na
2e,8e,1e

chlorine atom, Cl
2e,8e,7e

+

−

sodium ion, Na⁺
2e,8e

chloride ion, Cl⁻
2e,8e,8e

Figure 2.17.2 ▶
Electron transfer between atoms of sodium and chlorine to form ions

Figure 2.17.1 ▲
Sodium burning in chlorine. An exothermic reaction. The standard enthalpy of formation of sodium chloride is −411 kJ mol⁻¹

Figure 2.17.2 is simplified in many ways. It ignores the fact that when sodium reacts with chlorine the metal starts as a solid crystal lattice while the non-metal consists of molecules. The diagram also offers no explanation of where the energy comes from to remove electrons from the sodium atom. The first ionisation enthalpy of sodium is $+ 496$ kJ mol^{-1}.

One of the many research problems tackled by the German physicist Max Born (1882 – 1970) was detailed analysis of the enthalpy changes in the formation of ionic compounds. His work, with that of Fritz Haber, gave rise to the Born–Haber cycle which is a thermochemical cycle for investigating the stability and bonding in compounds of metals with non-metals.

Lattice enthalpies

The lattice enthalpy is usually defined as the standard enthalpy change when one mole of an ionic compound is formed from free gaseous ions. For sodium chloride the lattice enthalpy is defined by:

$$Na^+(g) + Cl^-(g) \longrightarrow NaCl(s) \qquad \Delta H^\ominus_{lattice} = -787 \text{ kJ mol}^{-1}$$

It is important to distinguish the lattice enthalpy from the standard enthalpy of formation of a compound which refers to the formation of a compound from its elements. For sodium chloride:

$$\Delta H^\ominus_f [NaCl(s)] = -411 \text{ kJ mol}^{-1}.$$

Definitions

The reverse of the lattice enthalpy is the **lattice dissociation enthalpy** when one mole of an ionic compound separates to give rise to free gaseous ions. This has the same magnitude but the opposite sign. So the lattice dissociation enthalpy for sodium chloride = $+787$ kJ mol^{-1}.

63

Figure 2.17.3 ▶
Lattice enthalpy is the energy which would be given out to the surroundings (red arrows) if 1 mol of a compound could form directly from free gaseous ions rushing together and assembling themselves into a crystal lattice (black arrows). A Born–Haber cycle applies Hess's law and makes it possible to calculate lattice enthalpies which cannot be measured experimentally

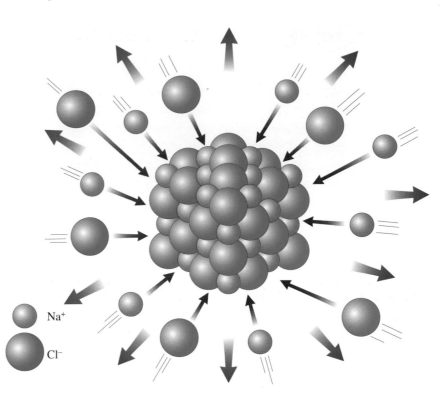

Na^+

Cl^-

Definitions

Electrostatic forces are the forces between charged particles. Opposite charges attract each other. Positive ions, for example, attract negative ions in ionic compounds. Like charges repel each other.

The size of the electrostatic force between two charges varies according to Coulomb's law. The bigger the charges the stronger the force – the force is proportional to the size of the charges, Q. The greater the distance, d, between the two charges the smaller the force – the force is inversely proportional to the distance squared.

electrostatic force $\propto \dfrac{Q_1 \times Q_2}{d^2}$

Born–Haber cycles

A Born–Haber cycle identifies all the enthalpy changes which contribute to the overall standard enthalpy of formation of a compound. Overall energy is required to create free gaseous ions from elements in their normal states. Energy is given out when these ions come together to form a crystal lattice.

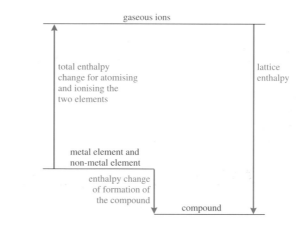

Figure 2.17.4 ▲
The main features of a Born–Haber cycle. Chemical reactions in laboratories do not normally involve gaseous atoms and ions. A Born–Haber cycle is a theoretical construct

A Born–Haber cycle is often set out as an enthalpy level diagram.

All the terms in a Born–Haber cycle can be measured experimentally except for the lattice enthalpy. Applying Hess's law makes it possible to calculate the one unknown term which is the lattice enthalpy.

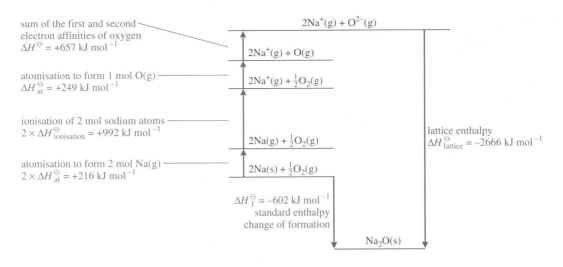

Figure 2.17.5 ▲
A Born–Haber cycle for sodium oxide

Definitions

Enthalpy change of atomisation is the enthalpy change to produce one mole of gaseous atoms from an element. The element at the start has to be in its standard state (normal stable state) at 298 K and 1 bar pressure.

The first **ionisation enthalpy** for an element is the energy needed to remove the first electron from one mole of gaseous atoms. Successive ionisation energies for the same element measure the energy needed to remove a second, third, fourth electrons and so on.

Electron affinity is the enthalpy change when gaseous atoms of an element gain electrons to become negative ions. This term is more precisely described as the **electron addition enthalpy**.

The first electron addition enthalpy of an element is the enthalpy change when one mole of gaseous atoms gains electrons to form one mole of gaseous ions with a single negative charge. These two equations define the first and second electron addition enthalpies for oxygen:

$$O(g) + e^- \longrightarrow O^-(g) \qquad \Delta H^\ominus = -141 \text{ kJ mol}^{-1}$$
$$O^-(g) + e^- \longrightarrow O^{2-}(g) \qquad \Delta H^\ominus = +798 \text{ kJ mol}^{-1}$$

The gain of the first electron is exothermic but adding the second electron to a negatively charged particle is an endothermic process. Overall, adding two electrons to a gaseous oxygen atom is endothermic.

The **lattice enthalpy** for an ionic solid is the standard enthalpy change when one mole of an ionic compound forms from free gaseous ions.

The **standard enthalpy of formation** of a compound refers to the formation of 1 mol of the compound from its elements in their standard states.

Test yourself D

1 Consider Figure 2.17.6, on page 66, which is a Born–Haber cycle for magnesium chloride.

   a) Identify the enthalpy changes $\Delta H^\ominus_1$, $\Delta H^\ominus_2$, $\Delta H^\ominus_3$, $\Delta H^\ominus_4$, $\Delta H^\ominus_5$, $\Delta H^\ominus_6$ and $\Delta H^\ominus_7$.

   b) Calculate the lattice enthalpy for magnesium chloride.

2 Show that a Born–Haber cycle is an application of Hess's law.

3 Draw up a Born–Haber cycle for sodium chloride using data from the reference section on pages 244 and 247. Which are the main enthalpy terms which determine the stability of sodium chloride?

4 Explain why the enthalpy change of atomisation of bromine is equal to half the sum of the enthalpy of vaporisation and the bond dissociation enthalpy.

Figure 2.17.6 ▶
A Born–Haber cycle for magnesium chloride

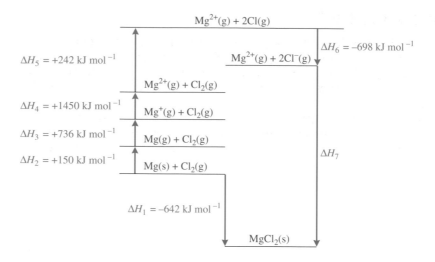

Stability of ionic compounds

A Born–Haber cycle can help to decide on the factors which determine the stability or instability of a compound. For a compound to be stable relative to its elements, the lattice enthalpy must be more negative than the total energy needed to produce gaseous ions from the elements.

Lattice energies give a measure of the strength of the ionic bonding in a crystal. Ionic bonding is the result of the electrostatic forces between the oppositely charged ions. So, for a given crystal structure, lattice enthalpies increase if:

- the charge on the ions increases
- the ions get smaller (and so are closer together).

Test yourself

5 a) Identify the enthalpy changes below which refer to the formation of the hypothetical compound MgCl. State which quantities can be found experimentally (directly or indirectly) and which have to be estimated using theory.

$$Mg^+(g) + Cl^-(g) \longrightarrow MgCl(s) \qquad \Delta H^\ominus = -753 \text{ kJ mol}^{-1}$$
$$\tfrac{1}{2}Cl_2(g) \longrightarrow Cl(g) \qquad \Delta H^\ominus = +122 \text{ kJ mol}^{-1}$$
$$Cl(g) + e^- \longrightarrow Cl^-(g) \qquad \Delta H^\ominus = -349 \text{ kJ mol}^{-1}$$
$$Mg(s) \longrightarrow Mg(g) \qquad \Delta H^\ominus = +156 \text{ kJ mol}^{-1}$$
$$Mg(g) \longrightarrow Mg^+(g) \qquad \Delta H^\ominus = +738 \text{ kJ mol}^{-1}$$

b) Use the data above to construct a Born–Haber cycle for the non-existent compound MgCl(s) and then calculate a value for the standard enthalpy of formation of MgCl(s).

c) Use your answer to **b)** and the standard enthalpy of formation of $MgCl_2(s)$ to calculate the standard enthalpy change for the reaction:

$$2MgCl(s) \longrightarrow MgCl_2(s) + Mg(s)$$
$$\Delta H_f^\ominus [MgCl_2(s)] = -642 \text{ kJ mol}^{-1}$$

d) Comment on the stability of MgCl(s) relative to $MgCl_2(s)$.

6 A Born–Haber cycle for the hypothetical compound $MgCl_3$, suggests that for this compound $\Delta H_f^\ominus [MgCl_3(s)] = +3950 \text{ kJ mol}^{-1}$ despite the fact that the estimated lattice enthalpy for the compound is $-5440 \text{ kJ mol}^{-1}$. Account for the large positive value of enthalpy change of this non-existent compound and comment on its stability relative to the free elements and to $MgCl_2(s)$.

7 How do your answers to questions 5 and 6 provide a theoretical justification of the octet rule as it applies to the formation of magnesium ions from magnesium atoms?

Ionic or covalent?

A Born–Haber cycle can also help to decide whether the bonding in a compound is truly ionic. The experimental lattice energy calculated from a Born–Haber cycle can be compared with the theoretical value calculated using the laws of electrostatics and assuming that the only bonding in the crystal is ionic.

With the help of the laws of electrostatics it is possible to calculate a theoretical value for the lattice enthalpy of an ionic crystal by summing the effects of all the attractions and repulsions between the ions in the crystal lattice.

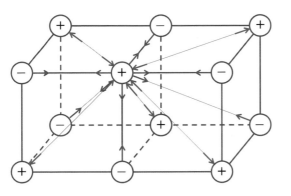

◄ *Figure 2.17.7*
Some of the many attractions and repulsions between ions which have to be taken into account when using the laws of electrostatics to calculate a theoretical value for the lattice enthalpy of an ionic crystal

Compound	Experimental lattice enthalpy from a Born–Haber cycle/ kJ mol⁻¹	Theoretical lattice enthalpy calculated assuming that the only bonding is ionic/ kJ mol⁻¹
NaCl	−788	−766
NaBr	−719	−731
NaI	−670	−686
KCl	−718	−692
KBr	−656	−667
KI	−615	−631
AgCl	−921	−769
ZnS	−3739	−3430

Figure 2.17.8 ▲

Pure ionic bonding arises solely from the electrostatic interactions between the ions in a crystal. The values in the table show that ionic bonding can largely account for the lattice enthalpies of sodium and potassium halides. There is good agreement between the theoretical and experimental values.

The lattice energy of silver chloride is greater than can be explained by ionic bonding. The bonding is intermediate between ionic and covalent. The bonding is stronger than predicted by a pure ionic model because of the contribution from covalent bonding. There is a similar discrepancy for zinc sulfide.

Test yourself

8 Here are four values for lattice enthalpy, in kJ mol⁻¹: 3791, 3299, 3054, 2725. The four ionic compounds with these four values are: BaO, MgO, BaS and MgS. Match the formulae with the values and justify your choice.

Physical Chemistry

Section two

Fajans' rules

Fajans' rules are a guide which helps to predict the extent to which the bonding in a compound will be ionic or intermediate between ionic and covalent. Kasimir Fajans (1887–1975) was a physical chemist who noted that ionic bonding is favoured if:

■ the charges on the ions are small (1+ or 2+, 1– or 2–)
■ the radius of the positive ions is large and the radius of the negative ions is small.

The way to apply Fajans' rules is to start by picturing ionic bonding between two atoms and then to consider the extent to which the positive metal ion will polarise the neighbouring negative ions giving rise to some degree of electron sharing (that is a degree of covalent bonding).

Figure 2.17.9 ▶
Ionic bonding with increasing degrees of electron sharing because the positive ion has polarised the neighbouring negative ion. Dotted circles show the unpolarised ions

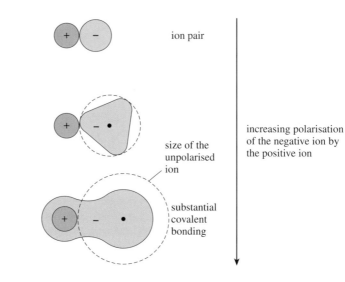

ion pair

increasing polarisation of the negative ion by the positive ion

size of the unpolarised ion

substantial covalent bonding

Test yourself

9 The lattice enthalpy of LiF is –1031 kJ mol^{-1} while the value for LiI is –759 kJ mol^{-1}.

a) Why is the lattice enthalpy of lithium iodide lower (less negative) than the lattice enthalpy of lithium fluoride?

b) For which compound would you expect the closer agreement between the Born–Haber cycle value and the theoretical value based on the ionic model?

The larger the negative ion and the larger its charge, the more polarisable it becomes. So iodide ions are more polarisable than fluoride ions. Fluorine, which has a small, singly charged fluoride ion, forms more ionic compounds than any other non-metal.

Enthalpy changes and covalent bonding

The stability of molecular compounds can be explored with the help of average bond enthalpies.

The overall enthalpy change for a reaction is the difference between the energy needed to break bonds in the reactants and the energy released as new bonds form in the products (see pages 64–65).

When investigating bond breaking, chemists distinguish between:

■ **bond dissociation enthalpies** – precise values for particular bonds
■ **average bond enthalpies** – mean values which are a useful approximate guide.

Note

Energy is needed to break covalent bonds. Bond breaking is endothermic so bond enthalpies are positive.

The bond dissociation enthalpy is the enthalpy change on breaking one mole of a particular covalent bond in a gaseous molecule. In molecules with two or more bonds between similar atoms, the energies needed to break successive bonds are not the same. In water, for example, the energy needed to break the first O—H bond in H—O—H(g) is 498 kJ mol^{-1} but the energy needed to break the second O—H bond in OH(g) is 428 kJ mol^{-1}.

Average bond enthalpies (or bond energies) are the average values of bond dissociation enthalpies used in approximate calculations to estimate enthalpy changes for reactions.

The mean values of bond enthalpies take into account the facts that:

■ the successive bond dissociation enthalpies are not the same in compounds such as water or methane
■ the bond dissociation enthalpy for a specific covalent bond varies slightly from one molecule to another.

> **Definition**
>
> **Dissociation** is a change in which a molecule splits into two or more smaller particles.

Worked example

Use average bond enthalpies to estimate the enthalpy of formation of hydrogen peroxide.

Notes on the method

Write out a diagram or an equation showing all the atoms and bonds in the molecules to make it easier to count the numbers of bond broken and formed.

Look up the mean bond energies in a book of data. The symbol E(O—H) stands for the average bond energy of a covalent bond between an oxygen atom and a hydrogen atom.

Answer

A diagram for the reaction:

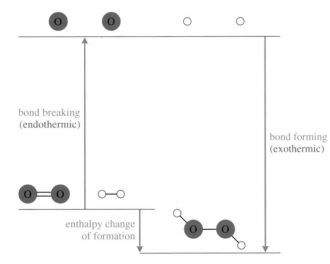

bond breaking (endothermic)

bond forming (exothermic)

enthalpy change of formation

◀ **Figure 2.17.10**
Diagram to show bond breaking and bond forming for the conversion of oxygen and hydrogen to hydrogen peroxide

The energy needed to break the bonds in the reactants

$$= E(O{=}O) \text{ kJ mol}^{-1} + E(H{-}H) \text{ kJ mol}^{-1}$$

$$= 498 \text{ kJ mol}^{-1} + 436 \text{ kJ mol}^{-1}$$

$$= 934 \text{ kJ mol}^{-1}$$

The energy given out as new bonds form to make the product

$$= E(O{-}O) \text{ kJ mol}^{-1} + 2E(O{-}H) \text{ kJ mol}^{-1}$$

$$= 144 \text{ kJ mol}^{-1} + 2 \times 464 \text{ kJ mol}^{-1}$$

$$= 1072 \text{ kJ mol}^{-1}$$

Less energy is needed to break bonds than is given out when bonds are formed so the reaction is exothermic and the enthalpy change is negative.

$$\Delta H^{\ominus} = +934 \text{ kJ mol}^{-1} -1072 \text{ kJ mol}^{-1} = -138 \text{ kJ mol}^{-1}$$

Physical Chemistry

Section two

Test yourself | D

10 Estimate the enthalpy change when 1 mol of ozone, O_3, forms from oxygen. Given that $E(O-O) = 302$ kJ mol^{-1} in ozone and that $E(O=O) = 498$ kJ mol^{-1} in oxygen.

11 Use the table of average bond enthalpies on page 247 to calculate the enthalpy change for the formation of these hydrides from their molecular elements. Comment on any patterns in the values.

   a) ammonia, $NH_3(g)$

   b) water, $H_2O(g)$

   c) hydrogen fluoride, $HF(g)$

12 Which is likely to give a more accurate answer – calculating the enthalpy change for a reaction

   a) from average bond energies, or

   b) from enthalpies of formation?

Test yourself

13 Comment on these observations:

   a) N_2O gas can relight a glowing splint. $\Delta H_f^{\ominus} [N_2O] = +82.0$ kJ mol^{-1}.

   b) Chlorine does not react with hot graphite. $\Delta H_f^{\ominus} [CCl_4] = -129.9$ kJ mol^{-1}.

   c) Plunging a red hot wire into a jar of hydrogen iodide gas produces purple fumes and a flame. $\Delta H_f^{\ominus} [HI] = +26.5$ kJ mol^{-1}.

Enthalpy changes and stability

Stability and the direction of change

Heat magnesium in a flame and, once the reaction starts, it burns brilliantly to form magnesium oxide, which is a very stable compound. Like many other exothermic reactions, once started, it tends to go. Many exothermic reactions just keep going once they have begun. In general, chemists expect that a reaction will go if it is exothermic.

For this reason, chemists expect compounds to be stable relative to their elements if their standard enthalpies of formation are negative. Therefore it is the reactions which give out energy to their surroundings that are the ones which happen. This ties in with the common experience that change happens in the direction in which energy is spread around and dissipated in the surroundings.

So the sign of $\Delta H^{\ominus}$ is a guide to chemical stability and to the likely direction of change, but it is not a totally reliable guide for three main reasons:

■ First, the direction of change may depend on the conditions of temperature and pressure. Ammonia is stable and does not tend to decompose into its elements at room temperature. $\Delta H_f^{\ominus} [NH_3] = -46.0$ kJ mol^{-1}. It becomes less and less stable as the temperature rises.

■ Second, a reaction may be highly exothermic and tend to go yet the rate of change may be so slow that the chemical appears inert. The decomposition of the solid sodium azide is highly exothermic and can be explosive but the solid is apparently stable at room temperature unless heated or given a sudden shock.

■ Third, there are examples of endothermic reactions which go spontaneously. So a reaction for which ΔH is positive can tend to happen. The decomposition of magnesium carbonate is endothermic but the reaction goes readily once the compound is hotter than 540 °C.

The stability of group 2 carbonates

All the carbonates of group 2 metals are stable at room temperature but they decompose on heating. When they decompose they split up into the metal oxide and carbon dioxide.

$$MgCO_3(s) \longrightarrow MgO(s) + CO_2(g)$$

Whenever chemists use the term 'stability' they are making comparisons. For the group 2 carbonates the question is which is more stable: the metal carbonate or a mixture of the metal oxide and carbon dioxide?

Most of the compounds of groups 1 and 2 elements are ionic. So chemists try to explain differences in the properties of the compounds of these elements in terms of two factors:

■ the charge on the metal ions, and
■ the size of the metal ions.

Group 2 carbonates are generally less stable than the corresponding group 1 compounds. This suggests that the larger the charge on the metal ion the less stable the compounds.

Down group 2 the carbonates become more stable. This suggests that the larger the metal ion the more stable the compound.

Figure 2.17.11 shows the temperatures at which the carbonates of group 2 metals begin to decompose. The figures confirm that magnesium carbonate is the least stable – it decomposes easily when heated with a Bunsen flame. Barium carbonate is the most stable.

Compound	T/°C	$\Delta H^{\ominus}_{\text{decomposition reaction}}$/kJ mol^{-1}
$MgCO_3$	540	+117
$CaCO_3$	900	+176
$SrCO_3$	1280	+238
$BaCO_3$	1360	+268

Figure 2.17.11 ▲
Temperature at which group 2 carbonates begin to decompose and the enthalpy change for the decomposition reaction

Chemists seek to explain the trend in thermal stability by analysing the energy changes.

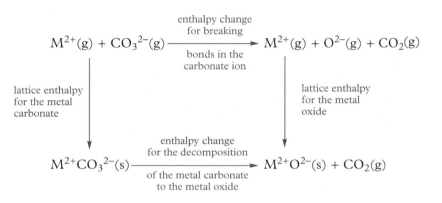

◀ **Figure 2.17.12**
An energy cycle for the decomposition of a group 2 carbonate

The key energy quantities in Figure 2.17.12 are:

■ the energy needed to break the carbonate ion into an oxide ion and carbon dioxide
■ the energy given out as the 2+ and 2− charges get closer together when the larger carbonate ions break up into smaller oxide ions and carbon dioxide. (This is the difference between the less negative lattice enthalpy of the carbonate and the more negative lattice enthalpy of the oxide.)

Figure 2.17.13 ▶
Decomposition of a group 2 carbonate. The smaller the metal ion, the less stable the compound. Energy is needed to break bonds in the carbonate ion. Energy is given out as the larger 2− carbonate ions change to smaller 2− oxide ions and the charges get closer together. This energy is proportionately greater when the metal ion is small

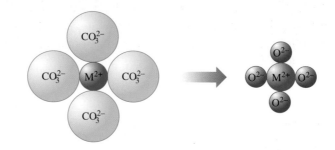

The key factor in explaining the trend in stability is the energy released as the ions get closer together. This energy is the difference between the lattice enthalpies of the carbonate (MCO_3) and the oxide (MO). The difference is greater when the metal ion is small than when it is large.

The lattice enthalpy for an ionic compound is inversely proportional to the sum of the ionic radii of the positive ion and the negative ion. Down group 2 lattice enthalpies decrease in magnitude (become less negative) from one compound to the next as the ionic radii of the metal ions increase.

For a large anion, such as the carbonate, the lattice enthalpy decreases (becomes less negative) only slowly as the metal ion radius increases down the group from Mg^{2+} to Ba^{2+}. For the smaller oxide ion, however, the magnitude of the lattice enthalpy decreases more quickly as the cation radius increases. As a result, the difference between the lattice enthalpies gets smaller down the group, so less energy is released as the lattice contracts when the compounds decompose and the carbonates become more stable.

Detailed analysis of the energy changes takes time, so chemists find it convenient to correlate the stability of compounds such as carbonates with the polarising power of the metal ions (see page 68). Generally the greater the polarising power of the metal ion the less stable the carbonate and the more easily it decomposes to the oxide.

Test yourself

14 a) Are the values of the enthalpy change for the decomposition of the group 2 carbonates given in Figure 2.17.11 a guide to the relative stability of these compounds?

b) Can the values of these enthalpies changes account for the fact that these compounds become less stable as the temperature rises?

15 Why is the lattice enthalpy of magnesium oxide more negative than the lattice enthalpy of barium oxide?

16 Why is the lattice enthalpy of magnesium oxide more negative than the lattice enthalpy of magnesium carbonate?

17 Why is the difference between the lattice enthalpies of the metal carbonates and oxides so significant in explaining the trend in the stability of group 2 carbonates?

2.18 The solubility of ionic crystals

Why do ionic crystals dissolve in water given that the ions in the lattice are strongly attracted to each other? Why is barium sulfate insoluble in water while magnesium sulfate is soluble? Why does sodium chloride dissolve in water despite the fact the process is endothermic? What, in general, are the factors which decide the extent to which an ionic salt dissolves in water? Chemists look for answers to questions of this kind by analysing the energy changes which take place as crystals dissolve.

◀ **Figure 2.18.1**
Crystallised salts in the Dead Sea in Israel

Energy changes during dissolving

An ionic compound such as sodium chloride does not dissolve in non-polar solvents such as hexane but it will dissolve in a polar solvent such as water. The overall enthalpy change is quite small:

$$NaCl(s) + aq \longrightarrow Na^+(aq) + Cl^-(aq) \qquad \Delta H^{\ominus}_{solution} = +3 \text{ kJ mol}^{-1}$$

In this equation '+ aq' is short for adding water.

Sodium chloride dissolves in water despite the fact that the process is endothermic. This is another example to show that the sign of ΔH is not a sure guide to whether or not a process will happen, especially when the magnitude of ΔH is small. What *is* surprising is that the overall enthalpy change is so small given that the energy needed to separate the ions from a sodium chloride crystal is +787 kJ mol^{-1} (– lattice enthalpy).

It not obvious why the charged ions in a crystal of sodium chloride separate and go into solution in water with only a small energy change. Where does the energy come from to overcome the attraction between the ions? The explanation is that the ions are so strongly hydrated by the polar water molecules that the sum of the enthalpy changes of hydration nearly balances the lattice energy.

Figure 2.18.2 ▶
Sodium and chloride ions are hydrated when they dissolve. Polar water molecules are attracted to both positive ions and negative ions. Here the bond between the ions and the polar water molecules is electrostatic attraction

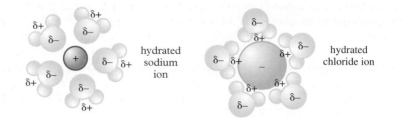

hydrated sodium ion

hydrated chloride ion

Definitions

The **enthalpy change of hydration** is the enthalpy change when one mole of gaseous ions is hydrated under standard conditions to produce a solution in which the concentration of ions is 1 mol dm^{-3}. For example, for sodium ions

$Na^+(g) + aq \longrightarrow Na^+(aq)$
$\Delta H^{\ominus}_{hydration} = -444 \text{ kJ mol}^{-1}$

The **enthalpy change of solution** for a compound is the enthalpy change when one mole of a substance dissolves in a stated amount of water under standard conditions.

$NaCl(s) + aq \longrightarrow$
$\quad Na^+(aq) + Cl^-(aq)$
$\Delta H^{\ominus}_{solution} = +3 \text{ kJ mol}^{-1}$

The enthalpy change of solution is the difference between the energy needed to separate the ions from the crystal lattice (– lattice enthalpy) and the energy given out as the ions are hydrated (sum of the hydration enthalpies).

$$\Delta H^{\ominus}_{solution} = \Delta H^{\ominus}_{hydration\ cation} + \Delta H^{\ominus}_{hydration\ anion} - \Delta H^{\ominus}_{lattice}$$

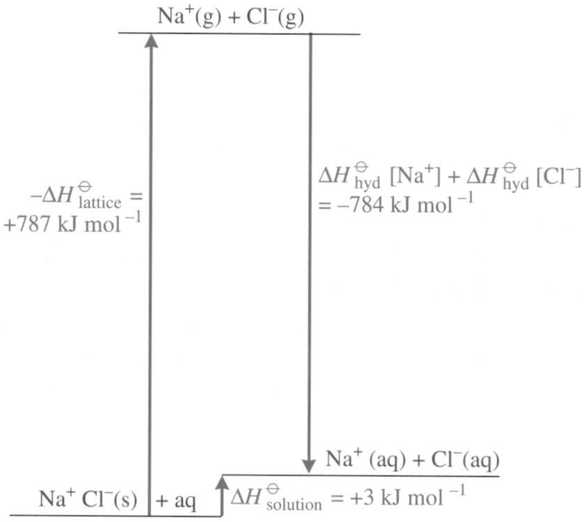

$Na^+(g) + Cl^-(g)$

$-\Delta H^{\ominus}_{lattice} = +787 \text{ kJ mol}^{-1}$

$\Delta H^{\ominus}_{hyd} [Na^+] + \Delta H^{\ominus}_{hyd} [Cl^-]$
$= -784 \text{ kJ mol}^{-1}$

$Na^+(aq) + Cl^-(aq)$

$Na^+ Cl^-(s) + aq$ $\quad \Delta H^{\ominus}_{solution} = +3 \text{ kJ mol}^{-1}$

Figure 2.18.3 ▲
Energy level diagram for sodium chloride dissolving in water. Note that the enthalpy change for going from the crystal to the gaseous ions is (– lattice energy)

Solubility trends

It is difficult to use enthalpy cycles to account for trends in solubilities of ionic compounds. One reason for this is that the enthalpy change of solution is a small difference between two large enthalpy changes. So even quite modest errors in estimating trends in the values can lead to large errors in the predicted differences between lattice and hydration enthalpies.

Test yourself   D

1 **a)** Calculate the enthalpies of solution of lithium fluoride and lithium iodide from the data on pages 247–248.

   b) Account for the relative values of the lattice enthalpies and hydration enthalpies of the two compounds in terms of ionic radii.

   c) To what extent, if at all, can your answers to **a)** explain the differences in the solubilities of the two compounds?
   (Solubilities: LiF = 5×10^{-5} mol in 100 g water; LiI = 1.21 mol in 100 g water.)

A further complication is that the lattice enthalpies and the hydration enthalpies tend to be affected in the same way by changes in the sizes of the ions and their charges. The smaller the ions and the larger the charges, the greater the lattice enthalpies but also the greater the hydration enthalpies.

Finally it is clearly not enough to consider the value of $\Delta H^{\ominus}_{solution}$. The sign and magnitude of the enthalpy change alone is not a reliable indicator of whether or not a solid will dissolve. Other factors have to be taken into account as explained in sections 2.19 and 2.20.

Solubilities of salts of group 2 metals

In a series of similar compounds with the same structure and ionic charges it is possible to suggest rationalisations for patterns of solubility despite the difficulty in using an analysis of enthalpy changes to make predictions.

Group 2 sulfates

In general, group 2 metal salts with large anions become less soluble down the group. This is illustrated by the sulfates (see Figures 2.18.4 and 2.18.5). Solubilities decrease from 1.8×10^{-1} mol per 100 g water for $MgSO_4$ to 9.4×10^{-7} mol per 100 g water for $BaSO_4$.

In a series of similar compounds lattice enthalpies vary with the sum of the ionic radii. Hydration enthalpies vary with the individual ionic radii.

In group 2 compounds with a large anion, such as the sulfate ion, the increasing size of the metal ion means that the hydration enthalpy decreases faster down the group than the lattice enthalpy. With a large anion the sum of the ionic radii is affected less in proportion because its value is largely determined by the large size of the sulfate ion. So the enthalpy change of solution becomes less exothermic down the group and the salts get less soluble.

Group 2 hydroxides

In general, the solubility of group 2 metal compounds with small anions become more soluble down the group. This is illustrated by the hydroxides. Magnesium hydroxide is the active ingredient in the antacid 'milk of magnesia'. It only dissolves very slightly, though enough to make the solution alkaline. Its solubility is 2.0×10^{-5} mol per 100 g water. Barium hydroxide is much more soluble and gives a strongly alkaline solution. Its solubility is 1.5×10^{-2} mol per 100 g water.

In group 2 compounds with a small anion, such as the hydroxide ion, the increasing size of the metal ion down the group has a much more significant effect on the sum of the ionic radii. With a small anion the lattice energy is much more sensitive to the size of the metal ion and it is the lattice enthalpy which decreases most as the metal ion size increases down the group. So the enthalpy change of solution becomes more exothermic and the salts get more soluble down the group.

Figure 2.18.4 ▲
Crystals of magnesium sulfate photographed through a microscope with polarised light. Magnesium sulfate is a soluble salt which occurs naturally as Epsom salts

Figure 2.18.5 ▲
Yellow prisms of barytes on haematite from Cumbria. Barytes is a mineral which is very insoluble in water. It consists of barium sulfate

Physical Chemistry

Section two

Test yourself

2   Suggest another example of a series of group 2 compounds with large anions which becomes *less* soluble down the group from Mg to Ba.

3   Suggest another example of a series of group 2 compounds with small anions which becomes *more* soluble down the group from Mg to Ba.

2.19 Entropy changes

Physical chemists have devised a range of ways for predicting the direction and extent of change. They use equilibrium constants, electrode potentials and enthalpy changes to explain why some reactions go while others do not. How are these quantities related and is there a more fundamental concept which links them all together? The answer is yes and the unifying concept is entropy.

Figure 2.19.1 ▲
Chemically the change of diamond to graphite is spontaneous. Graphite is more stable than diamond. Fortunately for owners of valuable jewellery the change is very very slow

Spontaneous changes

A spontaneous reaction is a reaction which tends to go without being driven by any external agency. Spontaneous reactions are the chemical equivalent of water flowing downhill. Any reaction which occurs naturally tends to happen spontaneously in this sense, even if it is very slow, just as water has a tendency to flow down a valley even when held up behind a dam. The chemical equivalent of a dam is a high activation energy for a reaction.

Figure 2.19.2 ▲
Metals such as magnesium, iron and aluminium react spontaneously with oxygen. They are ingredients of fireworks fuelled by the spontaneous reactions between sulfur, carbon and potassium nitrate in gunpowder

In practice, chemists also use the word *spontaneous* in its everyday sense to describe reactions which not only tend to go but also go fast on mixing the reactants at room temperature. Here is a typical example:

> 'The hydrides of silicon catch fire spontaneously in air, unlike methane which has an ignition temperature of about 500 °C.'

The reaction of methane with oxygen is also spontaneous in the thermodynamic sense, even at room temperature. However the activation energy for the reaction is so high that nothing happens until the gas is heated with a flame.

The fact that some endothermic processes are spontaneous shows the limitation of using the enthalpy change to decide the likely direction of change. One example is the reaction of citric acid with sodium hydrogencarbonate. The mixture fizzes vigorously while getting colder and colder. A more colourful example is the reaction of sulfur dichloride oxide with the dark red

Figure 2.19.3 ▲
Add citric acid crystals to a solution of sodium hydrogencarbonate and there is a vigorous reaction. The temperature falls. This is a spontaneous endothermic reaction

crystals of cobalt(II) chloride. The reaction looks violent and quantities of fuming hydrogen chloride gas bubble from the mixture while a thermometer dipping into the chemicals shows that the temperature falls sharply.

Entropy

Entropy, S, is a thermochemical quantity which makes it possible to decide whether or not a reaction is spontaneous. Change happens in the direction which leads to a total increase in entropy. When considering chemical reactions it is essential to calculate the total entropy change in two parts: the entropy change of the system and the entropy change of the surroundings.

$$\Delta S_{total} = \Delta S_{system} + \Delta S_{surroundings}$$

The entropy of a system measures the number of ways, W, of arranging the molecules and sharing out the energy between the molecules.

The relationship between S and W was derived by Ludwig Boltzman (1844 – 1906), the Austrian physicist who was the first person to explain the laws of thermodynamics in terms of the behaviour of atoms and molecules in motion.

$$S = k\ln W$$

where S is the entropy of the system, k the Boltzman constant and $\ln W$ the natural logarithm of the number of ways of arranging the particles and energy in the system.

Chemists sometimes describe entropy as a measure of disorder or randomness. These descriptions have to be interpreted with care because the disorder refers not only to the arrangement of the particles in space but much more significantly to the numbers of ways of distributing the energy of the system across all the available energy levels.

Standard molar entropies

Nothing seems to be happening in a closed flask of a gas kept at a constant temperature. Kinetic theory tells us, however, that the molecules are in rapid motion colliding with each other and with the walls of the container. The gas stays the same while constantly rearranging the molecules and redistributing the energy. The 'number of ways', W, is large for a gas. Gases generally have higher entropies than comparable liquids which have higher entropies than similar solids.

1 Classify these changes as spontaneous/fast, spontaneous/slow or not spontaneous:

a) ice melting at 5 °C

b) ammonia condensing to a liquid at room temperature

c) charcoal reacting with oxygen at room temperature

d) water splitting up into hydrogen and oxygen at room temperature

e) sodium reacting with water at room temperature.

Physical Chemistry

Section Two

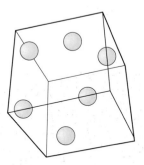

Figure 2.19.6 ▲
Particles of a gas. Gases have higher standard molar entropies than comparable liquids because the atoms or molecules are not only free to move but also widely spaced. There are even more ways of distributing the particles and energy. The standard molar entropy of argon is higher than that of mercury. As with liquids, molecules with more atoms have even higher standard molar entropies because they can vibrate, rotate and arrange themselves in more ways

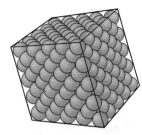

Figure 2.19.4 ▲
Structure of a solid. Solids have relatively low values for standard molar entropies. In diamond the carbon atoms are held firmly in place by strong, highly directional covalent bonds. The standard molar entropy of diamond is low. Lead has a higher value for its standard molar entropy because metallic bonds are not directional. The heavier, larger atoms can vibrate more freely and share out their energy in more ways than carbon atoms in diamond

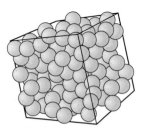

Figure 2.19.5 ▲
Structure of a liquid. In general, liquids have higher standard molar entropies than comparable solids because the atoms or molecules are free to move. There are many more ways of distributing the particles and energy. The standard molar entropy of mercury is higher than that of lead. Molecules with more atoms have higher standard molar entropies because they can vibrate, rotate and arrange themselves in yet more ways

solids	$S^{\ominus}$/J K^{-1} mol^{-1}	liquids	$S^{\ominus}$/J K^{-1} mol^{-1}	gases	$S^{\ominus}$/J K^{-1} mol^{-1}
carbon (diamond)	2.4	mercury	76.0	argon	154.7
magnesium oxide	26.9	water	69.9	ammonia	192.4
copper	33.2	ethanol	160.7	carbon dioxide	213.6
lead	64.8	benzene	173.3	propane	269.9

Figure 2.19.7 ▲

Definition

Standard molar entropy, $S^{\ominus}$ is the entropy per mole for a substance under standard conditions. Chemists use values for standard molar entropies to calculate entropy changes and so to predict the direction and extent of chemical change.

The units for standard molar entropy are joules per Kelvin per mole (J K^{-1} mol^{-1}). Note that the units are joules and not kilojoules.

Test yourself

2 Which substance in each pair is expected to have the higher standard molar entropy at 298 K?

a) $Br_2(l)$, $Br_2(g)$

b) $H_2O(s)$, $H_2O(l)$

c) $HF(g)$, $NH_3(g)$

d) $CH_4(g)$, $C_2H_6(g)$

e) $NaCl(s)$, $NaCl(aq)$

CD-ROM

The entropy change of the system

Tables of standard molar entropies make it possible to calculate the entropy change of the system of chemicals during a reaction.

$$\Delta S^{\ominus}_{system} = \begin{array}{c}\text{the sum of the standard}\\\text{molar entropies of}\\\text{the products}\end{array} - \begin{array}{c}\text{the sum of the standard}\\\text{molar entropies of}\\\text{the reactants}\end{array}$$

Worked example

Calculate the entropy change for the system, $\Delta S^{\ominus}_{system}$, for the synthesis of ammonia from nitrogen and hydrogen. Comment on the value.

Notes on the method

Write the balanced equation for the reaction.

Look up the standard molar entropies on pages 243–244 taking careful note of the units.

Answer

$N_2(g) + 3H_2(g) \longrightarrow 2NH_3(g)$

Sum of the standard molar entropies of the products = $2S^{\ominus}$ [NH$_3$(g)]

= 2×192.4 J K^{-1} mol^{-1} = 384.8 J K^{-1} mol^{-1}

Sum of the standard molar entropies of the reactants

= $S^{\ominus}$ [N$_2$(g)] + $3S^{\ominus}$ [H$_2$(g)]

= 191.6 J K^{-1} mol^{-1} + (3 × 130.6 J K^{-1} mol^{-1}) = 583.4 J K^{-1} mol^{-1}

$\Delta S^{\ominus}_{system}$ = 384.8 J K^{-1} mol^{-1} − 583.4 J K^{-1} mol^{-1}

= −198.6 J K^{-1} mol^{-1}

This shows that the entropy of the system decreases when nitrogen and hydrogen combine to form hydrogen. This is not surprising since the change halves the number of molecules so the amount of gas decreases.

Test yourself

3 Without doing any calculations predict whether the entropy of the system will increase or decrease as a result of these changes:

a) $KCl(s) + aq \longrightarrow KCl(aq)$

b) $H_2O(l) \longrightarrow H_2O(g)$

c) $Mg(s) + Cl_2(g) \longrightarrow MgCl_2(s)$

d) $N_2O_4(g) \longrightarrow 2NO_2(g)$

e) $NaHCO_3(s) + HCl(aq) \longrightarrow NaCl(aq) + H_2O(l) + CO_2(g)$

The entropy change of the surroundings

It is not enough to consider only the entropy of the system. What matters is the total entropy change, which is the sum of the entropy changes of the system and the entropy change in the surroundings.

It turns out that the entropy change of the surroundings during a chemical reaction is determined by the size of the enthalpy change, ΔH, and the temperature, T. The relationship is:

$$\Delta S_{surroundings} = \frac{-\Delta H}{T}$$

The minus sign is included because the entropy change is bigger the more energy is transferred to the surroundings. For an exothermic reaction, which transfers energy to the surroundings, ΔH is negative, so $-\Delta H$ is positive.

What this relationship shows is that the more energy transferred to the surroundings by an exothermic process, the larger the increase in the entropy of the surroundings. It also shows that, for a given quantity of energy, the increase in entropy is greater when the surroundings are cool than when they are hot. Adding energy to molecules in a cool system has a proportionately greater effect on the number of ways of distributing matter and energy than adding the same quantity of energy to a system that is already very hot.

Bearing in mind that a reaction is only spontaneous if the total entropy change, ΔS_{total}, is positive:

■ Most exothermic reactions tend to go because at about room temperature the value of $-\Delta H/T$ is much larger and more positive than ΔS_{system} which means that ΔS_{total} is positive.
■ An endothermic reaction can be spontaneous so long as the increase in the entropy of the system is greater than the decrease in the entropy of the surroundings.
■ A reaction which does not tend to go at room temperature may become spontaneous as the temperature rises because $\Delta S_{surroundings}$ decreases in magnitude as T increases.

Test yourself

4 Calculate the entropy change for the system, $\Delta S_{system}^{\ominus}$, for the catalytic reaction of ammonia with oxygen to form NO and steam. Comment on the value.

Physical Chemistry

Section two

Test yourself

5 Consider the reaction of magnesium with oxygen:

$$2Mg(s) + O_2(g) \longrightarrow 2MgO(s) \quad \Delta H^{\ominus} = -601.7 \text{ kJ mol}^{-1} \quad \Delta S_{system}^{\ominus} = -216.6 \text{ J K}^{-1} \text{ mol}^{-1}$$

a) Why does the entropy of the system decrease?

b) Show why the reaction of magnesium with oxygen is spontaneous at 298 K despite the decrease in the entropy.

6 Consider the decomposition of magnesium carbonate:

$$MgCO_3(s) \longrightarrow MgO(s) + CO_2(g) \quad \Delta H^{\ominus} = +117 \text{ kJ mol}^{-1} \quad \Delta S_{system}^{\ominus} = +168 \text{ J K}^{-1} \text{ mol}^{-1}$$

a) Why does the entropy of the system increase?

b) Show that the reaction is not spontaneous at room temperature (298 K).

c) Assuming that ΔH and ΔS_{system} do not vary with temperature, estimate the temperature at which the decomposition becomes spontaneous.

2.20 Free energy

A chemical change is spontaneous if the total entropy change is positive. There is no doubt about this. The problem is that using entropy to decide on the direction change involves three steps: working out the entropy change of the system, working out the entropy change of the surroundings and then putting the two together to calculate the total entropy change. This can be laborious and chemists are grateful to the American physicist Willard Gibbs who discovered an easier way of unifying all that chemists know about predicting the extent and direction of change.

Free energy and entropy

Willard Gibbs (1839 – 1903) was the first to define the thermochemical quantity free energy. The symbol for a free energy change is ΔG. If ΔG is negative, the reaction is spontaneous and tends to go.

If a reaction is spontaneous, that means that it tends to go without being pushed. In other words the reaction is feasible – even if happens very very slowly. In this sense the reaction between hydrogen and oxygen is feasible at room temperature but, because of a high activation energy, the reaction is so slow that it does not happen. Put a match to the mixture of gases and the reaction goes with a bang.

The advantage of $\Delta G^{\ominus}$ values for chemists is that tables of standard free energies of formation can be used to calculate the standard free energy change for any reaction. The calculations follow exactly the same steps as the calculations for standard enthalpy changes for reactions from standard enthalpies of formation.

The quantity 'free energy' is closely related to the idea of entropy and can be thought of as the 'total entropy change' in disguise. Willard Gibbs defined free energy as:

$$\Delta G^{\ominus} = -T\Delta S^{\ominus}_{\text{total}}$$

Given that:

$$\Delta S^{\ominus}_{\text{total}} = \Delta S^{\ominus}_{\text{system}} + \Delta S^{\ominus}_{\text{surroundings}}$$

And that for a change at constant temperature and pressure:

$$\Delta S^{\ominus}_{\text{surroundings}} = \frac{-\Delta H^{\ominus}}{T}$$

It follows that:

$$\Delta S^{\ominus}_{\text{total}} = \Delta S^{\ominus}_{\text{system}} + \left(\frac{-\Delta H^{\ominus}}{T}\right)$$

Hence $-T\Delta S^{\ominus}_{\text{total}} = -T\Delta S^{\ominus}_{\text{system}} + \Delta H^{\ominus}$

From Gibb's definition this becomes: $\Delta G^{\ominus} = \Delta H^{\ominus} - T\Delta S^{\ominus}_{\text{system}}$

The great advantage of this equation is that all the terms refer to the system and so it is no longer necessary to calculate changes in the surroundings. Given that this is the case the equation is usually written, as here, with the understanding that the entropy change is $\Delta S^{\ominus}_{\text{system}}$

$$\Delta G^{\ominus} = \Delta H^{\ominus} - T\Delta S^{\ominus}$$

Figure 2.20.1 summarises the implications of this important relationship.

Enthalpy change	Entropy change of the system	Is the reaction feasible?
exothermic ($\Delta H^\ominus$ negative)	increase ($\Delta S^\ominus$ positive)	yes, $\Delta G^\ominus$ is negative
exothermic ($\Delta H^\ominus$ negative)	decrease ($\Delta S^\ominus$ negative)	yes, if the number value of $\Delta H^\ominus$ is greater than the magnitude of $T\Delta S^\ominus$
endothermic (ΔH° positive)	increase ($\Delta S^\ominus$ positive)	yes, if the magnitude of $T\Delta S^\ominus$ is greater than the number value of $\Delta H^\ominus$
endothermic ($\Delta H^\ominus$ positive)	decrease ($\Delta S^\ominus$ negative)	no, $\Delta G^\ominus$ is positive

Figure 2.20.1 ▲

The possibilities listed in Figure 2.20.1 show why chemists sometimes say that the feasibility of a reaction depends on the balance between the enthalpy changes and the entropy changes for the process.

Figure 2.20.1 also shows that a change that is not feasible at one temperature may become feasible if the temperature is higher. Generally the values of $\Delta H^\ominus$ and $\Delta S^\ominus$ do not change markedly with temperature and so it is possible to estimate the temperature at which a reaction, which is not feasible at room temperature, becomes spontaneous at a higher temperature.

Full data is not available for all reactions so it is not always possible to calculate the free energy change. Often the $T\Delta S^\ominus_{system}$ term is relatively small compared to the enthalpy change so that: $\Delta G^\ominus \approx \Delta H^\ominus$.

This means that chemists can often use the sign of $\Delta H^\ominus$ as a guide to feasibility. This can be misleading if the magnitude of the entropy change for the reaction system is large. Also the approximation becomes less justified at higher temperatures when T is bigger and so the magnitude of $T\Delta S^\ominus_{system}$ is bigger.

Test yourself

1 Can an exothermic reaction which is not feasible at room temperature become feasible at a higher temperature if the entropy change for the reaction is negative?

2 Consider the reaction

   $BaCO_3(s) \longrightarrow BaO(s) + CO_2(g)$ $\Delta H^\ominus = +268$ kJ mol^{-1} $\Delta S^\ominus = +168$ J K^{-1} mol^{-1}

   a) Calculate $\Delta G^\ominus$ for the reaction at 298 K.

   b) Is the reaction spontaneous at room temperature?

   c) At what temperature does the reaction become feasible?

   d) Compare your answers to the answers to question 6 in section in section 2.19 on page 79. Comment on the similarities and differences.

3 a) Calculate $\Delta S^\ominus$ and $\Delta H^\ominus$ for the synthesis of methanol from carbon dioxide and hydrogen taking the values from the tables on pages 243 and 244.

   b) Work out the temperature at which the synthesis ceases to be feasible.

How far and in which direction?

Gibbs's concept of free energy makes it possible to unify all the means that chemists use to predict the extent and direction of change.

For many reactions, such as acid–base reactions, the easiest guide to the direction and extent of change is the equilibrium constant, K_c. It is possible to show that the standard free energy change is related to the value of the equilibrium constant: $\Delta G^\ominus \propto -\ln K_c$.

For redox reactions, standard electrode potentials offer an alternative way of deciding whether or not reactions tend to go (see section 2.14).

Physical Chemistry

Section two

The free energy change for a redox reaction and cell e.m.f. are closely related: $\Delta G^{\ominus} \propto - E^{\ominus}_{cell}$. So if the cell e.m.f. is positive the redox reaction in the cell will tend to go.

What this shows is that $\Delta G^{\ominus}$, $E^{\ominus}_{cell}$ and K_c values can all answer the same questions for a reaction:

- Will the reaction go?
- How far will it go?

Figure 2.20.2 shows how the three methods of predicting the direction and extent of change are related.

	$\Delta G^{\ominus}$/kJ mol^{-1}	$E^{\ominus}_{cell}$/V	K_c (units depend on the reaction)
Reaction goes to completion	more negative than -60	more positive than $+0.6$	$> 10^{10}$
Equilibrium with more products than reactants	≈ -10	$\approx +0.1$	$\approx 10^2$
Equilibrium with more reactants than products	$\approx +10$	≈ -0.1	$\approx 10^{-2}$
Reaction does not go	more positive than $+60$	more negative than -0.6	$< 10^{-10}$

Figure 2.20.2 ▲

In practice chemists use the quantity which is most convenient to measure. Knowing the value of one of the three quantities it is possible to calculate the other two. It is always important to bear in mind, however, that even if a reaction is spontaneous it may be very slow.

The stability of compounds

Compounds are stable if they do not tend to decompose into their elements or into other compounds. A compound which is stable at room temperature and pressure may become more or less stable as conditions change.

Chemists often use standard enthalpy changes as an indicator of stability. Strictly they should use standard free energy, $\Delta G^{\ominus}_f$, values but in many cases $\Delta G^{\ominus}_f \approx \Delta H^{\ominus}_f$.

Inert chemicals

A chemical is inert if it has no tendency to react under given circumstances. The lighter noble gases, helium and neon live up to the original name for group 8. They are inert towards all other reagents.

Nitrogen is a relatively unreactive gas which can be used to create an 'inert atmosphere' free of oxygen (which is much more reactive). However nitrogen is not inert in all circumstances. It reacts, for example, with hydrogen in the Haber process to form ammonia and with oxygen at high temperatures to form nitrogen oxides.

Sometimes there is no tendency for a reaction to go because the reactants are stable. This occurs if the free energy change for the reaction is positive.

Test yourself

4 Suggest examples of reactions to illustrate each of the four possibilities in Figure 2.20.3.

$\Delta G^{\ominus} \approx \Delta H^{\ominus}$	Activation energy	Change observed	
positive	high	no reaction	reactants stable relative to products
negative	high	no reaction	reactants unstable relative to products but kinetically inert
positive	low	no reaction	reactants stable relative to products
negative	low	fast reaction	reactants unstable relative to products

Figure 2.20.3 ▲

Sometimes there is no reaction even though thermochemistry suggests that it should go. The free energy change is negative therefore the change is feasible. However, high activation energy means that the rate of reaction is very, very slow.

A compound such as the gas N_2O, for example, is thermodynamically unstable. $\Delta H_f^{\ominus}(= + 82 \text{ kJ mol}^{-1})$ and $\Delta G_f^{\ominus}(= + 104 \text{ kJ mol}^{-1})$ are both positive. The decomposition reaction is exothermic. The compound tends to decompose into its elements but the rate is very slow under normal conditions.

Chemists sometimes say that N_2O is kinetically 'stable'. It is clearer to use a different word and to refer to the kinetic inertness of N_2O.

Examples of kinetic inertness are:

■ a mixture of hydrogen and oxygen at room temperature
■ a solution of hydrogen peroxide in the absence of a catalyst
■ aluminium metal in dilute hydrochloric acid.

'Kinetic stability' is the term often used for 'kinetic inertness'. It helps, however, to make a sharp distinction between the two quite different types of explanation. For clarity, chemists refer to:

■ systems with no tendency to react as 'stable' and
■ systems which should react but do not do so for a rate (kinetic) reason as 'inert'.

Definition

Kinetic inertness is a term used when a reaction does not go even though the reaction appears to be feasible. The reaction tends to go according to the equilibrium constant, or the standard electrode potentials, or the free energy change, yet nothing happens. There is no change because the rate of reaction is too slow to be noticeable. There is a barrier preventing change – usually a high activation energy. The compound or mixture is inert.

Test yourself

5 Draw a reaction profile to show the energy changes from reactants to products for reactants which are thermodynamically unstable relative to the products but kinetically inert.

Physical Chemistry

Section two

Section two study guide

Review

This guidance will help you to organise your notes and revision. Check the terms and topics against the specification you are studying. You will find that some topics are not required for your course.

Key terms

Show that you know the meaning of these terms by giving examples. Consider writing the key term on one side of an index card and the meaning of the term with an example on the other side. Then you can easily test yourself when revising. Alternatively use a computer data base with fields for the key term, the definition and the example. Test yourself with the help of reports which show just one field at a time.

- Rate of reaction
- Rate equation
- Order of reaction
- Rate constant
- Half-life
- Activation energy
- Rate determining step
- Mechanism of a reaction
- Homogeneous equilibrium
- Heterogeneous equilibrium
- Phase
- Partial pressure
- Mole fraction
- Brønsted acid
- Brønsted base
- pH
- K_w and pK_w
- K_a and pK_a
- Electrochemical cell
- Reference electrode
- Hydrogen electrode
- Salt bridge
- Disproportionation reaction
- Standard enthalpy change of atomisation
- Ionisation enthalpy
- Electron affinity (or enthalpy change for electron addition)
- Lattice enthalpy
- Standard enthalpy change of formation
- Enthalpy change of hydration
- Enthalpy change of solution
- Thermodynamic stability
- Kinetic stability (or inertness)
- Entropy
- Free energy
- Spontaneous change
- Feasible reaction

Symbols and conventions

Make sure that you understand the conventions which chemists use when working with physical quantities, writing rate equations and equilibrium constants, describing cells and working with thermochemical quantities. Illustrate your notes with examples.

- Conventions for writing rate equations and showing the reaction orders.
- Conventions for writing the expressions for K_c and K_p for homogeneous and heterogeneous equilibria.
- Conventions for representing electrochemical cells and standard electrode potentials.
- Conventions for representing Born–Haber cycles.
- The standard conditions for defining thermochemical quantities.

Patterns and principles

Use tables, charts, concept maps or mind maps to summarise key ideas. Brighten your notes with colour to make them memorable.

- The effect of temperature changes on the rate constant for a reaction (with reference to the collision theory and the Arrhenius equation).
- The effects of catalysts on reaction rates and on systems at equilibrium.
- The application of the equilibrium law to explain the behaviour of indicators in acid–base titrations and the action of buffer solutions.
- The explanation of the corrosion of iron as an electrochemical change.

Predictions

Use examples to show that you can apply chemical principles to make predictions.

- Use the equilibrium law to predict the effects of changing pressure or concentration on a system at equilibrium.
- Predict the effect of changing the temperature on a system at equilibrium given the value of the enthalpy change for the reaction.
- Predict the shape of the pH curve during acid–base titrations in all combinations of strong and weak acids.
- Predict the direction and likely extent of change for redox reactions given cell e.m.f.s, or values for standard electrode potentials.
- Predict the direction of spontaneous change given data for calculating $\Delta S^{\ominus}_{total}$ or $\Delta G^{\ominus}$ (while understanding that the predictions can say nothing about the rate of change).

Laboratory techniques

Use labelled diagrams to illustrate and describe these practical procedures

- Measuring reaction rates.
- Measuring enthalpies of neutralisation.
- Selecting a suitable indicator for an acid–base titration and justifying the choice
- Setting up electrochemical cells, measuring their e.m.f.s and hence determining standard electrode potentials.

Interpretation of data

Give your own worked examples, with the help of the Test Yourself and examination questions, to show that you can do the following from given data.

- Work out rates of reaction and half-lives from concentration–time graphs.
- Recognise rate–concentration graphs for zero, first and second order reactions.
- Use the initial rate method to deduce the order of a reaction with respect to a reactant.
- Calculate the rate constant from the rate equation given experimental results.
- Suggest a reaction mechanism consistent with a given rate equation.
- Calculate values for equilibrium constants from given data for homogeneous and heterogeneous systems (K_c or K_p).
- Calculate the pH of a strong acid given its molar concentration (and vice versa).

- Use K_w or pK_w to calculate the pH of a solution of a strong base.
- Calculate the pH of a weak acid given its molar concentration and the value of K_a (and vice versa).
- Determine the value of K_a for a weak acid from the pH curve during a titration of the acid with a strong base.
- Calculate the pH of a buffer solution given the composition and the value of K_a for the weak acid.
- Calculate an enthalpy of neutralisation from experimental results and account for the relative values for different combinations of strong and weak acids and bases.
- Determine $E^\ominus_{cell}$ values from standard electrode potentials.
- Draw a Born–Haber cycle given data and use it to calculate an unknown term in the cycle.
- Calculate free energy changes using the relationship $\Delta G^\ominus = \Delta H^\ominus - T\Delta S^\ominus$
- Calculate the entropy change given appropriate data using the relationship $\Delta S^\ominus_{total} = \Delta S^\ominus_{system} + \Delta S^\ominus_{surroundings}$

Key skills

Problem solving

Planning, carrying out and interpreting a laboratory investigation calls for problem solving skills, as long as you are not told how to set about the task. You will have to make your own decisions about the procedure, and you will have to work independently using your initiative. This could be an investigation of reaction rates, energy changes or reversible reactions.

Application of number

Quantitative practical work to investigate rates of reaction, equilibrium constants or electrode potentials will give you opportunities to develop and practice the individual skills you need for the application of number key skill. This includes: selecting data from a large data set, working to the right number of significant figures, making measurements in appropriate units, using standard form, choosing the appropriate method of calculating the result, carrying out multi-stage calculations and using formulae, plus checking to identify mistakes and experimental uncertainty.

Information technology

You can use spreadsheets or special software to explore the factors which affect rates of reaction or systems at equilibrium. Kinetics software allows you to simulate kinetics experiments and practice interpreting the results.

Section *three*

Inorganic Chemistry

Contents

3.1 Inorganic chemistry in action

Inorganic chemistry is the study of the chemical elements and their compounds especially their oxides, chlorides and hydrides. In inorganic chemistry it is crucial to understand that when an element reacts and combines with other elements it can form different chemical species with different properties. An element may be essential to life as one chemical species. As another it may be a deadly poison.

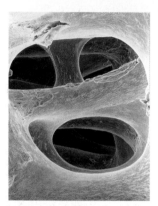

Figure 3.1.1 ▲
A false-colour scanning electron micrograph showing the honeycomb structure of the mineralised inner part of a human bone. In living tissue the cavities contain bone marrow

Elements for health

Fourteen metals and about ten non-metals are essential to good health. The most abundant metal in the human body is calcium which, as the mineral calcium phosphate, makes our bones hard and tough.

Next, in terms of abundance, are the elements potassium and sodium. The human diet must include the chlorides of potassium and sodium. The ions of these metals have many important roles in the body, but they are especially vital because the conduction of impulses along nerves depends on potassium and sodium ions crossing cell membranes.

Other metals are also important but in smaller amounts. Several of them are d-block elements including iron, zinc, manganese and cobalt. Zinc ions, for example, are present in many enzymes. Zinc ions also have a crucial part to play as cells translate the genetic code of DNA molecules into proteins.

Elements for medicine

There is growing interest in the use of metal compounds in medicine, stimulated by the great success of cisplatin as a treatment for cancer. Doctors apply ointments containing silver compounds when treating patients suffering from severe burns. Several complex, soluble gold salts are the basis for treatment of patients with arthritis.

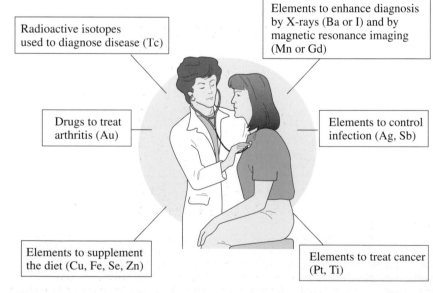

Radioactive isotopes used to diagnose disease (Tc)

Elements to enhance diagnosis by X-rays (Ba or I) and by magnetic resonance imaging (Mn or Gd)

Drugs to treat arthritis (Au)

Elements to control infection (Ag, Sb)

Elements to supplement the diet (Cu, Fe, Se, Zn)

Elements to treat cancer (Pt, Ti)

Figure 3.1.2 ▲
Uses of elements in medicine. It is compounds rather than the free elements which are used

Oxides to brighten our lives

Titanium is not only valuable as a metal. Its oxide, titanium(IV) oxide, is a brilliant white pigment. The main use of the pigment is in paint but it is also a surface coating for paper, a filler in plastics and an ingredient of cosmetics and toothpaste. The oxide makes a good white pigment because it has a very high refractive index and absorbs almost no light in the visible part of the spectrum. When ground to a fine powder it is both intensely white and very opaque so as a pigment it has excellent covering power.

Aluminium oxide is another decorative oxide. When pure, this oxide is a white solid with a very high melting point. It also exists naturally as a group of minerals called corundum. Corundum is very hard which makes it an important abrasive. Some gemstones are varieties of corundum including rubies, which are red because of the presence of some chromium(III) ions in place of aluminium(III) ions in the crystal structure. Sapphires are blue because some of the aluminium ions are replaced by a mixture of titanium(IV) and cobalt(II) or iron(II) ions.

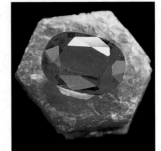

Figure 3.1.3 ▲
A cut and polished ruby on top of an uncut ruby. Rubies consist of aluminium oxide containing traces of chromium oxide

Elements for greener chemistry

One of the new frontiers in chemistry is the development of new catalysts, especially catalysts which work at low temperatures and are highly selective. Modern catalysts allow industry to manufacture chemicals using less energy while creating much less hazardous waste.

Transition metal atoms lie at the heart of many catalysts. In the early 1960s a group of industrial chemists developed a new process for making ethanoic acid by combining methanol and carbon monoxide with the help of a catalyst based on cobalt in the presence of iodine. A few years later the process was improved by replacing the cobalt with the metal rhodium which lies below cobalt in the periodic table. Using rhodium the yield of unwanted by-products fell below 1%.

Then, in the late 1990s, the chemical industry found an even more efficient catalyst based, this time, on the metal iridium which is the element below rhodium in the periodic table. Finding a way to make this better catalyst effective depended on a detailed understanding of the mechanism of the reaction. This knowledge allowed the research team to identify the additives which can control the sequence of bond breaking and bond forming during the steps of the reaction.

Figure 3.1.4 ▲
A member of the Chinese opera wearing stage make up. Pigments used in make up include chromium(III) oxide (green), iron oxides (reds and yellows) and manganese salts (violet)

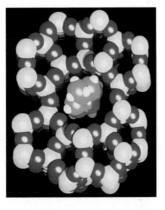

Figure 3.1.5 ▲
A computer graphic showing a molecule of an aromatic hydrocarbon in a cavity of a zeolite catalyst. Note that the carbon atoms are green. Zeolites are sodium-aluminium silicates in which the three-dimensional structures of the silicon and oxygen atoms form tunnels or cavities into which ions and small molecules can fit. Synthetic zeolites make excellent heterogeneous catalysts because they can be developed with active sites to favour the required products by acting on molecules with particular shapes and sizes

3.2 Metal ions in water

The oceans, which cover four fifths of the Earth's surface, contain many dissolved salts and can be a source of elements such as bromine, iodine and magnesium. The water which evaporates from the oceans and cycles in the natural environment helps to weather rocks and release nutrients into solution for plant growth. Water makes up about two-thirds of the human body and an even higher percentage of some animals and plants. So theories to explain the behaviour of ions in solution play an essential part of our understanding of environmental chemistry and biochemistry as well as helping chemists to control reactions in laboratories and industry.

Figure 3.2.1 ▶
The slopes of volcanoes can be immensely fertile thanks to the soluble ions in the soil produced by the weathering of igneous rocks

Hydration

Sodium chloride dissolves easily to make water salty; this hardly seems to present a chemical problem. But the ionic bonding between the sodium and chloride ions is strong with a lattice energy of -787 kJ mol^{-1} (see pages 63–64). So from a chemical point of view it is hard to see why the charged ions in a crystal of sodium chloride separate and go into solution in water with only a small energy change. Chemists now understand that the ions are so strongly hydrated by the polar water molecules that the sum of the enthalpy changes of hydration nearly balances the lattice energy (see pages 73–74).

Water molecules are polar, therefore they are attracted to both positive sodium ions and negative chloride ions. Here the bond between the ions and the polar water molecules is simply electrostatic attraction. Crystallising the solution produces an anhydrous salt.

Figure 3.2.2 ▶
Sodium and chloride ions leaving a crystal lattice and becoming hydrated as they dissolve in water. Here the bond between the ions and the polar water molecules is electrostatic attraction

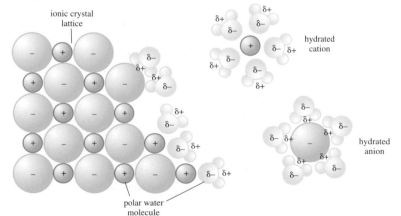

With other metal ions, especially the ions of p-block and d-block metals, the bonding between the metal ions and the water molecules is stronger. Dative covalent bonds hold the water molecules to the metal ions. As a result, the metal ions are hydrated in crystals as well as in solution in the form of compounds such as $FeSO_4.7H_2O$ and $CoCl_2.6H_2O$.

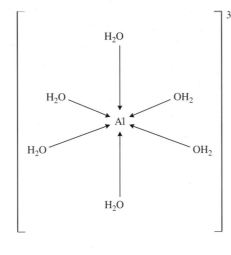

an octahedron with 8 faces

◄ *Figure 3.2.3*
A hydrated aluminium ion with six water molecules forming dative covalent bonds to the central metal atom. The ion has an octahedral shape

Hydrolysis

It is not possible to make crystals of anhydrous aluminium chloride by heating a solution of the salt to dryness. Instead the products are aluminium oxide and hydrogen chloride gas.

$$2AlCl_3(aq) + 3H_2O(l) \longrightarrow Al_2O_3(s) + 6HCl(g)$$

This is an example of hydrolysis. The water breaks the polar covalent bond between the aluminium and chlorine atoms.

Even without heating there is a reaction between hydrated aluminium ions and water in solution. A solution of aluminium chloride is acidic.

The hydrated aluminium(III) ion is an acid. The aluminium ion, because of its polarising power, draws electrons towards itself. The electrons in the water molecules are pulled towards the highly polarising Al^{3+} ion making it easier for the water molecules linked to the aluminium ion to give away protons.

$$[Al(H_2O)_6]^{3+} + H_2O \rightleftharpoons [Al(H_2O)_5OH]^{2+} + H_3O^+$$

$$[Al(H_2O)_5OH]^{2+} + H_2O \rightleftharpoons [Al(H_2O)_4(OH)_2]^+ + H_3O^+$$

The hydrated aluminium ion is as strong an acid as ethanoic acid (see pages 37–40). The hydrated ion gives protons to water molecules forming oxonium ions. The solution is acidic enough to release carbon dioxide when added to sodium carbonate.

Adding a base such as hydroxide ions, to a solution of aluminium ions removes a third proton to produce an uncharged complex. The uncharged complex is much less soluble in water and forms a white jelly-like precipitate called hydrated aluminium hydroxide, often written as $Al(OH)_3$.

$$[Al(H_2O)_4(OH)_2]^+ + OH^- \rightleftharpoons [Al(H_2O)_3(OH)_3](s)$$

Adding excess alkali removes yet another proton. A negatively charged ion is produced which is soluble in water, causing the precipitate to redissolve. This demonstrates the amphoteric properties of aluminium hydroxide.

$$[Al(H_2O)_3(OH)_3](s) + OH^- \rightleftharpoons [Al(H_2O)_2(OH)_4]^- (aq)$$

aluminate(III) ion, $Al(OH)_4^-$

A **hydrolysis** reaction uses water to split a compound apart. A change of pH is a symptom of partial hydrolysis of a salt in solution. Hydrolysis changes the pH of a solution of a salt by altering the concentrations of $H^+(aq)$ and $OH^-(aq)$ ions.

Chemists use the term **polarising power** to describe the extent to which a positive ion is able to distort the electron cloud around a neighbouring molecule or negative ion. The larger the charge on a positive ion and the smaller its size the greater its polarising power.

CD-ROM

Inorganic Chemistry

Section three

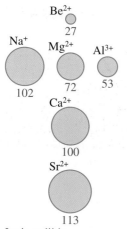

Be^{2+}
27

Na$^+$ Mg^{2+} Al^{3+}

102 72 53

Ca^{2+}
100

Sr^{2+}
113

Ionic radii in
picometres, pm
1 pm = 1×10^{-12} m

Figure 3.2.4 ▲
Relative sizes and charges of metal ions

All these changes are reversed by adding a solution of a strong acid, such as hydrochloric acid, which turns the OH$^-$ ions in the aluminate(III) ion back to water molecules.

The hydrated ions in M^{2+} (where M stands for a metal) salts are much less acidic than in M^{3+} ions because of the smaller polarising power of the M^{2+} metal ions.

Figure 3.2.5 ▲
A laterite cliff. Intense chemical weathering of igneous rocks in the tropics removes the silicate minerals and leaves behind a mixture of the hydroxides of iron(III) and aluminium creating deep lateritic soils. Where the iron content is low, the mineral becomes bauxite which is the ore of aluminium

Definitions

A **precipitation reaction** produces an insoluble product from soluble chemicals in solution. A common example is the formation of an insoluble salt on mixing solutions of two soluble salts.

Ionic equations describe chemical changes by only showing the reacting ions in solution. These equations leave out the spectator ions which remain in solution unchanged.

An **amphoteric oxide** or hydroxide dissolves in both acids and alkalis. Aluminium hydroxide is amphoteric.

Test yourself

1 Write equations to show what happens, step by step, on adding hydrochloric acid to a solution of aluminate(III) ions.

2 a) Write an ionic equation for the reaction between solutions of magnesium chloride and sodium carbonate.

   b) Explain why it is possible to use a solution of sodium carbonate to precipitate magnesium carbonate but impossible to precipitate aluminium carbonate in a similar way.

3 Weathering breaks down the metal silicates. The sodium and potassium ions released by weathering end up as salts in the sea. The aluminium(III) and iron(III) ions end up as deposits of bauxite, clay and other minerals. Account for the differences in the fates of these metal ions in the environment.

4 Comment on the significance of these three features of water molecules for the solvent properties of water: the molecules are polar, they have lone pairs of electrons and they can form hydrogen bonds.

5 How do you account for the fact that the salts with the hydrated aluminium(III) and the aluminate(III) ions are soluble but that aluminium hydroxide is insoluble?

6 What is the trend in the polarising power of the metal ions:

   a) across period 3 from Na$^+$ to Al^{3+}

   b) down group 2 from Be^{2+} to Sr^{2+}?

3.3 Complex formation

The haem groups in haemoglobin molecules are complexes of iron(II) ions with porphyrin rings. Lone pairs of electrons on nitrogen atoms form dative bonds with the metal ions. This is just one of very many example of complex associations between metal ions and organic molecules which are vital to biochemistry.

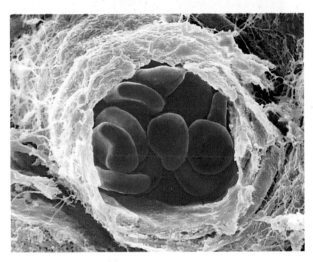

Figure 3.3.1 ▲
Red blood cells in a branch of a small artery. Many complexes of d-block elements are brightly coloured. Haemoglobin in red blood cells is bright red when combined with oxygen. Carbon monoxide is toxic because it binds to the iron in the haem group more strongly than oxygen and irreversibly. Death is inevitable if most of the haemoglobin in a person's blood has combined with carbon monoxide

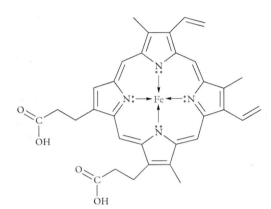

Figure 3.3.2 ▲
A haem group. There are four haem groups in each haemoglobin molecule. A reversible reaction between the haem groups and oxygen molecules allows haemoglobin to pick up oxygen in the lungs and release it to cells in body tissues. Each haem group is attached to a polypeptide chain (see page 201)

Complex ions

A complex ion consists of a central metal ion linked to a number of molecules or ions. The surrounding molecules or ions are ligands which use their lone-pair of electrons to form dative covalent bonds with the metal ion. The number of ligands in a complex ion is typically two, four or six.

Hydrated metal ions are the commonest complexes. The hydrated aluminium, iron(III) and copper(II) ions are all examples of aqua complexes.

Chemists have an alternative name for dative covalent bonds which they often prefer to use when describing complex ions. The alternative name is 'co-ordinate bond' which also gives rise to the name 'co-ordination compound' for materials containing complexes.

Co-ordination compounds contain complexes which may be cations, anions or neutral molecules. Examples of co-ordination compounds include:

- $K_3[Fe(CN)_6]$ containing the negatively charged complex ion $[Fe(CN)_6]^{3-}$

- $[Ni(NH_3)_6]Cl_2$ with the positively charged complex ion $[Ni(NH_3)_6]^{2+}$

- $Ni(CO)_4$ a neutral complex between nickel atoms and carbon monoxide molecules.

Definition

Ligands are the molecules or ions bound to the central metal ion in a complex ion.

Definition

The **co-ordination number** of a metal ion in a complex is the number of dative covalent (co-ordinate) bonds formed with the metal ion by the surrounding ligands.

Figure 3.3.3 ▶
Crystals of some co-ordination compounds: hydrated $NiSO_4$, $FeSO_4$, $CoCl_2$, $CuSO_4$, $Cr_2(SO_4)_3$ plus $K_3[Fe(CN)_6]$

Figure 3.3.4 ▶
The blue pigment in this picture is Prussian blue which consists of iron(III) hexacyanoferrate(II) (George Romney, 1763)

Figure 3.3.5 ▶
Naming of complex ions. Note that the overall charge on the complex ion is the sum of the charges on the metal ion and the ligands

Names of complex ions

The systematic names of complex ions show:

- ■ first the number of ligands, di-, tri-, tetra-, penta-, hexa-
- ■ then the type of ligands (in alphabetical order if there is more than one type of ligand), such as aqua- for water molecules, ammine- for ammonia, chloro- for chloride ions and cyano- for cyanide ions, CN^-
- ■ next the identity of the central metal atom in a form which shows whether or not the ion is a cation or an anion
  - for cations (and uncharged complexes) the metal name is normal, such as silver, iron or copper
  - for anions the metal name ends in –ate and often has an old-fashioned style such as argentate for silver, ferrate for iron and cuprate for copper
- ■ finally the oxidation number of the metal.

Examples:

- ■ $[Ag(NH_3)_2]^+$ is the colourless diamminesilver(I) ion
- ■ $[Cu(H_2O)_6]^{2+}$ is the pale blue hexaaquacopper(II) ion
- ■ $[CuCl_4]^{2-}$ is the yellow tetrachlorocuprate(II) ion
- ■ $[Fe(CN)_6]^{3-}$ is the yellow hexacyanoferrate(III) ion.

There are two common visible signs of a reaction to form a new complex ion:

- ■ a colour change, or
- ■ an insoluble solid dissolving.

A very familiar example of a colour change is seen on adding excess ammonia solution to copper(II) sulfate solution. This produces a deep-blue solution containing copper(II) ions complexed with ammonia molecules.

The test for chloride ions is an example of an insoluble salt dissolving as a result of a complex forming. Adding silver nitrate to a solution of chloride ions produces a white precipitate. The precipitate redissolves on adding ammonia solution as the silver ions in the silver chloride form the diamminesilver(I) ion.

Section three **Inorganic Chemistry**

Test yourself

Test yourself

1 What is the oxidation state of the metal ion in these complex ions:

$$[NiCl_4]^{2-}, [Ag(NH_3)_2]^+, [Ag(S_2O_3)_2]^{3-}, [Fe(H_2O)_6]^{3+}, [Fe(CN)_6]^{4-}?$$

CD-ROM

2 Write ionic equations for the reactions used to test for the presence of chloride ions in solution.

3 The fixer used to remove unexposed and undeveloped silver bromide from photographic film contains thiosulfate ions. A silver ion forms a complex with two thiosulfate ions as the silver bromide dissolves. Write an equation for this reaction.

4 Tollen's reagent contains the diamminesilver(I) ion. How is this reagent used in organic chemistry (see pages 174–175)? What is the result of a positive test? Why are the silver ions present as a complex in the reagent?

Shapes of complex ions

The shapes of complex ions depend on the number of ligands around the central metal ion. There is no simple rule for predicting the shapes of complexes from their formulae. Typically complexes with:

- ▪ six smaller ligands, such as H_2O and NH_3, are octahedral as in $[Mn(H_2O)_6]^{2+}$
- ▪ four larger ligands, such as Cl^-, are usually tetrahedral as in $[CuCl_4]^{2-}$
- ▪ two ligands are linear as in $[Ag(NH_3)_2]^+$, $[Ag(S_2O_3)_2]^{3-}$.

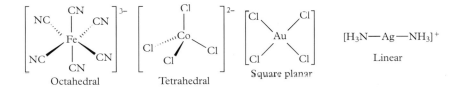

Octahedral   Tetrahedral   Square planar $[H_3N-Ag-NH_3]^+$ Linear

▲ **Figure 3.3.6**
Shapes of complex ions

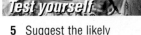

Figure 3.3.7 ▲
A coffee pot electroplated with silver. Experience has shown that for good results the silver should be present in the plating solution as a complex ion. The usual complex is a cyanide complex, $[Ag(CN)_2]^-$

Cisplatin is an anticancer drug consisting of a complex of platinum, ammonia molecules and chloride ions, $Pt(NH_3)_2Cl_2$. The ion is planar and can exist as *cis* and *trans* isomers (see page 147). The *cis* isomer inhibits cell division by cross linking the two chains in the DNA double helix. The complex does not, however, prevent cell growth. This makes it a useful treatment for cancer. Unfortunately cisplatin is toxic and has unpleasant side effects. Clinical trials have led to the discovery of other platinum complexes which are also effective.

Types of ligand

Some ligands use a single lone-pair of electrons to form one bond with the central metal atom. These are all monodentate (or 'one-toothed') ligands. Examples of molecules which can act as monodentate ligands are water and ammonia. Examples of ions which act as monodentate ligands are hydroxide ions, chloride ions and cyanide ions.

Some ligands have more than one lone-pair of electrons which can form dative bonds to the metal ion. Bidentate ligands, for example, form two dative covalent bonds with metal ions in complexes. Examples of bidentate (or 'two-toothed') ligands are: 1,2-diaminoethane, amino acids (see page 197) and the ethanedioate ion.

Test yourself

Test yourself

5 Suggest the likely shape of these complex ions: $[Ag(CN)_2]^-$, $[Cr(NH_3)_6]^{3+}$, $[NiCl_4]^{2-}$.

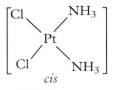

cis

Figure 3.3.8 ▲
Structure of cisplatin $Pt(NH_3)_2Cl_2$. This is the cis *isomer with both chloride ions on the same side of the planar complex.*

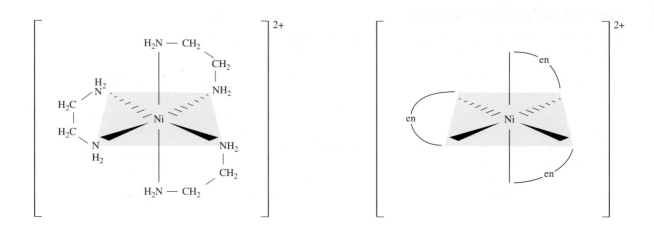

Figure 3.3.9 ▲
Representations of a complex formed by the bidentate ligand, 1,2-diaminoethane. Note the use of 'en' as an abbreviation for the ligand

Note

Monodentate ligands may also be called unidentate ligands.

Note

Edta crystals consist of the disodium salt of **e**thylene **d**iaminetetraacetic **a**cid, or as it is now called 1,2-bis [bis(carboxymethyl)amino] ethane. Chemists sometimes use the abbreviation Na_2H_2Y for the salt where Y represents the 4– ion.

Definition

Chelates are complex ions in which each ligand molecule or ion forms more than one dative covalent bond with the central metal ion. Chelates are formed by bidentate and polydentate ligands such as edta. The term chelate comes from the Greek word for a crab's claw reflecting the claw-like way in which chelating ligands grab hold of metal ions. Powerful chelating agents trap metal ions and effectively isolate them in solution.

Particularly impressive is the hexadentate ligand **edta** which forms six bonds with the central metal ion in a complex. Edta is the common abbreviation for an ion which binds so firmly with metal ions that it holds them in solution and effectively makes them chemically inactive. Edta is added to salad dressing to trap traces of metal ions which otherwise would catalyse the oxidation of oils. Edta is included in bathroom cleaners to help remove scale by dissolving deposits of calcium carbonate left by hard water. Edta can be used to treat lead poisoning by forming such a stable complex with the metal ions that they can be excreted through the kidneys.

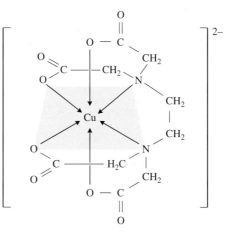

Figure 3.3.10 ▲
Complex ion formed by edta ions with a metal ion. The negative ion in the salt can fold itself round metal ions so that four oxygen atoms and two nitrogen atoms are able to form co-ordinate bonds with the metal ion. It is a hexadentate ligand

Test yourself

6 Draw a diagram to represent the complex formed between a Cr^{3+} ion and three ethanedioate ions.

7 **a)** Place in order of stability (with the most stable first):
      $[Ni(NH_3)_6]^{2+}$, $[Ni(en)_3]^{2+}$, $[Ni\text{-}edta]^{2-}$.

  b) Suggest a way of using equilibrium constants, K_c, to compare the stability of complex ions (see page 24).

Ligand substitution reactions

Complex ions often react by swapping one ligand for another. These ligand substitution reactions are usually reversible.

$$[Cu(H_2O)_6]^{2+}(aq) + 4NH_3(aq) \rightleftharpoons [Cu(NH_3)_4(H_2O)_2]^{2+}(aq) + 4H_2O(l)$$

pale blue   deep blue

The ligands NH_3 and H_2O are similar in size and are uncharged. Exchange reactions can take place between these ligands without a change in the co-ordination number of the metal ion.

The chloride ion is larger than uncharged ligands, such as water, so that fewer chloride ions can fit round a central metal ion. As a result, ligand exchange involves a change in co-ordination number.

$$[Cu(H_2O)_6]^{2+}(aq) + 4Cl^-(aq) \rightleftharpoons [CuCl_4]^{2-}(aq) + 6H_2O(l)$$

pale blue   yellow

◄ **Figure 3.3.11**
Reversible ligand exchange reaction of copper(II) ions. Note that in this example substitution is incomplete. The aqua complex is pale blue but the ammine complex is a very deep blue

◄ **Figure 3.3.12**
Ligand exchange with change of co-ordination number. The aqua complex is pale blue but the chloro complex is yellow

Test yourself

8 Write equations for these ligand substitution reactions:

a) the hexaaquacobalt(II) ion with ammonia molecules to form the hexaamminecobalt(II) ion

b) the hexaamminecobalt(II) ion with chloride ions to form the tetrachlorocobaltate(II) ion

c) the hexaaquairon(II) ion with cyanide ions to form the hexacyanoferrate(II) ion.

9 Account for these changes with the help of equations:
adding ammonia solution to a pale blue solution of hydrated copper(II) ions produces a pale blue precipitate of the hydrated hydroxide; on adding more ammonia solution the precipitate dissolves to give a deep blue solution.

Complex-forming titration

Edta forms such stable complex ions with metal ions that it can be used to measure the concentration of metal ions in solution. The procedure for these complex-forming titrations is the same as for any other titration.

The titration determines the volume of a standard solution of the complex-forming reagent needed to react exactly with a measured volume of the unknown solution of a metal ions. To find the end-point, the analyst adds an indicator which forms a coloured but unstable complex with the metal ion. An added buffer solution makes sure that the pH of the solution is in the right range for the indicator to be effective.

A suitable indicator for edta titrations is Eriochrome black T which is normally blue in solution at pH 10. On adding the indicator to the solution of metal ions at the start of a titration it produces a wine–red complex.

Edta forms a more stable complex with metal ions than the indicator. So as edta solution is run in from the burette, it takes the metal ions from the indicator. At the end-point all the metal ions have been complexed by titration with edta. The last drop of edta leaves no metal ions to form the red complex with the indicator. The indicator colour reverts to its normal blue colour.

[metal-indicator complex] + edta $\rightleftharpoons$ [metal-edta complex] + free indicator

red   blue

Figure 3.3.13 ▲
The colour of Eriochrome black T when complexed with calcium ions and the colour of the free indicator at pH10

10 a) Water hardness is often related to calcium as calcium carbonate. By convention, hardness values are quoted in mg dm^{-3} $CaCO_3$. Calculate the hardness of water in mg dm^{-3} $CaCO_3$ for the sample tested in the worked example.

b) Show that a concentration of 1 mg dm^{-3} $CaCO_3$ is equivalent to 1 ppm (parts per million) $CaCO_3$.

11 An analyst carried out an edta titration to measure the mass of zinc ions in a capsule which people swallow to supplement the zinc in their diet. The analyst dissolved one capsule in water and titrated it with 0.0200 mol dm^{-3} edta solution in the presence of Eriochrome black T and an alkaline buffer solution. The average volume of edta solution required to react with the zinc in one capsule was 11.45 cm^3. Calculate the mass of zinc in one capsule.

Worked example

An alkaline buffer and a few drops of Eriochrome black T indicator were added to 100.0 cm^3 of a hard water sample containing calcium ions. In the titration 22.8 cm^3 of 0.0100 mol dm^{-3} edta were run in from a burette until the indicator changed from red to blue. What was the concentration of the calcium ions?

Notes on the method

Always start by writing the equation for the reaction.

Remember to convert volumes in cm^3 to volumes in dm^3 by dividing by 1000.

In any titration there is one unknown – in this case the unknown concentration of the calcium ions, c_A (in mol dm^{-3}).

Answer

The equation for the reaction is:

$$Ca^{2+}(aq) + edta^{4-}(aq) \longrightarrow [Ca.edta]^{2-}(aq)$$

Volume of hard water in the flask $= \dfrac{100.0}{1000}$ dm^3 = 0.1 dm^3

So the amount of calcium ions in the flask = (0.1 dm$^3 \times c_A$) mol

Volume of edta from the burette $= \dfrac{22.8}{1000}$ dm^3 = 0.0228 dm^3

Concentration of edta = 0.0100 mol dm^{-3}

Amount of edta from the burette = 0.0228 dm$^3 \times$ 0.0100 mol dm^{-3}

The equation shows that 1 mol edta reacts with 1 mol calcium ions. So the amount of edta added from the burette at the end point must be the same as the amount of calcium ions in the flask.

Therefore: (0.1 dm$^3 \times c_A$) mol = 0.0228 dm$^3 \times$ 0.0100 mol dm^{-3}

Hence: $c_A = \dfrac{0.0228 \text{ dm}^3 \times 0.0100 \text{ mol dm}^{-3}}{0.1 \text{ dm}^3}$ = 0.00228 mol dm^{-3}

Concentration of calcium ions in hard water = 0.00228 mol dm^{-3}

3.4 Acid–base reactions

The first classification of acids was based on their properties: their sharp taste, effect on indicators and their characteristic reactions. Scientists, however, were not content simply to observe, classify and make generalisations. Chemists wanted to understand why acids have similar properties and for this they needed theories to explain the behaviour of these compounds. Different theories have led to changing explanations and to three related definitions.

The Arrhenius theory

Svante Arrhenius (1859 – 1927) suggested a theoretical explanation for the similarities between acids in aqueous solution. He was a Swedish physical chemist who pioneered the theory of ions in solution. He explained the behaviour of acids in terms of hydrogen ions. What acids have in common, according to his theory, is that they produce hydrogen ions when they dissolve in water.

$$HCl(g) + aq \longrightarrow H^+(aq) + Cl^-(aq)$$

Figure 3.4.1 ▲
A geologist using dilute hydrochloric acid to identify limestone

The Brønsted–Lowry theory

The normal definition of an acid generally used today is based on the Brønsted–Lowry theory which defines an acid as a molecule or ion which can give away a hydrogen ion to something else. Johannes Brønsted (1879 – 1947) was a Danish physical chemist. He published his theory in 1923 at the same time as Thomas Lowry (1874 – 1936) of the University of Cambridge but the two worked independently.

According to this theory, acids are hydrogen ion (proton) donors. This definition is more general than the Arrhenius theory because it covers reactions which happen without water and in non-aqueous solvents. According to this theory hydrogen chloride molecules give hydrogen ions to water molecules when they dissolve in water producing hydrated hydrogen ions called oxonium ions. The water accepts the proton from the acid thus acting as a *base*.

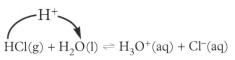

$$HCl(g) + H_2O(l) \rightleftharpoons H_3O^+(aq) + Cl^-(aq)$$

The Brønsted–Lowry theory remains the most popular definition of acids and bases and underlies the treatment of acid–base equilibria in section 2.10. Chemists use this theory to explain pH and buffer solutions.

Test yourself

1 Give one example for each of the characteristic reactions of acids with: metals, metal oxides, metal hydroxides and metal carbonates.

2 Use the Arrhenius theory to explain why dilute aqueous solutions of hydrochloric, sulfuric and nitric acids have similar properties.

CD-ROM

Test yourself

3 Which of the examples you have chosen to answer in question 1 are acid–base reactions according to the Brønsted–Lowry theory?

4 Show that the reaction of the hydrated aluminium(III) ion with hydroxide ions is an acid–base reaction according to the Brønsted–Lowry theory (see page 91).

5 Identify the Brønsted–Lowry acid and the Brønsted–Lowry base in the reaction taking place in Figure 3.4.1.

Lewis acids and bases

Gilbert Lewis (1875 – 1946) was the US physical chemist who clarified theories of chemical bonding. It was Lewis who developed the idea that atoms gain or lose outer electrons to create 'octets' either by transferring or sharing electrons. This led to the dot-and-cross diagrams to describe the ionic and covalent bonding in simple compounds. Electron pairs played an important part in Lewis's thinking and led him to suggest an even more general definition of acids and bases.

A Lewis acid is a molecule or ion which can form a bond by accepting a pair of electrons. The formation of an oxonium ion is a special case of a Lewis acid–base reaction between the proton (a Lewis acid) and a water molecule (a Lewis base). So in this theory it is the proton rather than the proton donor which is an acid.

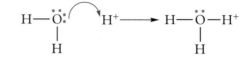

Figure 3.4.2 ▶
Formation of an oxonium ion. The Lewis base, water, forms a dative covalent bond with a proton which is the Lewis acid

This much wider definition of acids and bases includes the formation of a complex ion within the family of acid–base reactions. The metal ion (a Lewis acid) accepts pairs of electrons from the ligands (Lewis bases).

$$6H_2O + Cu^{2+} \longrightarrow [Cu(H_2O)_6]^{2+}$$

Lewis base   Lewis acid   aqua complex (hydrated ion)

Figure 3.4.3 ▶
Complex formation as a Lewis acid-base reaction

A Lewis base is a molecule or ion which can form a bond by donating a pair of electrons. This means that any Brønsted–Lowry base is also a Lewis base; but the Lewis definition is broader because it refers to any reaction in which the base provides a pair of electrons to form a co-ordinate bond.

The Lewis acid–base theory also covers the reactions between acidic oxides and basic oxides which do not involve proton transfer.

$$CaO + SO_3 \longrightarrow CaSO_4$$

basic oxide   acidic oxide   salt

Figure 3.4.4 ▶
The reaction of a basic metal oxide with an acidic non-metal oxide to form a salt

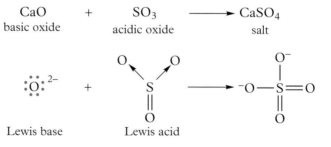

Lewis base   Lewis acid

Aluminium chloride, $AlCl_3$, is a Lewis acid. This accounts for its use as a catalyst in a Friedel Crafts reaction (see page 164).

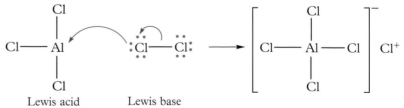

Figure 3.4.5 ▶
Aluminium chloride acting as a Lewis acid

Lewis acid   Lewis base
(electron pair acceptor)   (electron pair donor)

Chemists normally use the Brønsted-Lowry theory. When they use the terms acid and base they mean 'proton donor' and 'proton acceptor'. To signal that they are using the wider Lewis definition they refer to 'Lewis acids' and 'Lewis bases'.

6 a) Draw dot-and-cross diagrams to illustrate the octet rule giving one example of ionic bonding and another of covalent bonding.

b) Draw a dot-and-cross diagram to illustrate an exception to the octet rule.

c) What is the link between the octet rule and the electron configurations of noble gases?

7 Ammonia reacts with boron trifluoride to form a single product. By drawing dot-and-cross diagrams, suggest a structure for the product and show that this is a reaction of a Lewis acid with a Lewis base.

8 Which of the examples you have chosen to answer in question 1 (on page 99) are acid–base reactions according to the Lewis theory?

Section three   Inorganic Chemistry

3.5 Redox reactions

Redox reactions are very important in the natural environment, in living things and in modern technology. It is not surprising that planet Earth, with its oxygen atmosphere, has an extensive redox chemistry.

Redox in action

Approximately a thousand billion (10^{12}) moles of oxygen are removed from the atmosphere every year to oxidise inorganic ions and molecules such as iron(II) ions from weathered rock and volcanic gases such as hydrogen sulfide, carbon monoxide and methane.

A series of redox reactions is involved in the metabolic pathways of respiration. These pathways produce the molecule adenosine triphosphate (ATP). ATP transfers the energy from oxidising food to movement, growth and all the other activities in living things which need a source of energy.

The voltages of chemical cells are all derived from the free energy of redox reactions as described in section 2.15. Industrial processes which use electrolysis to make products such as chlorine and aluminium also involve redox reactions.

Test yourself

1 Describe, in terms of gain or loss of oxygen, the redox reactions in a blast furnace which help to extract iron from iron(III) oxide.

2 Describe, in terms of gain or loss of electrons, the redox reactions in these examples:

a) The changes at the electrodes during the manufacture of chlorine and sodium hydroxide from sodium chloride solution in a membrane cell.

b) The purification of copper metal by the electrolysis of copper(II) sulfate with copper electrodes.

3 Describe, in terms of changes of oxidation states, the redox reaction of concentrated sulfuric acid with potassium bromide.

Figure 3.5.1 ▲
Volcanoes release reducing gases into the atmosphere where they react with oxygen

Definitions of redox

Descriptions and theories of oxidation and reduction have developed over the years. As a result there have been several definitions of redox reactions. Even now chemists do not use just one definition but pick the one that most suits their purposes in a given context. Whatever the definition, oxidation and reduction always go together.

Oxidation originally meant combination with oxygen but the term now covers all reactions in which atoms or ions lose electrons. Chemists have extended the definition of oxidation to cover molecules as well as ions by defining oxidation as a change which makes the oxidation number of an element more positive, or less negative.

Similarly, reduction originally meant removal of oxygen but the term now covers all reactions in which atoms, molecules or ions gain electrons. Defining reduction as a change which makes the oxidation number of an element more negative or less positive extends this idea to cover molecules as well as ions.

Oxidation states

A p-block or d-block element typically forms a series of two or more compounds in different oxidation states. Displaying the compounds of an element on an oxidation state diagram provides a 'map' of its chemistry.

Oxidation number rules

1. The oxidation number of uncombined elements is zero.

2. In simple ions the oxidation number of the element is the charge on the ion.

3. The sum of the oxidation numbers in a neutral compound is zero.

4. The sum of the oxidation numbers for an ion is the charge on the ion.

5. Some elements have fixed oxidation numbers in all their compounds

Metals		**Non-metals**	
group 1 metals (e.g. Li, Na, K)	+1	hydrogen (except in metal hydrides, H^-)	+1
group 2 metals (e.g. Mg, Ca, Ba)	+2	fluorine	−1
Aluminium	+3	oxygen (except in peroxides, O_2^{2-}, and compounds with fluorine)	−2
		chlorine (except in compounds with oxygen and fluorine)	−1

◀ **Figure 3.5.2**
Rules for determining oxidation numbers. Oxidation number rules also apply, in principle, in organic chemistry but it is often easier to use the older description of oxidation as either the addition of oxygen to a molecule or the removal of hydrogen (see page 102)

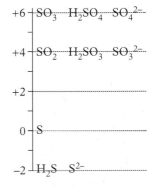

Figure 3.5.3 ▲
The oxidation states of sulfur

Test yourself

4. What is the oxidation number of:

   a) iodine in KIO_3
   b) nitrogen in the nitrate ion, NO_3^-
   c) oxygen in F_2O
   d) sodium in NaH
   e) oxygen in sodium peroxide, Na_2O_2?

5. Give examples of the sulfur compounds shown in Figure 3.5.3 which can:

   a) act as an oxidising agent or as a reducing agent depending on the conditions
   b) only act as an oxidising agent
   c) only act as a reducing agent.

6. Draw a chart to show the main oxidation states of these elements:

   a) nitrogen
   b) chlorine.

7. Are these elements oxidised or reduced in these conversions:

   a) magnesium to magnesium sulfate
   b) iodine to aluminium iodide
   c) hydrogen to lithium hydride
   d) iodine to iodine monochloride, ICl?

Inorganic oxidising agents

Oxidising agents are chemical reagents which can oxidise other atoms, molecules or ions by taking electrons away from them. Common oxidising agents are oxygen, chlorine, bromine, hydrogen peroxide, the manganate(VII) ion in potassium manganate(VII), and the dichromate(VI) ion in potassium or sodium dichromate(VI).

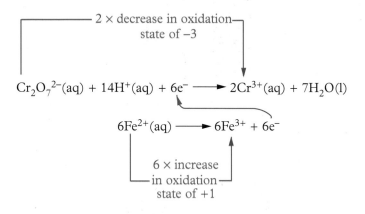

Figure 3.5.4 ▶

Dichromate(VI) ions acting as an oxidising agent by taking electrons from iron(II) ions in acid solution. An oxidising agent is itself reduced when it reacts

Some reagents change colour when they are oxidised which makes them useful for detecting oxidising agents. In particular a colourless solution of iodide ions is oxidised to iodine which turns the solution to a yellow–brown colour.

$$2I^-(aq) \rightarrow I_2(aq) + 2e^-$$

electrons taken by the oxidising agent

This can be a very sensitive test if starch is also present because starch forms an intense blue–black colour with iodine. Moistened starch–iodide paper is a version of this test which can detect oxidising gases such as chlorine and bromine vapour.

Inorganic reducing agents

Reducing agents are chemical reagents which can reduce other atoms molecules or ions by giving electrons to them. Common reducing agents are metals such as zinc, iron and tin often with acid, also sulfur dioxide, iron(II) ions and iodide ions.

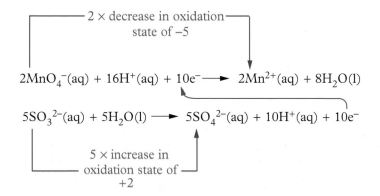

Figure 3.5.5 ▲

Sulfite ions acting as a reducing agent by giving electrons to manganate(VII) ions. A reducing agent is itself oxidised when it reacts

Some reagents change colour when they are reduced which makes them useful for detecting reducing agents (see Figures 3.5.6 and 3.5.7).

The **tri-iodide ion** forms when iodine dissolves in aqueous potassium iodide.

$$I_2(s) + I^-(aq) \rightleftharpoons I_3^-(aq)$$

Iodine is only very slightly soluble in water. It dissolves in potassium iodide solution because it forms $I_3^-(aq)$. A reagent labelled 'iodine solution' is normally $I_2(s)$ in KI(aq). The $I_3^-(aq)$ ion is a yellow-brown colour which explains why aqueous iodine looks quite different from a solution of iodine in a non-polar solvent such as hexane.

Test yourself

8 Write half-equations (see page 52) to show what happens when these molecules or ions act as oxidising agents:

a) iron(III) ions

b) bromine molecules

c) hydrogen peroxide.

9 Explain the use of moist starch–iodide paper to test for chlorine gas.

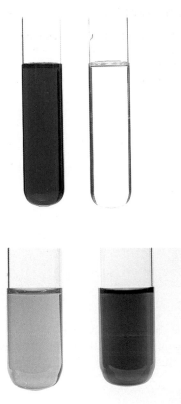

◀ *Figure 3.5.6*
A test for reducing agents. The test: add a solution of potassium manganate(VII) acidified with dilute sulfuric acid. The result: purple solution turns colourless. Purple MnO_4^- *ions are reduced to very pale pink* Mn^{2+} *ions*

◀ *Figure 3.5.7*
A test for reducing agents. The test: add potassium dichromate(VI) solution acidified with dilute sulfuric acid. The result: orange solution turns green. Orange $Cr_2O_4^{2-}$ *ions reduced to green* Cr^{3+} *ions*

Equations for redox equations

Oxidation numbers help to balance redox equations because the total decrease in oxidation number for the element reduced must equal the total increase in oxidation number for the element oxidised. This is illustrated here for the oxidation of iron(II) ions by manganate(VII) ions in acid solutions.

Step 1: Write down the formulae for the atoms, molecules and ions involved in the reaction.

$$MnO_4^- + H^+ + Fe^{2+} \longrightarrow Mn^{2+} + H_2O + Fe^{3+}$$

Step 2: Identify the elements which change in oxidation number and the extent of change.

$$\overbrace{MnO_4^- + H^+ + Fe^{2+} \longrightarrow Mn^{2+}}^{\text{change of } -5} + H_2O + Fe^{3+}$$

change of +1

Step 3: Balance the equation so that the decrease in oxidation number of one element equals the total increase of the other element.

In this example the decrease of −5 in the oxidation number of manganese is balanced by five iron(II) ions each increasing their oxidation number by +1.

$$MnO_4^- + H^+ + 5Fe^{2+} \longrightarrow Mn^{2+} + H_2O + 5Fe^{3+}$$

Step 4: Balance for oxygen and hydrogen.

In this example the four oxygen ions from the manganate(VII) ion join with eight hydrogen ions to form four water molecules.

$$MnO_4^- + 8H^+ + 5Fe^{2+} \longrightarrow Mn^{2+} + 4H_2O + 5Fe^{3+}$$

Test yourself

10 Write half-equations (see page 52) to show what happens when these molecules or ions act as reducing agents:

a) zinc atoms

b) iodide ions

c) sulfur dioxide.

Step 5: In an ionic equation, check that the + and − charges balance and add state symbols.

The net charge on the left is now 17+ which is the same as the net charge on the right.

$$MnO_4^-(aq) + 8H^+(aq) + 5Fe^{2+}(aq) \longrightarrow Mn^{2+}(aq) + 4H_2O(l) + 5Fe^{3+}(aq)$$

Test yourself

11 Write a balanced equation for each of these redox reactions:

a) iodate(V) ions with iodide ions in acid solution to form iodine

b) hydrogen peroxide with iron(II) to form water and iron(III) ions

c) dichromate(VI) ions in acid solution with sulfite ions to form chromium(III) ions and sulfate ions

d) manganese(IV) oxide with concentrated hydrochloric acid to form chlorine gas and manganese(II) ions

e) copper metal with nitrate ions in nitric acid to form copper(II) nitrate and nitrogen dioxide gas.

Redox titrations

A redox titration determines the concentration of a solution of an oxidising agent or of a reducing agent. A titration measures the volume of a standard solution of oxidising agent or reducing agent needed to react exactly with a measured volume of the unknown solution. The procedure only gives accurate results if the reaction is rapid and is exactly described by the chemical equation.

Iodine–thiosulfate titrations

Iodine–thiosulfate titrations measure amounts of oxidising agents. The method is based on the fact that oxidising agents convert iodide ions to iodine quantitatively.

$$2I^-(aq) \longrightarrow I_2(aq) + 2e^-$$

Among the oxidising agents which do this are iron(III) ions, copper(II) ions, chlorate(I) ions, dichromate(VI) ions in acid, iodate(V) ions in acid and manganate(VII) ions in acid. The iodine stays in solution in excess potassium iodide forming a yellow–brown colour.

The analyst titrates the iodine with a standard solution of sodium thiosulfate which reduces iodine molecules back to iodide ions. This also happens quantitatively exactly as in the equation.

$$I_2(aq) + 2S_2O_3^{2-}(aq) \longrightarrow 2I^-(aq) + S_4O_6^{2-}(aq)$$

The greater the amount of oxidising agent added, the more iodine is formed and so more thiosulfate is needed from a burette to react with it. On adding thiosulfate from a burette, the colour of the iodine gets paler. Near the end-point the solution is a very pale yellow. Adding a little soluble starch solution as an indicator near the end point gives a sharp colour change from blue–black to colourless.

Definition

Thio compounds are sulfur compounds with similar formulae and structures to equivalent oxygen compounds.

The thiosulfate ion, $S_2O_3^{2-}$, for example, can be thought of as a sulfate ion, SO_4^{2-}, with one oxygen atom replaced by a sulfur atom.

Worked example

A standard solution of potassium dichromate(VI) was prepared by dissolving 1.029 g in water and making the solution up to 250 cm³. Then 25.0 cm³ samples of the standard solution were added to excess potassium iodide and dilute sulfuric acid. In each case the iodine formed was titrated with the solution of sodium thiosulfate from a burette. The average volume of sodium thiosulfate solution needed to decolourise the blue iodine–starch colour at the end-point was 20.20 cm³. Calculate the concentration of the solution of sodium thiosulfate standardised in this way.

Notes on the method

First find the concentration of the standard solution.

Write the equations and work out the amount in moles of $S_2O_3^{2-}$ equivalent to 1 mol $Cr_2O_7^{2-}$.

There is no need to consider the amounts of iodine in the calculations.

Calculate the amount of potassium dichromate(VI) in a 25.0 cm³ (0.025 dm³) sample. Hence calculate the amount of sodium thiosulfate required to react with the iodine produced by this amount of the oxidising agent.

Answer

Molar mass of potassium dichromate(VI), $K_2Cr_2O_7 = 294$ g mol⁻¹.

Amount of $K_2Cr_2O_7$ in the standard solution $= \dfrac{1.029 \text{ g}}{294.0 \text{ g mol}^{-1}}$

The volume of the standard solution = 0.25 dm³

The concentration of the standard solution $= \left(\dfrac{1.029 \text{ g}}{294 \text{ g mol}^{-1}}\right) \div 0.25 \text{ dm}^3$

$= 0.014 \text{ mol dm}^{-3}$

The equations for producing iodine:

$$Cr_2O_7^{2-}(aq) + 6I^-(aq) + 14H^+(aq) \longrightarrow 2Cr^{3+}(aq) + 3I_2(aq) + 7H_2O$$

The equation for the reaction during the titration:

$$I_2(aq) + S_2O_3^{2-}(aq) \longrightarrow 2I^-(aq) + S_4O_6^{2-}(aq)$$

So 1 mol $Cr_2O_7^{2-}$ produces 3 mol I_2 which reacts with 6 mol $S_2O_3^{2-}$.

Overall 6 mol $S_2O_3^{2-}$ react with the iodine formed by 1 mol $Cr_2O_7^{2-}$.

Amount of $Cr_2O_7^{2-}(aq)$ in the flask at the start

$$= 0.025 \text{ dm}^3 \times 0.014 \text{ mol dm}^{-3}$$

So the amount of thiosulfate in 20.20 cm³ (= 0.0202 dm³) solution

$$= 6 \times 0.025 \text{ dm}^3 \times 0.014 \text{ mol dm}^{-3}$$

So the concentration of the sodium thiosulfate solution

$$= (6 \times 0.025 \text{ dm}^3 \times 0.014 \text{ mol dm}^{-3}) \div 0.0202 \text{ dm}^3$$

$$= 0.104 \text{ mol dm}^{-3}$$

Definitions

Chemists use the term **standard solution** to describe a solution with an accurately known concentration. The direct method for preparing a standard solution is to dissolve a weighed sample of a primary standard in water and to make the solution up to a definite volume in a graduated flask.

A **primary standard** is a chemical which can be weighed out accurately to make up a standard solution. A primary standard must:

■ be very pure

■ not gain or lose mass when exposed to the air

■ have a relatively high molar mass so weighing errors are minimised

■ react exactly as described by the chemical equation.

Primary standards for redox titrations include: potassium dichromate(VI) and potassium iodate(V).

Inorganic Chemistry

Section three

Potassium manganate(VII) titrations

Potassium manganate(VII) is an important reagent in redox titrations because it will oxidise many reducing agents in acid conditions. The reactions go according to their equations which makes them suitable for quantitative work.

$$MnO_4^- + 8H^+(aq) + 5e^- \longrightarrow Mn^{2+}(aq) + 4H_2O(l)$$

No indicator is required for a manganate(VII) titration. On adding the solution from a burette the manganate(VII) rapidly changes from purple to colourless (because the colour of the Mn^{2+} ion is so pale). At the end–point it only takes the slightest excess of manganate(VII) to give a permanent red–purple colour.

Test yourself · D

12 What volume of a solution of 0.02 mol dm^{-3} potassium manganate(VII) solution is required to oxidise 20 cm^3 of:

a) 0.010 mol dm^{-3} iron(II) sulfate solution

b) 0.080 mol dm^{-3} sodium sulfite solution

c) 0.200 mol dm^{-3} hydrogen peroxide solution.

13 A 10 cm^3 sample of domestic bleach was diluted to 100 cm^3 in a graduated flask. Then 10 cm^3 volumes of the diluted bleach were mixed with excess potassium iodide acidified with ethanoic acid. Each mixture was titrated with 0.150 mol dm^{-3} sodium thiosulfate solution using starch to detect the end-point. The mean titre was 20.55 cm^3. Calculate the concentration of chlorine in the undiluted bleach.

14 The iron minerals in a 1.340 g sample of iron ore were dissolved in acid. The solution was titrated with 0.020 mol dm^{-3} potassium manganate(VII) solution. The titre was 26.75 cm^3. Calculate the percentage by mass of iron in the ore.

15 A 0.275 g sample of a metal alloy containing copper was dissolved in acid and then diluted with water. An excess of potassium iodide was added. The copper(II) reacted quantitatively with the iodide ions to form a precipitate of copper(I) iodide together with iodine. In a titration, the iodine formed reacted with 22.50 cm^3 of 0.140 mol dm^{-3} sodium thiosulfate solution. Calculate the percentage by mass of copper in the metal sample.

3.6 The periodic table

The modern arrangement of elements in the periodic table reflects the underlying electron configurations of the atoms. Explaining the chemical behaviour of elements and their compounds in terms of electron configuration is a central theme of inorganic chemistry.

CD-ROM

The building-up principle

The division of the periodic table into s, p, d and f blocks reflects an underlying principle which governs how electrons fill the atomic orbitals in atoms. The principle states that the electron configurations of atoms build up according to a set of rules. The three rules are that:

- electrons go into the orbital at the lowest available energy level
- each orbital can only contain at most two electrons (with opposite spins)
- where there are two or more orbitals at the same energy, they fill singly before the electrons pair up.

▼ *Figure 3.6.1*
The arrangement of the elements in the periodic table reflects the order in which electrons fill atomic orbitals. The chemistry of an element is largely determined by the number and arrangement of electrons in its outer shell

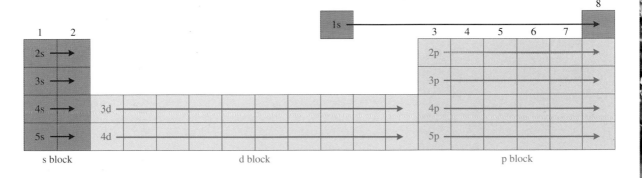

The s-block elements are the elements in groups 1 and 2. For these elements the last electron added to the atomic structure goes into the s-orbital in the outer shell. All the elements in the s-block are reactive metals.

The p-block elements are the elements in groups 3, 4, 5, 6, 7 and 8 in the periodic table. For these elements the last electron added to the atomic structure goes into one of the three p-orbitals in the outer shell.

◄ *Figure 3.6.2*
Electron configurations of an s-block element and a p-block element. The arrows represent electrons. Arrows pointing in opposite directions indicate that where there are two electrons in the same orbital they have opposite spins

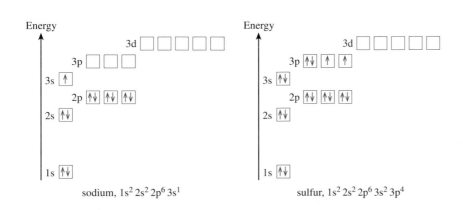

sodium, $1s^2 2s^2 2p^6 3s^1$ sulfur, $1s^2 2s^2 2p^6 3s^2 3p^4$

The d-block elements are the elements in the three horizontal rows of elements in periods 4, 5 and 6 for which the last electron added to the atomic structure goes into a d-orbital. In period 4, the d-block elements run from scandium ($1s^2 2s^2 2p^6 3s^2 3p^6 3d^1 4s^2$) to zinc ($1s^2 2s^2 2p^6 3s^2 3p^6 3d^{10} 4s^2$)

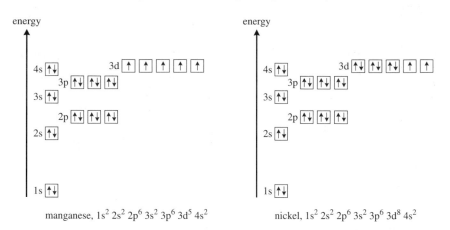

manganese, $1s^2\ 2s^2\ 2p^6\ 3s^2\ 3p^6\ 3d^5\ 4s^2$ nickel, $1s^2\ 2s^2\ 2p^6\ 3s^2\ 3p^6\ 3d^8\ 4s^2$

Figure 3.6.3 ▲
Electron configurations of two d-block elements as free atoms. Note that orbitals fill singly before the electrons start to pair up

The changes in the properties across a series of d-block elements are much less marked than the big changes across a p-block series. This is because from one element to the next, as the proton number of the nucleus increases by one, the extra electron goes into the inner d-sub-shell.

In the d-block series in period 4, the outer shell is always the 4s orbital which fills before the 3d starts to fill because it is at a slightly lower energy. Nevertheless the 4s electrons are the first electrons to take part in bonding and chemical changes because they are the outer electrons.

The chemistry of an atom is largely determined by its outer electrons because they are the first to get involved in reactions. So the elements Sc to Zn in period 4 are similar in many ways (see sections 3.11 and 3.12).

Patterns in the periodic table

Chemists look for trends and repeating patterns in the periodic table. These give chemists an overview which means that they can avoid having to learn many unrelated facts but instead can make sense of the chemistry of one element in the context of related elements.

The third horizontal row of elements in the periodic table starts with sodium and ends with argon. The changes in properties from Na to Ar across period 3 show up periodic patterns particularly clearly.

Definition

A **periodic pattern** repeats itself at more or less regular intervals. The most obvious repetition from one period to the next in the table is from metals on the left to non-metals on the right.

Test yourself

1 Write out the electron configurations of these atoms: oxygen, fluorine, magnesium, titanium, cobalt.

2 a) Show that these ions and atoms have the same electron configuration: N^{3-}, O^{2-}, F^-, Ne, Na^+, Mg^{2+}.

  b) Account for the trend in radius of the ions and atoms in a).

Figure 3.6.4 ▼
Patterns and trends in the periodic table

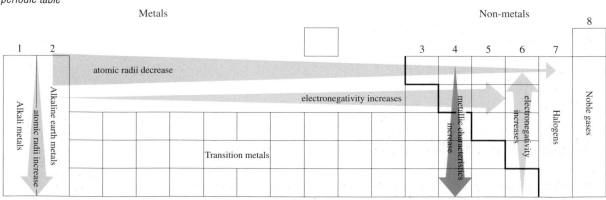

◄ Figure 3.6.5
The periodic variation of atomic radii of the elements Li to Ca.
1 pm = 10^{-3} nm = 10^{-12} m

It is not only the properties of the elements which show repeating patterns (see section 3.7) but the properties of the oxides (see section 3.8) and the chlorides (see see section 3.9) do so as well.

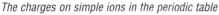

Figure 3.6.6 ▲
The charges on simple ions in the periodic table

In each s-block or p-block group of the periodic table the element in period 3 is often regarded as the typical element in the group. Chemists then generalise about trends down the group by comparing the elements in lower periods with the period 3 element. The first member or each group (usually an element in period 2) is often exceptional in some way.

Figure 3.6.7 ▲
Beryllium is a member of group 2 but it has some exceptional properties which make it useful as a material which is light and strong but also very resistant to corrosion

Definitions

Chemists often use the term **trend** to describe the way in which a property increases or decreases along a series of elements or compounds. In the periodic table the term can describe the variations of a property down a group or across a period.

The inner electrons of an atom reduce the pull of the nucleus on the electrons in the outer shell of an atom. Thanks to this **shielding** the electrons in the outer shell are attracted by an 'effective nuclear charge' which is less than the full charge on the nucleus.

Test yourself

3 Identify and describe trends and periodic patterns in Figures 3.6.5 and 3.6.6.

4 Explain in terms of electron configurations and shielding why the general trend across period 3 is for ionisation energies to rise while down group 2 the trend is for them to fall.

5 a) Give examples from the chemistries of beryllium and fluorine to show that the chemistry of a period 2 element can be exceptional when compared to the other elements in its group.

   b) What is it about the atomic structures of period 2 elements which means that they have some properties which are exceptional?

3.7 Period 3: the elements

The changes in properties from Na to Ar across period 3 show up particularly clearly the trends in the properties of the elements. Chemists seek to explain the patterns in terms of the electron configurations of the atoms as well as the structure and bonding of the elements

Aluminium is a shiny metal. A very thin layer of aluminium oxide on the surface of the metal protects it from corrosion. It consists of a giant structure of aluminium atoms held together by metallic bonding. The metal melts at 660 °C, boils at 2350 °C and is a good conductor of electricity. Electron configuration: $1s^2 2s^2 2p^6 3s^2 3p^1$.

Chlorine is a yellow–green gas which consists of diatomic molecules, Cl_2. It belongs to the halogen group of elements. The bonding between the atoms is covalent. Chlorine melts at −101 °C and boils at −34 °C and does not conduct electricity. Electron configuration: $1s^2 2s^2 2p^6 3s^2 3p^5$.

Sodium is a shiny metal in the group of alkali metals. It tarnishes quickly when exposed to the air. The element consists of a giant structure of sodium atoms held together by metallic bonding. The metal melts at 98 °C, boils at 900 °C and is a good conductor of electricity. Electron configuration: $1s^2 2s^2 2p^6 3s^1$.

Phosphorus can exists in various forms. The white allotrope consists of P_4 molecules. The more stable and less reactive red form has a giant structure. Bonding between the atoms in all forms is covalent. White phosphorus melts at 44 °C and boils at 280 °C and does not conduct electricity. Electron configuration: $1s^2 2s^2 2p^6 3s^2 3p^3$.

Magnesium is a silvery white metal which belongs to the group of alkaline–earth metals. It is harder and denser than sodium. The element consists of a giant structure of magnesium atoms held together by metallic bonding. The metal melts at 650 °C, boils at 1100 °C and is a good conductor of electricity. Electron configuration: $1s^2 2s^2 2p^6 3s^2$.

Sulfur is a yellow, crystalline solid. It is a non-metal. There are two solid allotropes with the same molecular structure but different crystal structures: rhombic and monoclinic sulfur. Both consist of S_8 molecules with the atoms in rings held together by covalent bonding. The stable form at room temperature is rhombic sulfur. The rhombic crystals melt at 113 °C; the liquid boils at 445 °C and does not conduct electricity. Electron configuration: $1s^2 2s^2 2p^6 3s^2 3p^4$.

Silicon is a blue–grey solid. The element consists of a giant structure of silicon atoms held together by covalent bonding. It melts at 1410 °C, boils at 2620 °C and is a semi-conductor. Chemically the element is classified as a non-metal but its physical properties mean that it can also be regarded as a metalloid. Electron configuration: $1s^2 2s^2 2p^6 3s^2 3p^2$.

Argon is a colourless, unreactive gas which belongs to the family of noble metals. The gas consists of single atoms. Argon melts at −189 °C and boils at −186 °C and does not conduct electricity. Electron configuration: $1s^2 2s^2 2p^6 3s^2 3p^6$.

▲ *Figure 3.7.1 The elements of period 3: properties, structure and bonding*

Structure and bonding of the elements

The three elements on the left of the period are metals. The elements to the right are non-metals. Chemists often classify silicon as a metalloid because it has some in-between properties. The metals, for example are all good conductors of electricity. The non-metals are electrical insulators. Silicon is a semi-conductor which is why the element plays such an important part in the modern electronics industry.

Reactions of the elements with water

During the reactions of reactive metals the atoms of the elements turn into positive ions. So metals give away electrons and are reducing agents. Most s-block metals can reduce water. For example, sodium floats on water reacting rapidly to produce hydrogen and a solution of sodium hydroxide.

Magnesium reacts slowly with cold water but it burns brightly on heating in steam to form magnesium oxide and hydrogen.

Aluminium does not react with water. A thin, tough and transparent layer of oxide on the surface of the solid protects the metal from reaction with water.

Apart from chlorine, none of the non-metals in period 3 react with water. Chlorine disproportionates (see page 134) when it reacts with water forming chloric(I) acid and hydrochloric acid. This reaction is reversible and does not go to completion.

Test yourself

1   a) Write equations for the reactions with water of:

   (i) sodium

   (ii) magnesium

   (iii) chlorine.

   b) Show that these are redox reactions.

Reactions of the elements with oxygen

All the elements in period 3 react directly with oxygen except for chlorine and argon. All three metals burn brightly on heating in oxygen forming solid oxides which are white. Sodium burns with a bright yellow flame forming an oxide, Na_2O, and some peroxide, Na_2O_2.

Firework makers include magnesium and aluminium powders in their formulations because both metals burn brightly giving off white light. Magnesium turns to MgO and aluminium to Al_2O_3.

The main source of silicon is silica sand and so burning the element in oxygen is not a useful reaction. Silicon does not react with oxygen at room temperature or on gentle heating. On strong heating it oxidises to the white solid SiO_2.

White phosphorus catches fire in oxygen at room temperature and burns very brightly to form a white smoke of the solid oxide with the molecular formula P_4O_{10}.

Sulfur burns on gentle heating with oxygen with an attractive blueish flame. The main product is the colourless gas sulfur dioxide.

Definition

Allotropes are different forms of the same element in the same physical state.

Test yourself **D**

2   What are the trends or patterns in the melting and boiling points of the elements in period 3?

3   Produce a summary chart to show the structure and bonding of the elements in period 3 based on the information in Figure 3.7.1.

4   Which properties of an element are useful indicators of whether it has a giant structure or is molecular? Suggest reasons.

5   Why is the boiling point of argon lower than the boiling point of chlorine?

6   Draw a dot-and-cross diagram to show the bonding in a molecule of chlorine.

7   Why do the solid metals conduct electricity while the solid non-metals are insulators?

Note

There are oxides of chlorine. They are unstable compounds and they cannot be formed by direct combination between the two elements.

Inorganic Chemistry

Section three

113

Figure 3.7.2 ▲
Sulfur burning in oxygen. Note the blue flame. The product is the gas sulfur dioxide

11 Plot a chart to show the pattern of oxidation states for the chlorides of period 3 elements.

12 Write equations for the reactions of these elements with chlorine:

a) aluminium

b) silicon

c) phosphorus with a limited amount of chlorine

d) phosphorus with excess chlorine.

13 a) Explain the purpose of the calcium chloride tube in the apparatus shown in Figure 3.7.3.

b) How would you modify the apparatus in Figure 3.7.3 to make a sample of silicon tetrachloride, $SiCl_4$?

8 Write equations for the reactions of these elements with oxygen:

a) sodium

b) phosphorus

9 Plot a chart to show the pattern of oxidation states for the oxides of period 3 elements.

10 Plot a bar chart to compare the standard enthalpy changes of formation of the oxides of the period 3 elements per mole of oxygen atoms. What trend to you notice?

Reactions of the elements with chlorine

CD-ROM

Apart from chlorine itself and argon, all the elements in period 3 react with chlorine on heating.

Sodium and magnesium both burn brightly in chlorine on heating but these are not useful reactions since the metals are extracted from their chlorides by electrolysis. The chlorides of both metals are white, crystalline solids.

The chemical industry uses the reaction of aluminium metal with chlorine to make aluminium chloride on a large scale. The reaction is highly exothermic. Manufacturers need the chloride as a catalyst in the Friedel–Crafts reaction (see page 164).

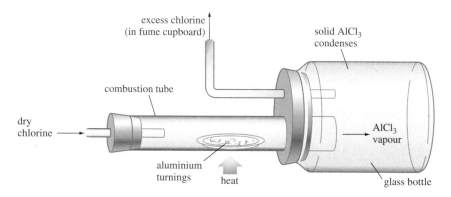

Figure 3.7.3 ▲
An apparatus for preparing aluminium chloride by direct combination of the elements. Aluminium chloride sublimes. It condenses as an off-white solid in the cool receiver

Hot silicon reacts with chlorine gas to form a colourless, liquid chloride $SiCl_4$. Phosphorus also reacts readily on warming with chlorine to form a colourless liquid chloride, PCl_3, but with excess chlorine this reacts further to form a solid with the empirical formula PCl_5.

3.8 Period 3: the oxides

The oxides of period 3 illustrate repeating patterns which can be seen more or less clearly in other periods. The oxides of reactive s-block metals are basic. The oxides of non-metals in the p-block are generally acidic. In between come the amphoteric oxides.

Metal oxides

The oxides of sodium and magnesium are ionic solids. The giant lattice of doubly charged magnesium and oxide ions gives rise to crystals with a high melting point. As a result magnesium oxide is a raw material for making refractory bricks to line furnaces.

The s-block metal oxides are basic which means that they react with acids to form salts and water. Magnesium oxide, for example, reacts with acids to produce magnesium salts.

$$MgO(s) + H_2SO_4(aq) \longrightarrow MgSO_4(aq) + H_2O(l)$$

Note that it is the oxide ion in the basic oxide which acts as the base by taking a hydrogen ion from the acid.

$$O^{2-} + 2H^+ \longrightarrow H_2O$$

Basic oxides which dissolve in water are called alkalis. The oxide ion, acting as a base, takes a hydrogen ion from water to form a hydroxide ion. Water turns sodium oxide into the strong alkali, sodium hydroxide.

$$Na_2O(s) + H_2O(l) \longrightarrow 2NaOH(aq)$$

Magnesium oxide reacts with water to form magnesium hydroxide which is slightly soluble. The antacid milk of magnesia is a suspension of magnesium hydroxide in water.

Aluminium oxide has a giant structure with strong bonding between the atoms so that it is both hard and strong. The pure oxide is colourless or white. Minerals consisting of this oxide include corundum and gemstones such as rubies and sapphires (see page 89).

Moving across the period from sodium to aluminium, the ions get smaller and the charges on the ions increase (see page 111). The smaller a positive ion and the larger its charge the greater the extent to which it tends to polarise a neighbouring negative ion (see page 68). So the order of the polarising power of these ions is: $Al^{3+} > Mg^{2+} > Na^+$.

The bonding in aluminium oxide is not purely ionic because the Al^{3+} ions distort the electron cloud round the oxide ions giving rise to significant electron sharing. So the bonding in this compound is intermediate between ionic and covalent bonding. Intermediate bonding gives rise to intermediate chemical properties. Aluminium oxide is amphoteric.

The anhydrous oxide is relatively inert to aqueous reagents so it is easier to demonstrate its amphoteric behaviour on a test tube scale with a freshly formed sample of the metal hydroxide.

Adding sodium hydroxide to a solution of aluminium ions produces a white precipitate of hydrated aluminium hydroxide (see page 91). The solid dissolves in acids to form salts (so acting as a base); it also dissolves in a solution of a strong alkali forming aluminate ions (thus acting as an acid).

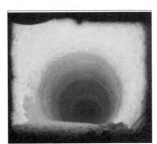

Figure 3.8.1 ▲
Magnesium oxide is an ionic ceramic with a very high melting point and so it is used in refractory bricks to line furnaces

Definition

An **amphoteric oxide** behaves both like a basic oxide and an acidic oxide.

Figure 3.8.2 ▶
Aluminium hydroxide dissolving in alkali and in acid. Note that this simpler description leaves out the hydrating water molecules (see also page 91)

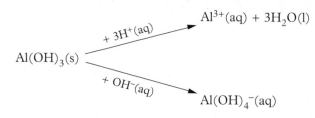

$$Al(OH)_3(s) \quad \begin{array}{c} {\scriptstyle + \, 3H^+(aq)} \nearrow \; Al^{3+}(aq) + 3H_2O(l) \\[1em] {\scriptstyle + \, OH^-(aq)} \searrow \; Al(OH)_4{}^-(aq) \end{array}$$

Figure 3.8.3 ▲
Crystals of quartz: a form of silicon(IV) oxide

Test yourself

1 Write equations for the reactions of:

a) sodium oxide with dilute hydrochloric acid

b) magnesium oxide with dilute nitric acid

c) magnesium oxide with water.

2 Show that when aluminium hydroxide dissolves in acids or in alkalis the reactions involve the addition or removal of protons.

3 a) Work out measures of the polarising power for each of these ions: Na^+, Mg^{2+}, Al^{3+}, Si^{4+}.

b) Explain why the bonding in silicon(IV) oxide is covalent.

Non-metal oxides

Silicon(IV) oxide, SiO_2, is a crystalline solid consisting of a giant structure of atoms held together by covalent bonding. The oxide is abundant in rocks such as quartz. Silicon(IV) oxide has a very high melting point. It melts at 1710 °C and turns to a viscous liquid. On cooling the liquid becomes a glass.

Amethyst is crystalline quartz coloured purple due to the presence of iron(III) ions. Sandstone and sand consist mainly of silica. Flint is a non-crystalline form of silica.

Silicon(IV) oxide is an acidic oxide which is insoluble in water. Silica reacts with basic metal oxides at high temperatures in furnaces during glass making and in steel making. Iron makers, for example, mix calcium oxide with the iron ore fed into blast furnace to remove silica and other impurities.

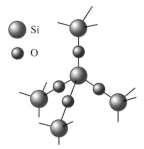
● Si
● O

Figure 3.8.4 ▲
Giant structure of quartz. Each Si atom is at the centre of a tetrahedron of oxygen atoms. The arrangement of silicon atoms is the same as the arrangement of silicon atoms in diamond but there is an oxygen atom between each silicon atom

$$CaO(s) + SiO_2(s) \longrightarrow CaSiO_3(l)$$

The calcium silicate produced is a liquid at the temperature of the furnace; it runs to the bottom where it floats on top of the molten iron as a slag which can be tapped off separately.

The oxides of phosphorus, sulfur and chlorine are molecular. All three elements form more than one oxide. The oxides all react with water and dissolve to give acidic solutions of oxoacids.

Acidic oxide	Oxoacid
phosphorus(V) oxide, $P_4O_{10}(s)$	phosphoric(V) acid, H_3PO_4
sulfur dioxide, $SO_2(g)$	sulfuric(IV) acid, H_2SO_3 [sulfurous acid]
sulfur trioxide, $SO_3(g)$	sulfuric(VI) acid, H_2SO_4 [sulfuric acid]
chlorine(I) oxide, $Cl_2O(g)$	chloric(I) acid, HOCl

Figure 3.8.5 ▲
Selected molecular oxides and oxoacids of non-metals in period 3

Note

If an oxide is insoluble in water it does not affect the pH.

	Ionic crystals			Covalent giant structure	Covalent molecular gases and solids		
$Na_2O(s)$	MgO(s)		$Al_2O_3(s)$	$SiO_2(s)$	$P_4O_{10}(s)$	$SO_3(s)$	$Cl_2O(g)$
$Na_2O_2(s)$					$P_4O_6(s)$	$SO_2(g)$	
	Basic		Amphoteric		Acidic		

Figure 3.8.7 ▲
Formulae, structures, bonding and acid-base character of oxides in period 3

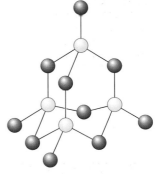

Figure 3.8.6 ▲
The structure of a molecule of phosphorus(v) oxide which is a white solid

Definitions

An **acidic oxide** is an oxide of a non-metal which reacts with water to form an acid. Some acidic oxides, such as SiO_2, are insoluble but they can be recognised because they react directly with basic oxides to form salts. Note that acidic oxides are not themselves acids as defined by the Brønsted–Lowry theory because they do not contain ionisable hydrogen atoms. They cannot act as proton donors.

The reaction of an acidic oxide with water produces an **oxoacid**. The systematic names for oxoacids are not widely used but they make it possible to work out the formulae of each oxoacid. The full systematic name for sulfurous acid is trioxosulfuric(IV) acid. Oxoacids ionise in solution by giving hydrogen ions to water molecules. Where an element forms two oxoacids, the one with element in the higher oxidation state is the stronger.

Test yourself   D

4 What is the empirical formula of phosphorus(v) oxide?

5 Write equations for the reactions of these oxides with water:

a) phosphorus(v) oxide

b) sulfur dioxide, $SO_2(g)$

c) sulfur trioxide, $SO_3(s)$.

6 Draw dot-and-cross diagrams for the bonding in the two oxides of sulfur and predict the shapes of the molecules.

7 a) Draw a dot-and-cross diagram to show the bonding in a molecule of Cl_2O and predict the shape of the molecule.

b) What does the value of the standard enthalpy change of formation of Cl_2O suggest about its stability? Does the value suggest that the compound is liable to explode?

8 Plot the melting points of the oxides of the period 3 elements against the atomic number of the elements and explain the pattern you observe in terms of structure and bonding.

9 Draw up a chart to show the pH values you would expect on shaking a small spatula measure of each of the solid oxides of period 3 elements with a little water in a test tube.

3.9 Period 3: the chlorides

There is a periodic pattern in the properties of chlorides which is clearly illustrated by the chlorides of period 3 elements. The chlorides of reactive s-block elements on the left of the table are ionic crystals which dissolve in water forming neutral solutions. The chlorides of non-metals in the p-block on the right consist of covalent molecules. They usually react vigorously with water. Aluminium chloride has bonding and properties which are intermediate between the two extremes.

Figure 3.9.1 ▶
Periodicity: formulae, structures, bonding and behaviour in water of chlorides in periods 2 and 3

Ionic crystals	Polar covalent giant structures	Covalent molecular gases and solids				
$LiCl(s)$	$BeCl_2(s)$	$BCl_3(g)$	$CCl_4(l)$	$NCl_3(l)$	$Cl_2O(g)$	$ClF(g)$
$NaCl(s)$	$MgCl_2(s)$	$AlCl_3(s)$	$SiCl_4(l)$	$PCl_5(s)$ $PCl_3(l)$	$S_2Cl_2(l)$	$Cl_2(g)$
Dissolve in water giving a neutral solution		Partially hydrolysed to give an acidic solution	Hydrolysed by water giving an acidic solution (unless inert as in CCl_4)			

Figure 3.9.2 ▲
Crystals of sodium chloride. The shape of the crystals reflects the cubic arrangement of the ions in the crystal structure

Metal chlorides

The chlorides of sodium and magnesium are white, crystalline solids with high melting points. They dissolve in water and as they do so the polar water molecules hydrate the ions (see page 73).

Aluminium atoms in $AlCl_3$ only have six electrons and have a strong tendency to accept two more. As a result aluminium chloride vapour contains Al_2Cl_6 dimers. Solid $AlCl_3$ has a layer lattice with bonding intermediate between ionic and covalent.

Aluminium chloride ($AlCl_3$) is an off-white solid which sublimes on heating and fumes in moist air because of hydrolysis to hydrogen chloride and the hydroxide.

A solution of aluminium chloride is acidic as a result of partial hydrolysis (see page 91).

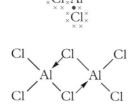

Figure 3.9.3 ▲
A dot-and-cross diagram for $AlCl_3$ and the bonding in an Al_2Cl_6 dimer present in the vapour of this compound. Lone pairs of electrons on chlorine atoms form dative bonds with aluminium atoms

Test yourself

1 Draw a dot-and-cross diagram to show the bonding in magnesium chloride.

2 Why does magnesium chloride have a higher melting point than sodium chloride?

3 Write half-equations to show the changes at the electrodes during the electrolysis of molten sodium chloride.

4 Explain why the bonding in aluminium chloride is not purely ionic (see page 68).

5 Write an equation for the reaction of hot aluminium chloride vapour with moisture in the air.

6 Explain why a solution of aluminium chloride is acidic. Why is this chloride partially hydrolysed in solution?

Non-metal chlorides

CD-ROM

Most non-metal chlorides are liquids consisting of molecules. The atoms in the molecules are held together by strong covalent bonds but the intermolecular forces are weak. The chlorides do not mix with water, but react with it. Hydrolysis splits the compounds into an oxoacid or hydrated oxide of the non-metal and hydrogen chloride.

$$SiCl_4(l) + 2H_2O(l) \longrightarrow SiO_2(s) + 4HCl(g)$$

Phosphorus(v) chloride vapour consists of PCl_5 molecules. The compound is a solid at room temperature. The solid consists of PCl_4^+ and PCl_6^- ions. It is the ionic bonding between the complex ions which makes this non-metal chloride a solid rather than a liquid. Hydrolysis converts phosphorus(v) chloride to phosphoric(v) acid and hydrochloric acid.

$$PCl_5(s) + 4H_2O(l) \longrightarrow H_3PO_4(s) + 5HCl(g)$$

Disulfur dichloride, S_2Cl_2, is a yellow liquid with a foul smell. Hydrolysis of this chloride produces a yellow precipitate of sulfur together with hydrochloric acid.

Figure 3.9.4 ▲
Samples of non-metal chlorides

Test yourself D

7 a) Draw dot-and-cross diagrams to show the bonding in these molecules: $SiCl_4$, PCl_3, PCl_5 and S_2Cl_2.

b) Predict the shapes of these molecules.

c) Which of these molecules are polar?

8 Write equation for the hydrolysis of PCl_3 to H_3PO_3.

9 Predict the shapes of the PCl_4^+ and PCl_6^- ions in solid PCl_5.

10 In what ways are the reactions of PCl_5 with water and with ethanol similar?

11 Suggest tests to show that the hydrolysis of non-metal chlorides produces hydrogen chloride.

12 Plot the melting points of the chlorides of period 3 elements against atomic number and explain the pattern you observe in terms of structure and bonding.

Figure 3.9.5 ▲
PCl_5 *reacting with water*

3.10 Group 4

This group of elements shows a trend from non-metals at the top to metals at the bottom. The similarities between the elements in this group are more subtle and harder to detect than the much more obvious likenesses between the elements in groups 2 or 7 of the periodic table.

CD-ROM

C
Si
Ge
Sn
Pb

The elements

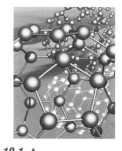

Figure 3.10.1 ▲
Carbon's remarkable ability to form stable chains and rings of atoms with single, double and triple bonds gives rise to organic chemistry

Figure 3.10.2 ▲
Silicon is the second most abundant element in the Earth's crust. Combined with oxygen it forms many minerals including silica, SiO_2, and silicates. Silicate minerals, based on giant structures of silicon and oxygen atoms, make up most of the crust of the Earth

Test yourself

1 To what extent, if at all, do the appearance and physical properties of group 4 elements illustrate the trend from non-metallic to metallic characteristics down the group?

Figure 3.10.3 ▲
Germanium is an element with some metallic and some non-metallic features. It is a metalloid. Germanium is a semi-conductor

Figure 3.10.5 ►
Lead is a greyish metal often seen on roofs of buildings. It resists corrosion and is malleable, so it can be bent to fit the space where roof tiles meet walls to create a waterproof flashing. Lead is a dense metal and is an effective shield to stop X-rays and other radiations. Lead melts at a relatively low temperature for a metal and, when alloyed with tin, produces solder with an even lower melting point

Figure 3.10.4 ▲
At room temperature tin has a metallic structure but below the transition temperature 13.2°C the stable form is grey tin which has the diamond structure

Structure and bonding in the elements

The two common allotropes of carbon (diamond and graphite) consist of covalent giant structures. Silicon and germanium have diamond-like giant structures. They are semi-conductors.

The room temperature allotrope of tin has a metallic structure. The low temperature allotrope, grey tin has the non-metallic diamond structure. Lead has a metallic structure.

The two metals, tin and lead, have lower melting points than the non-metals with giant structures.

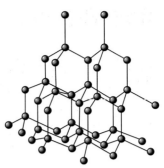

Figure 3.10.6 ▲
A fragment of the giant structure of a diamond

◄ **Figure 3.10.7**
Tin is a shiny metal with a long history; it was mined in Cornwall from the Roman times until 1998 when the last mine closed. Tin is valued as an ingredient of a range of alloys including pewter (with *Sb*), solder (with *Pb*) and bronze (with *Cu*). The main use of tin today is for coating the steel for tin cans. The layer of tin stops the iron corroding

Oxidation states

All the elements of group 4 have four electrons in their outer shell. The characteristic oxidation states of the group are +4 and +2. The +2 state becomes more important down the group and is the more stable state of lead.

carbon, C	$[He]2s^22p^2$
silicon, Si	$[Ne]3s^23p^2$
germanium, Ge	$[Ar]3d^{10}4s^23p^2$
tin, Sn	$[Kr]4d^{10}5s^23p^2$
lead, Pb	$[Xe]4f^{14}5d^{10}6s^23p^2$

Figure 3.10.8 ▲
Abbreviated electron configurations of group 4 elements

In group 4 (as in groups 3 and 5) of the periodic table, the oxidation state which is 2 below the highest state becomes increasingly stable in elements nearer the bottom of the group. In group 4, for example the main oxidation state for carbon and silicon is +4. The +2 oxidation state becomes important in the chemistry of tin and lead. In the chemistry of lead the +2 state is even more stable relative to the +4 state than it is in the chemistry of tin.

What this means is that two of the four electrons in the outer shell become less available for bonding down the group. Chemists use the term 'inert-pair' effect' to describe the increasing importance of the lower oxidation state down the group. This 'effect' is better regarded as a reminder of a trend rather than an explanation.

Test yourself D

2 **a)** What is the trend in the melting points of the elements down group 4, from carbon to lead?

b) How can you account for this trend in terms of structure and bonding?

3 Write out the full electron configurations of silicon and tin.

4 **a)** What is the trend in the first ionisation energies of the group 4 elements down the group?

b) Account for the trend in terms of the electronic structures of the atoms.

c) What is the significance of this trend for the chemistry of the elements?

The oxides

The +4 state

Carbon forms two oxides both of which are colourless, molecular gases with no smell. Carbon monoxide, CO, is a neutral oxide. Carbon dioxide is an acidic oxide. It dissolves in water forming carbonic acid which is a weak acid.

$$CO_2(g) + H_2O(l) \longrightarrow H_2CO_3(aq)$$

Carbon dioxide is strongly absorbed by alkalis such as potassium hydroxide.

The oxide SiO_2 is also an acidic oxide though relatively unreactive because it has a giant structure. So CO_2 and SiO_2 are good examples of the generalisation that the group 4 compounds in the +4 state behave more like the compounds of non-metals.

SnO_2 and PbO_2 are amphoteric with a bias towards acidic oxide behaviour.

The +2 state

The compounds of tin and lead in the +2 state (SnO and PbO) are more typical of metallic compounds. The bonding in these compounds is highly polar though not purely ionic. They are amphoteric oxides but with a bias towards basic oxide properties.

Tin(II) oxide dissolves in concentrated hydrochloric acid to give a solution of tin(II) chloride which is a reducing agent.

Figure 3.10.9 ▶
Samples of the oxides of lead: PbO, Pb$_3$O$_4$, PbO$_2$. Lead oxides can be used as pigments in paint but, like all lead compounds, they are highly toxic

Lead oxides react with acids to form salts. There are two soluble lead(II) compounds: the nitrate and the ethanoate. Lead ions are colourless in solution.

Adding alkali to a solution containing lead(II) ions gives a white precipitate of $Pb(OH)_2$ which dissolves in excess forming the $[Pb(OH)_4]^{2-}$ ion in solution, showing that the hydroxide, like the oxide, is amphoteric.

The +2 state is the more stable state for lead. So lead(IV) oxide is a strong oxidising agent. Lead(IV) oxide can oxidise chloride ions to chlorine gas.

The chlorides

The +4 state

The +4 chlorides are molecular liquids which, with the exception of CCl_4, are rapidly hydrolysed by water. Again, this shows that the compounds in the +4 state generally behave more like the compounds of non-metals. $SiCl_4$, for example, is rapidly hydrolysed by water to hydrated silica and hydrogen chloride.

$$SiCl_4(l) + 2H_2O(l) \longrightarrow SiO_2(s) + 4HCl(g)$$

Test yourself ▷ D

5 Write balanced equations for the reactions of:

a) carbon dioxide with aqueous potassium hydroxide

b) tin(II) oxide with concentrated hydrochloric acid

c) lead(IV) oxide with concentrated hydrochloric acid

d) lead(II) oxide with sodium hydroxide.

6 Use standard electrode potentials to show:
a) that tin(II) ions in acid solution can reduce iron(III) to iron(II) ions b) that lead(IV) oxide can oxidise chloride ions to chlorine under acid conditions.

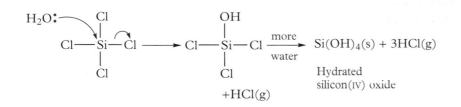

$Si(OH)_4(s) + 3HCl(g)$

Hydrated
silicon(IV) oxide

$+HCl(g)$

◀ **Figure 3.10.10**
A possible mechanism for the hydrolysis of silicon tetrachloride. A lone pair on a water molecule attacks the δ+ end of the polar Si–Cl bond. Some chemists argue that this mechanism is possible with silicon because there are empty 3d-orbitals in silicon atoms which can accept an electron pair during this first step

The tetrachloride of carbon (tetrachloromethane) is a colourless liquid with tetrahedral molecules, CCl_4. Unlike most other non-metal chlorides it is not hydrolysed by water or by alkalis. A popular explanation for the inertness of CCl_4 to hydrolysis suggests that it is because there are no d-orbitals in the second shell. As a result, when a water molecule attacks a CCl_4 molecule the carbon atom does not have an empty orbital to accept the incoming electron pair (see Figure 3.10.10). The problem with this explanation is that it suggests that the hydrolysis of halogenoalkanes by a similar mechanism (see page 157) should also be impossible.

An alternative explanation argues that the difference in the behaviour of CCl_4 and $SiCl_4$ with water is a result of the much smaller size of a carbon atom compared to a silicon atom. As a result, it is impossible for water molecules to attack a carbon atom when it is surrounded by four relatively large chlorine atoms in CCl_4 and is strongly bonded to them.

The +2 state

The compounds of tin and lead in the +2 state are more typical of metallic compounds. The +2 chlorides are white solids. $PbCl_2$ is an ionic solid.

The +4 state is more stable for tin, so that tin(II) compounds are reducing agents. Tin(II) chloride reduces iron(III) to iron(II) for example.

The +2 state is the more stable state for lead and so lead(II) chloride is not a reducing agent.

Test yourself

7 How can you account for the fact that CCl_4 is a liquid which mixes freely with a hydrocarbon solvent such as hexane but is immiscible with water?

8 How would you demonstrate that lead(II) chloride is an ionic solid?

9 What do chemists mean when they say that the hydrolysis of CCl_4 is feasible (spontaneous) but does not happen because the compound is kinetically stable (inert)? What energy quantities do chemists refer to when justifying explanations of this kind?

10 What are the chemical similarities between the elements in group 4 that make them a chemical family in the same group?

11 What explanations can you suggest for the observation that group 4 compounds in the +4 state generally behave more like the compounds of non-metals, while the compounds of tin and lead in the +2 state are more typical of metallic compounds?

12 Generally the properties of the element in period 3 is more characteristic of the rest of the group than the chemistry of the element in period 2. To what extent is this the case in group 4?

13 What similarities are there, if any, between the trends in properties down group 4 and the trends down groups 2 and 7?

3.11 Period 4: d-block elements

The d-block elements are vital to life and bring colour to our lives. They are also materials of great engineering and industrial importance. Chemically these elements are interesting because they are much more similar to each other than might be expected following a study of the elements in period 3. Across the ten elements from scandium to zinc in period 4 the similarities are as striking as the differences. Chemists explain the characteristics of these elements in terms of the electron configurations of their atoms.

Figure 3.11.1 ▲
Window in the Robie House, designed by Frank Lloyd Wright in 1909. The commonest colourants in stained glass are the oxides of d-block elements

Figure 3.11.2 ▶
Electron configurations of d-block elements as free atoms. Note that orbitals fill singly before the electrons start to pair up. Note too that the configurations of chromium and copper do not fit the general pattern

Figure 3.11.4 ▶
Specimens of d-block metals

CD-ROM

Electron configurations

In period 4, the d-block elements run from scandium ($1s^2 2s^2 2p^6 3s^2 3p^6 3d^1 4s^2$) to zinc ($1s^2 2s^2 2p^6 3s^2 3p^6 3d^{10} 4s^2$). As shown in Figure 3.6.3 on page 110, the 4s orbital fills before the 3d orbitals. So the last electron added to the atomic structure goes into a d-orbital.

Along this series the number of protons in the nucleus increases by one from one element to the next. The added electrons go into an inner d-sub shell. The outer electrons are always in the 4s. This means that the changes in the properties across the series are much less marked than the big changes across a p-block series such as Al to Ar.

		3d					4s			3d					4s
Sc	[Ar]	↑					↑↓	Fe	[Ar]	↑↓	↑	↑	↑	↑	↑↓
Ti	[Ar]	↑	↑				↑↓	Co	[Ar]	↑↓	↑↓	↑	↑	↑	↑↓
V	[Ar]	↑	↑	↑			↑↓	Ni	[Ar]	↑↓	↑↓	↑↓	↑	↑	↑↓
Cr	[Ar]	↑	↑	↑	↑	↑	↑	Cu	[Ar]	↑↓	↑↓	↑↓	↑↓	↑↓	↑
Mn	[Ar]	↑	↑	↑	↑	↑	↑↓	Zn	[Ar]	↑↓	↑↓	↑↓	↑↓	↑↓	↑↓

The chemistry of an atom is to a large extent determined by its outer electrons because they are the first to get involved in reactions. So the elements Sc to Zn in period 4 are similar in many ways.

The metals

All the d-block elements are metals with useful properties for engineering and construction. Most have high melting points. A plot of physical properties against proton number often has two peaks corresponding to the half filling and then filling of the d-shell.

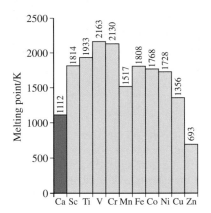

◀ *Figure 3.11.4*
Plot of melting point against proton number for the elements Ca to Zn

Oxidation states

Most of the d-block elements in period 4 show more than one oxidation state in their compounds.

				+7					
			+6						
		+5		+6					
	+4	+4		+4					
+3	+3	+3	+3	+3	+3	+3			
	+2	+2	+2	+2	+2	+2	+2	+2	+2
								+1	
Sc	Ti	V	Cr	Mn	Fe	Co	Ni	Cu	Zn

Figure 3.11.5 ▲
Main oxidation states of the elements scandium to zinc with the commoner states in red. Note that scandium and zinc only form ions in one oxidation state. The other metals can all exist in more than one state, though the higher states of nickel are very unusual

The elements at each end of the series, scandium and zinc, give rise to only one oxidation state. The elements near the middle of the series have the greatest range of states.

Most of the elements form compounds in the +2 state corresponding to the use of both 4s electrons in bonding. Across the series the +2 state becomes more stable relative to the +3 state.

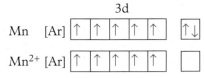

Figure 3.11.6 ▲
The electron configurations of a manganese atom and a manganese(II) ion

From scandium to manganese the highest oxidation state corresponds to the total number of electrons in the 3d and 4s energy levels. The higher oxidation states never exist as simple ions. Typically they arise in compounds in which the metal is covalently bonded to an electronegative atom, especially oxygen as in the dichromate(VI), $Cr_2O_7^{2-}$, and manganate(VII), MnO_4^-, ions.

Inorganic Chemistry — *Section three*

Test yourself D

1 Write out the full electron configuration of:
   a) an iron atom
   b) an iron(II) ion
   c) an iron(III) ion.

2 Compare the rise in the first ionisation energy from scandium to zinc with the rise from sodium to argon. Account for the differences in terms of the electron configurations of the atoms.

3 Suggest three generalisations about d-block element oxidation states based on Figure 3.11.5.

4 Give examples of compounds of:
   a) chromium in the +3 and + 6 states
   b) manganese in the +2, +4 and +7 states
   c) iron in the +2 and +3 states
   d) copper in the +1 and + 2 states.

5 Use a table of standard electrode potentials to show that in moving from chromium to iron the +2 state becomes more stable and the +3 state less stable.

Complex formation

The ions of most d-block elements exist as complex ions in solution (see section 3.3) and some oxidation states are only stable in the form of complex ions (see also section 3.12 for examples of complexes formed by some d-block metals).

The d-block elements can form a wide variety of complexes because the ions have partially filled d-orbitals as well as 4p and 4d orbitals which can accept electron pairs from the ligands.

Figure 3.11.7 ▶
A diagram to show the electron configuration of copper in the $[Cu(H_2O)_6]^{2+}$ ion

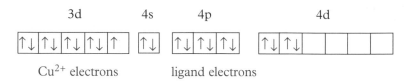

Some of the vital enzymes in the human body are complexes between proteins and d-block element ions. One example is a cobalt complex which plays a key part in the activity of vitamin B_{12}.

Test yourself

6 Suggest the likely shape of these complex ions:

a) $CuCl_4^{2-}$

b) $Fe(H_2O)_6^{3+}$

Colour of d-block element compounds

Coloured ions

In many instances coloured compounds get their colour by absorbing some of the radiation in the visible region of the electromagnetic spectrum with wavelengths between 400 nm and 700 nm.

Colour of compound	Wavelength absorbed/nm	Colour of light absorbed
greenish yellow	400–430	violet
yellow to orange	430–490	blue
red	490–510	blue-green
purple	510–530	green
violet	530–560	yellow-green
blue	560–590	yellow
greenish blue	590–610	orange
blue-green to green	610–700	red

Figure 3.11.8 ▲
A table with complementary colours in left and right-hand columns. The colour of a compound is the complementary colour to the light it absorbs

It is the electrons in coloured compounds which absorb radiation as they jump from their normal state to a higher excited state. According to quantum theory there is a fixed relationship between the size of the energy jump and the wavelength of the radiation absorbed. In many compounds the jumps are so big that they absorb in the ultraviolet part of the spectrum. These compounds are colourless.

Colour in d-block metal ions arises from electronic transitions between d-orbitals. In a free atom or ion the five d-orbitals are at the same energy level. The five orbitals are not the same shape and they split into two groups with different energies when a d-block element ion is surrounded by molecules or ions in a complex ion. This helps to account for the colour of complex ions.

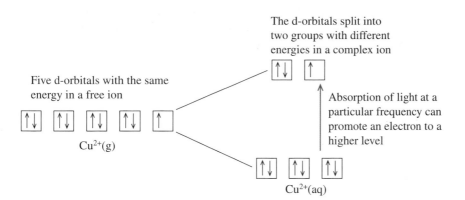

Definition

Quantum theory states that radiation is emitted or absorbed in discrete amounts called energy quanta. Quanta have energy $\Delta E = h\nu$ where h is Planck's constant and ν is the frequency of the radiation.

◄ **Figure 3.11.9**
Energy difference between d-orbitals in a transition metal complex ion makes possible electronic transitions from a lower level to a higher level, so the ion absorbs light with a particular frequency

If all the d-orbitals are full, or empty, there is no possibility of electronic transitions between them.

The colour of a d-block metal complex ion depends on:

- the metal
- the oxidation state of the metal
- the ligand
- the co-ordination number.

Test yourself

7 Explain why Zn^{2+}, Cu^+ ions and Sc^{3+} ions are colourless by writing out their electron configurations.

8 Are these colours the result of a change of oxidation state, a change in ligand, a change in co-ordination number or a combination of more than one of these changes?

a) $[Cr(H_2O)_6]^{2+} \longrightarrow [Cr(H_2O)_6]^{3+}$
    blue   red

b) $[Cu(H_2O)_6]^{2+} \longrightarrow [Cu(NH_3)_4(H_2O)_2]^{2+}$
    pale blue   deep blue

c) $[Co(H_2O)_6]^{2+} \longrightarrow [CoCl_4]^{2-}$
    pink   blue

d) $[Fe(H_2O)_6]^{2+} \longrightarrow [Fe(H_2O)_5OH]^{3+}$
    pale green   yellow

Colorimetry

A colorimeter can be used to measure the concentrations of chemicals which are themselves coloured or which produce a colour when mixed with a suitable reagent (see Figure 2.2.2 on page 9). This makes it possible to determine the formula of a complex ion. Figure 3.11.10 shows the results of measuring the absorbance of a series of mixtures of 0.01 mol dm^{-3} $Cu^{2+}(aq)$ ions and 0.01 mol dm^{-3} edta(aq).

Inorganic Chemistry   **Section three**

Figure 3.11.10 ▶
Plot of the results of measuring the absorbance of ten mixtures of 0.01 mol dm^{-3} Cu^{2+}(aq) ions and 0.01 mol dm^{-3} edta(aq)

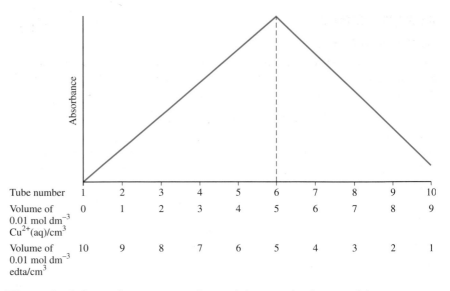

Tube number	1	2	3	4	5	6	7	8	9	10
Volume of 0.01 mol dm^{-3} Cu^{2+}(aq)/cm^3	0	1	2	3	4	5	6	7	8	9
Volume of 0.01 mol dm^{-3} edta/cm^3	10	9	8	7	6	5	4	3	2	1

The peak of absorption corresponds to mixing equal volumes of the two solutions which both have the same molar concentration. This shows that 1 mol Cu^{2+}(aq) forms a complex with 1 mol edta(aq) (for a diagram see Figure 3.3.10 on page 96).

Spectrophotometry

The concentration of metal ions in solution can be determined more precisely using a spectrophotometer. The procedure is very similar to the use of an ultraviolet spectrometer in organic chemistry (see page 230). Instead of a filter a spectrophotometer has a diffraction grating which can select a specific wavelength for measuring the absorbance of radiation. The instrument can scan across a range of wavelengths to produce an absorption spectrum.

Figure 3.11.11 ▶
Absorption spectrum of four d-block element ions.

Test yourself

9 Describe, in outline, how you would calibrate a spectrophotometer to measure concentrations of iron(III) ions in the range 2.00 × 10^{-5} mol dm^{-3} to 3.00 × 10^{-5} mol dm^{-3}.

10 With the help of Figure 3.11.8, predict the colours of the ions from the spectra in Figure 3.11.11 and match the spectra with these ions: Cr$_2$O$_7$$^{2-}$(aq), Co^{2+}(aq), Ni^{2+}(aq) and Cu^{2+}(aq).

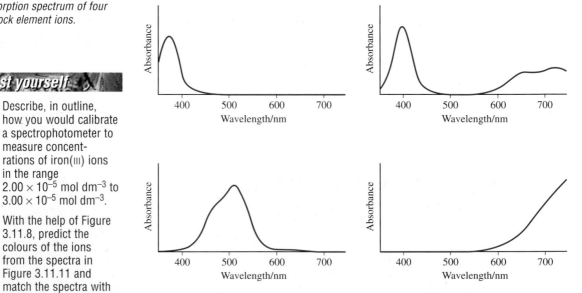

Chemists use complex ions and spectrophotometry to determine the concentration of metal ions. Iron(III) ions, for example, are not intensely coloured in dilute solutions. Adding a solution of thiocyanate ions produces an intense red complex.

Catalytic activity

Homogeneous catalysis

Transition metal ions can be effective homogeneous catalysts because they can gain and lose electrons as they change from one oxidation state to another. The oxidation of iodide ions by peroxodisulfate(VI) ions, for example, is very slow.

$$S_2O_8^{2-}(aq) + 2I^-(aq) \longrightarrow 2SO_4^{2-}(aq) + I_2(aq)$$

The reaction is catalysed by iron(III) ions in the solution. A possible mechanism is that iron(III) is reduced to iron(II) as it oxidises iodide ions to iodine. The $S_2O_8^{2-}$(aq) ions oxidise the iron(II) back to iron(III) ready to oxidise some more of the iodide ions, and so on.

Sometimes one of the products of a reaction can act as a catalyst for the process. This is autocatalysis. An autocatalytic reaction starts slowly but then speeds up as the catalytic product starts to form. Mn^{2+} ions catalyse the oxidation of ethanedioate ions by manganate(VII) ions in acid solution. This is also an example of homogeneous catalysis.

$$2MnO_4^-(aq) + 5C_2O_4^{2-}(aq) + 16H^+(aq) \longrightarrow 10CO_2(g) + 2Mn^{2+}(aq) + 8H_2O(l)$$

Test yourself

11 a) Suggest methods of speeding up the reaction of MnO_4^-(aq) ions with $C_2O_4^{2-}$(aq) ions at the start before the reaction has got underway.

b) What would you expect to see on adding a solution of potassium manganate(VII) to an acidified solution of potassium ethanedioate

(i) at the start, and

(ii) as the reaction begins to speed up?

Heterogeneous catalysis

Heterogeneous catalysts are used in large scale continuous processes. For example, during the manufacture of ammonia in the Haber process, nitrogen and hydrogen gas flow through a reactor containing small lumps of iron.

Platinum metal alloyed with other metals such as rhodium is an important catalyst. It is used to oxidise ammonia during nitric acid manufacture. It is also used in catalytic converters. In a catalytic converter the expensive catalyst is finely divided and supported on the surface of a ceramic to increase the surface area in contact with the exhaust gases.

Impurities in the reactants can poison catalysts causing them to become less effective. Carbon monoxide poisons the iron catalyst used in the Haber process. Lead compounds in the exhaust from a car engine poison catalytic converters so lead-free petrol must be used.

Definition

A **homogeneous catalyst** is a catalyst which is in the same phase as the reactants. Typically the reactants and the catalyst are dissolved in the same solution.

Definition

A **heterogeneous catalyst** is a catalyst which is in a different phase from the reactants. Generally a heterogeneous catalyst is a solid while the reactants are gases or in solution in a solvent. The advantage of heterogeneous catalysts is that there is no difficulty in separating the products from the catalyst.

◀ *Figure 3.11.12*
A catalytic converter is an example of heterogeneous catalysis. The surface of the metal catalyst absorbs the pollutants NO and CO, where they react to form CO_2 and N_2. Note that carbon atoms are shown green in this computer graphic

Inorganic Chemistry

Section three

Heterogeneous catalysts work by adsorbing reactants at active sites on the surface of the solid. Nickel acts as a catalyst for the addition of hydrogen to C=C in unsaturated compounds by adsorbing hydrogen molecules which probably split up into single atoms held on the surface of the metal crystals.

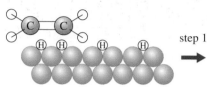 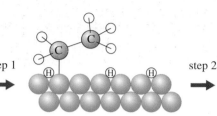

step 1 → step 2 →

Ethene adsorbed on catalyst surface where hydrogen gas is also adsorbed as single atoms

Ethene adds one hydrogen atom and the CH₃CH• radical is attached to the surface

After adding a second hydrogen atom the hydrocarbon, now ethane, escapes from the surface

Figure 3.11.13 ▲
Possible mechanism for the hydrogenation of an alkene in the presence of a nickel catalyst. The reaction takes place on the surface of the metal which adsorbs hydrogen and splits the molecules into atoms

If a metal is to be a good catalyst for a hydrogenation reaction it must not adsorb the hydrogen *so* strongly that the hydrogen atoms becomes unreactive. This happens with tungsten. Equally, if adsorption is too weak there will not be enough adsorbed atoms for the reaction to go at a useful rate, as is the case with silver. The strength of adsorption must have a suitable intermediate value. Suitable metals are nickel, platinum and palladium.

Test yourself

12 Explain the advantage of using heterogeneous catalysts in industrial processes.

13 Which catalyst is used to convert SO_2 to SO_3 during the manufacture of sulfuric acid?

3.12 Transition metals

The chemical reactions of vanadium, chromium, manganese, iron, cobalt, nickel and copper provide many examples of the characteristic properties of transition metals. Transition element chemistry is colourful because of the range of oxidation states and complex ions. Transition metals matter because their properties are fundamental not only to life but also to much modern technology.

Characteristics

Transition metals are d-block elements which have partially filled d-energy levels in one or more of their oxidation states. This definition excludes zinc $[Ar]3d^{10}4s^2$, because the 3d orbitals are full in its atoms and also in its compounds in the +2 state. Losing two electrons gives an ion Zn^{2+} with the electron configuration $[Ar]3d^{10}$ in which all the d-energy levels are still full. Some chemists include oxidation state zero in the definition and so count scandium as a transition metal. Others exclude scandium, $[Ar]3d^14s^2$, because it only forms compounds in the +3 state and it loses all its three outer electrons when it forms a 3+ ion.

Transition metals share a number of common features. They:

- are metals with useful mechanical properties and with high melting points
- form compounds in more than one oxidation state
- form coloured compounds
- form a variety of complex ions
- act as catalysts either as the elements or as compounds.

Vanadium

Vanadium is a d-block metal with the electron configuration $[Ar]3d^34s^2$. The metal is used to make alloy steels which are strong and tough, making them suitable for machine tools and engine parts. Vanadium is a typical transition metal: it forms coloured compounds in several oxidation states, it forms complex ions and it has compounds which can be used as catalysts such as vanadium(V) oxide in sulfuric acid manufacture.

In solution vanadium forms ions in the +2, +3, +4 and +5 oxidation states. The +5 state is available as the yellow solid ammonium vanadate(V).

◄ *Figure 3.12.1*
Oxidation states of vanadium showing the colours of the ions: +5, +4, +3 and +2 oxidation states

Standard electrode potentials help to identify reagents which can reduce vanadium(V) to the succession of lower states (see page 57).

Iodide ions reduce vanadium(V) to vanadium(IV).

Figure 3.12.2 ▶
Half-equations and electrode potentials for the reduction of vanadium(V) to vanadium(IV). Green arrows show the direction of change

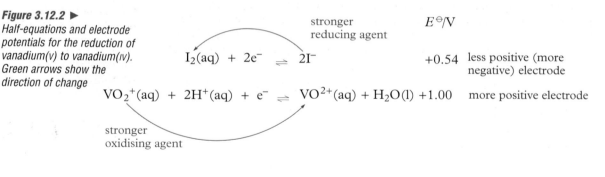

	$E^{\ominus}/V$	
$I_2(aq) + 2e^- \rightleftharpoons 2I^-$	+0.54	less positive (more negative) electrode
$VO_2^+(aq) + 2H^+(aq) + e^- \rightleftharpoons VO^{2+}(aq) + H_2O(l)$	+1.00	more positive electrode

stronger reducing agent

stronger oxidising agent

Copper metal reduces vanadium(V) to vanadium(III).

Zinc in acid will reduce vanadium(V) all the way to vanadium(II). The $V^{2+}(aq)$ ion is a strong reducing agent.

Figure 3.12.3 ▶
Half-equations and electrode potentials for the reduction of vanadium(III) to vanadium(II). Green arrows show the direction of change

stronger reducing agent

$Zn^{2+}(aq) + 2e^- \rightleftharpoons Zn(s)$	$E^{\ominus} = -0.76\,V$	more negative electrode
$V^{3+}(aq) + e^- \rightleftharpoons V^{2+}(aq)$	$E^{\ominus} = -0.26\,V$	less negative (more positive) electrode

stronger oxidising agent

Test yourself

1 Write out the full electron configuration of a vanadium atom and of a V^{3+} ion.

2 What is the oxidation state of vanadium in VO^{2+}?

3 Write an ionic equation for the reduction of vanadium(V) to vanadium(II) by zinc.

Chromium

CD-ROM

Chromium is a hard, silvery d-block metal with the electron configuration $[Ar]3d^5 4s^1$. This electron configuration is an exception to the normal $[Ar]3d^x 4s^2$ pattern for the first series of d-block elements. Energetically it is more favourable to have one electron in each d-orbital and thus to half-fill the d-sub shell.

In solution chromium forms ions in the +2, +3 and +6 oxidation states

+6 — CrO_4^{2-} (yellow), $Cr_2O_7^{2-}$ (orange)

+4 --

— Cr^{3+}

+2 — Cr^{2+} ----------------------------

0 --

Figure 3.12.4 ▶
Oxidation states of chromium: CrO_4^{2-}, $Cr_2O_7^{2-}$, Cr^{3+} and Cr^{2+}

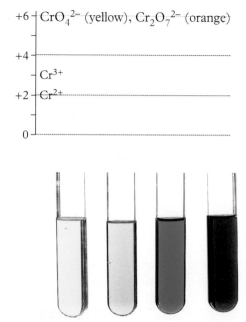

There is an equilibrium between yellow chromate(VI) ions and orange dichromate(VI) ions in aqueous solution. The position of the equilibrium depends on the pH. In acid the hydrogen ion concentration is high so the solution is orange because dichromate ions predominate. Adding alkali removes hydrogen ions and turns the solution yellow as chromate ions form. These shifts in the position of equilibrium are as predicted by Le Chatelier's principle. Note that this is not a redox reaction. Chromium is in the +6 state on both sides of the equation.

$$2CrO_4^{2-}(aq) + 2H^+(aq) \longrightarrow Cr_2O_7^{2-}(aq) + H_2O(l)$$

Potassium dichromate(VI) is used as a primary standard in redox titrations because it is a powerful oxidising agent in acid solution and reacts quantitatively according to the balanced equation (see page 107).

$$Cr_2O_7^{2-}(aq) + 14H^+(aq) + 6e^- \longrightarrow 2Cr^{3+}(aq) + 7H_2O(l)$$

Chromium(III) ions in solution exist as aquo complexes: $[Cr(H_2O)_6]^{3+}$. The hydrated ions are acidic and behave in a very similar way to aluminium(III) ions.

The removal of successive protons from the hydrated chromium(III) ion, $[Cr(H_2O)_6]^{3+}$, produces a neutral complex, $[Cr(H_2O)_3(OH)_3]$, which is sparingly soluble and precipitates. The removal of further protons produces a negative ion, $[Cr(OH)_6]^{3-}$ and the ionic compound redissolves. This is possible because of the high polarising power of the small, highly charged chromium(III) ions.

Under alkaline conditions hydrogen peroxide oxidises chromium(III) to chromium(VI).

$$H_2O_2(aq) + 2e^- \longrightarrow 2OH^-(l)$$

Zinc reduces a green solution of chromium(III) ions to a blue solution of chromium(II) ions. Chromium(II) is a powerful reducing agent and is quickly converted to chromium(III) by oxygen in the air.

Manganese

 *CD-ROM*

Manganese is a hard, grey, brittle d-block metal with the electron configuration $[Ar]3d^54s^2$.

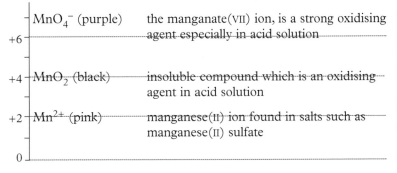

MnO_4^- (purple) — the manganate(VII) ion, is a strong oxidising agent especially in acid solution

MnO_2 (black) — insoluble compound which is an oxidising agent in acid solution

Mn^{2+} (pink) — manganese(II) ion found in salts such as manganese(II) sulfate

+6, +4, +2, 0

Figure 3.12.5 ▲
The main oxidation states of manganese: $KMnO_4$, MnO_2 and $MnSO_4$

Definition

Deprotonation describes the removal of a proton from a molecule or ion. Adding aqueous hydroxide ions to a solution of chromium(III) ions leads to a series of deprotonation reactions.

Test yourself

4 Write an equation for the oxidation of iron(II) to iron(III) ions by dichromate(VI) ions in acid solution.

5 Given an example of the use of dichromate(VI) as an oxidising agent in organic chemistry.

6 Explain the use of paper moistened with dichromate(VI) solution in the test for sulfur dioxide gas.

Inorganic Chemistry

Section three

Definition

A **disproportionation reaction** is a reaction in which the same element both increases and decreases its oxidation number.

Potassium manganate(VII) consists of greyish–black crystals. It is used as a powerful oxidising agent. Potassium manganate(VII) is an important reagent in redox titrations because it will oxidise many reducing agents in acid conditions. The reactions go according to their equations which makes them suitable for quantitative work.

$$MnO_4^-(aq) + 8H^+(aq) + 5e^- \longrightarrow Mn^{2+}(aq) + 4H_2O(l)$$

Under special conditions it is also possible to produce solutions with red manganese(III) ions or green manganate(VI) ions. Manganate(VI) is stable in alkaline solution but disproportionates to manganate(VII) and manganese(IV) oxide on adding acid.

Manganese(IV) oxide is a catalyst for the decomposition of solutions hydrogen peroxide. Hydrogen peroxide decomposes slowly on standing in the dark, but much more rapidly in the present of a catalyst. This is another disproportionation reaction with oxygen starting in the −1 state and ending up in the −2 and 0 states. Hydrogen peroxide both oxidises and reduces itself.

$$2H_2O_2(aq) \longrightarrow 2H_2O(l) + O_2(g)$$

Test yourself

7 a) Write the balanced equation for the oxidation of hydrogen peroxide to oxygen by potassium manganate(VII).

b) Why is no indicator required for a manganate(VII) titration?

c) Calculate the concentration of a solution of hydrogen peroxide given that in a titration with 25 cm³ of 0.020 mol dm⁻³ potassium manganate(VII) the titre was 28.5 cm³.

8 Write a balanced equation for the disproportionation of green manganate(VI) ions, MnO_4^{2-}, in acid.

9 Adding aqueous sodium hydroxide to a solution of manganese(II) ions produces an off-white precipitate which gradually turns brown in air. What are the chemical changes underlying these observations?

Iron

CD-ROM

Iron is another d-block metal which shows the characteristic behaviour of the transition metals. Its electron configuration is: $[Ar]3d^6 4s^2$.

Iron can exist in more than one oxidation state. Iron dissolves in dilute acids to form iron(II) salts. The aqueous iron(II) ion is pale green. Oxidising agents, including oxygen in the air, chlorine or potassium manganate(VII), all oxidise iron(II) to iron(III). The aqueous iron(III) ion is yellow.

$$Fe^{2+}(aq) \longrightarrow Fe^{3+}(aq) + e^-$$

Iron ions form a variety complexes in both the +2 and the +3 states. The hexaaquo complex of the iron(III) ion, $Fe(H_2O)_6^{3+}$, is acidic in solution for the same reason that the aluminium(III) and chromium(III) ions are acidic (see page 91). The hydrated iron(III) iron is sufficiently acidic to react with carbonate ions to form carbon dioxide.

Adding aqueous sodium hydroxide to solutions of iron salts produces precipitates which can help to distinguish iron(II) from iron(III) compounds. Iron(II) hydroxide appears as a greenish precipitate while iron(III) hydroxide is a rusty browny–red colour.

A very sensitive test for iron(III) is to add a dilute solution of thiocyanate ions, SCN⁻. The solution turns to a deep blood–red colour due to the formation of iron complexes such as $[Fe(H_2O)_5(SCN)]^{2+}$. The SCN⁻ ligand takes the place of a water molecule in the aquo complex.

An example of an iron(III) complex with a bidentate ligand is $[Fe(C_2O_4)_3]^{3-}$ in which the ligand is the ethanedioate ion.

Figure 3.12.6 ▲
Solutions containing iron in the +2 and +3 states

Definition

Steels are alloys of iron with carbon and with other metals. Mild steel contains about 0.2 per cent carbon. Alloy steels consist of iron with small amounts of carbon together with up to 50% of one or more of these metals: aluminium, chromium, cobalt, manganese, molybdenum, nickel, titanium, tungsten, and vanadium.

Inorganic Chemistry

Section three

10 Show how deprotonation reactions lead to the formation of iron(III) hydroxide on adding sodium hydroxide to a solution of hydrated iron(III) ions.

11 Suggest a reason why a solution of hydrated iron(III) ions is more acidic than a solution of hydrated iron(II) ions.

12 Explain why it is possible to precipitate an insoluble carbonate from a solution of iron(II) ions but not from a solution of iron(III) ions.

13 Write an equation for the ligand exchange reaction between hydrated iron(III) ions and thiocyanate ions.

14 Show that the suggested mechanism for the use of iron(III) ions to catalyse the oxidation of iodide ions by peroxodisulfate(VI) ions is consistent with the electrode potentials for the half-equations involved (see pages 249–250).

15 Give an example of iron metal acting as a heterogeneous catalyst in an industrial process.

16 Produce a summary table to show that iron illustrates all the characteristic properties of transition elements.

Figure 3.12.7 ▲
Filter paper soaked in pink cobalt(II) chloride solution and dried in an oven until it is blue can be used to test for the presence of water

Cobalt   *CD-ROM*

Cobalt is a hard, silvery d-block metal which is less reactive than iron. It has the electron configuration $[Ar]3d^7 4s^2$.

Cobalt is an ingredient of alloy steels. One of them is the ferromagnetic alloy, Alnico which is used to make permanent magnets.

In solution cobalt forms ions in the +2 and +3 oxidation states. Cobalt(II) is the more stable state. Anhydrous cobalt(II) chloride is blue but it turns pink on adding water as the cobalt ions are hydrated to the $[Co(H_2O)_6]^{2+}$ ion.

A dilute solution of cobalt chloride is pink because the cobalt(II) ions are hydrated. A concentrated solution of the salt is blue. A dilute solution also turns blue on adding concentrated hydrochloric acid. The colour change is due to a ligand exchange reaction. Chloride ions replace the water molecules. The reaction is reversible.

17 Explain the chemical basis for the test illustrated in Figure 3.12.7.

18 Explain why raising the concentration of chloride ions by adding concentrated hydrochloric acid changes the colour of a dilute solution of cobalt(II) chloride solution.

$$[Co(H_2O)_6]^{2+}(aq) + 4Cl^-(aq) \rightleftharpoons [CoCl_4]^{2-}(aq) + 6H_2O \ (aq)$$

It is normally very difficult to oxidise aqueous cobalt(II) to cobalt(III) but the reaction goes readily if the cobalt(II) ions are complexed with ammonia molecules. The Co(III) complex with ammonia is more stable than the Co(II) complex. The value for the standard electrode potential shows that hydrated Co(III) is a stronger oxidising agent than potassium manganate(VII) in acid solution.

$$[Co(H_2O)_6]^{3+}(aq) + e^- \rightleftharpoons [Co(H_2O)_6]^{2+}(aq) \qquad E^\ominus = +1.82 \text{ V}$$

When the two states are complexed with ammonia the standard electrode potential shifts to a value that shows that the Co(III) state is much more stable. Cobalt(II) is now a reducing state and can be oxidised to Co(III) by oxygen or hydrogen peroxide.

19 Show that the standard electrode potentials reflect the greater stability of the +3 state relative to the +2 state when cobalt ions are complexed with ammonia molecules.

$$[Co(NH_3)_6]^{3+}(aq) + e^- \rightleftharpoons [Co(NH_3)_6]^{2+}(aq) \qquad E^\ominus = +0.10 \text{ V}$$

$$H_2O_2(aq) + 2H^+(aq) + 2e^- \rightleftharpoons 2H_2O(l) \qquad E^\ominus = +1.77 \text{ V}$$

Figure 3.12.8 ▲
A solution of nickel(II) sulfate

Nickel

Nickel is a hard, greyish but shiny d-block metal with the electron configuration $[Ar]3d^84s^2$. Nickel is relatively unreactive so it is used to make spatulas and crucibles. Nickel is a constituent of many alloys including some alloy steels and the ferromagnetic alloy Alnico in permanent magnets.

The common oxidation state of nickel is +2. Nickel(II) salts, such as the sulfate $NiSO_4$, are green. Other oxidation states only arise in special circumstances.

Copper

Copper is a ductile metal with a familiar coppery colour. It has the electron configuration $[Ar]3d^{10}4s^1$. This electron configuration is an exception to the normal $[Ar]3d^x4s^2$ pattern for the first series of d-block elements. Energetically it is more favourable to half-fill the d-sub-shell and leave only one electron in the 4s.

Copper is relatively unreactive. It corrodes very slowly in moist air. It is not attacked by dilute non-oxidising acids. Copper is also a good conductor of electricity. The metal is widely used in electricity cables and for domestic water pipes. Copper's mechanical properties are enhanced by making alloys such as brass and bronze.

Copper forms compounds in the +1 and +2 states. Under normal conditions copper(II) is the stable state in aqueous solution. Copper(I) disproportionates in aqueous solution. So when Cu_2O dissolves in dilute sulfuric acid the products are copper(II) sulfate and copper.

$$2Cu^+(aq) \rightleftharpoons Cu^{2+}(aq) + Cu(s)$$

The equilibrium lies well to the right. In the presence of water, copper(I) can exist as very insoluble compounds such as Cu_2O, CuI or $CuCl$. Iodide ions, for example, reduce copper(II) to copper(I) ions which immediately precipitate with more iodide ions as white copper(I) iodide.

$$2Cu^{2+}(aq) + 4I^-(aq) \longrightarrow 2CuI(s) + I_2(s)$$

Copper(I) can exist in aqueous solution as stable complexes such as $[Cu(NH_3)_2]^+$ or $[Cu(CN)_4]^{3-}$.

Fehling's reagent contains a deep blue copper(II) complex in alkali. The reagent is used to distinguish aldehydes from ketones. An aldehyde reduces the reagent on heating to copper(I) oxide. The blue colour first turns greenish then disappears as a reddish-brown precipitate forms (see Figure 4.7.13 on page 175).

(see Figure 4.7.13 on page 175).

Test yourself

20 Given an example of nickel acting as a catalyst.

21 What shape would you expect for the $[Ni(H_2O)_6]^{2+}$ and the $[NiCl_4]^{2-}$ ions?

Definitions

Brass is an alloy of copper (60–80%) and zinc (20–40%). Brass is easily worked, has an attractive gold colour and does not corrode.

Bronze is an alloy of copper with up to 12% tin. Bronze is a strong, hardwearing alloy with good resistance to corrosion. It is used to make gear wheels, bearings, propellers for ships, statues and coins.

Test yourself
D

22 Use standard electrode potentials to show that copper(I) is expected to disproportionate in solution under standard conditions.

23 Give an example of copper acting as a catalyst.

24 Describe the ligand exchange reactions when hydrated copper(II) ions react with:

a) ammonia solution

b) chloride ions.

25 Explain why the copper(II) ions must be present as a complex ion in Fehling's solution.

Review

This guidance will help you to organise your notes and revision. Check the terms and topics against the specification you are studying. You will find that some topics are not required for your course.

Key terms

Show that you know the meaning of these terms by giving examples. Consider writing the key term on one side of an index card and the meaning of the term with an example on the other side. Then you can easily test yourself when revising. Alternatively use a computer data base with fields for the key term, the definition and the example. Test yourself with the help of reports which show just one field at a time.

- Acidic oxide
- Basic oxide
- Amphoteric oxide
- Hydrolysis
- Protonation and deprotonation
- Inert-pair effect
- d-block element
- Transition element
- Ligand
- Co-ordination number

- Monodentate ligand
- Bidentate ligand
- Hexadentate ligand
- Chelate
- Ligand substitution
- Polarising power
- Heterogeneous catalyst
- Homogeneous catalyst
- Autocatalysis
- Absorption spectrum

Symbols and conventions

Make sure that you understand the conventions which chemists use when using symbols and writing equations. Illustrate your notes with examples.

- Use of 'arrows in boxes' to represent the electron configurations of d-block element atoms and ions.
- The distinction which chemists make between d-block elements and transition elements based on electron configurations.
- Rules for naming complex ions.
- Diagrammatic methods for representing the shapes of complex ions.
- Use of oxidation numbers to balance redox equations.

Facts, patterns and principles

Use tables, charts, concept maps or mind maps to summarise key ideas. Brighten your notes with colour to make them memorable.

- The variation across period 3 in the chemical properties of the elements, illustrated by their reactions with oxygen, chlorine and water.
- The properties of the oxides of the elements in period 3 and the interpretation of these properties in terms of structure and bonding.
- The properties of the chlorides of the elements in period 3 and the interpretation of these properties in terms of structure and bonding.
- The similarities and trends in the properties of the elements, oxides and chlorides in group 4.
- The characteristic properties of transition elements illustrated with examples from specific elements.

- The chemistry of complex ions including their formulae, shapes and ligand exchange reactions.
- Interpretation of the chemistry of selected d-block elements and compounds in terms of electron configuration.
- The behaviour of hydrated metal ions in aqueous solutions and especially their reactions with alkalis.

Predictions

- Predict the colour of a transition metal complex ion given its absorption spectrum.
- Use standard electrode potentials to predict the expected direction of change for reactions involving transition metal ions.

Laboratory techniques

Draw diagrams to show these key steps in practical procedures.

- Carrying out a redox titration with potassium manganate(VII).
- Using a titration to estimate the concentration of a solution of an oxidising agent with potassium iodide and sodium thiosulfate.
- Use of a colorimeter or spectrometer to determine the formula of a complex ion.

Calculations

Give your own worked examples, with the help of the Test Yourself and examination questions to show that you can carry out calculations to work out the following from given data.

- Calculations to interpret the results of titrations to estimate the concentration of oxidising or reducing agents.

Chemical applications

Find examples to illustrate the uses of specified elements and compounds.

- Examples of the uses of complex ions.
- Describe and explain examples of the uses of heterogeneous and homogeneous catalysts.

Key skills

Communication

Finding out about the sources and uses of complex ions or of catalysts requires you to select and read information from a range of sources. You can bring the information together as a coherent report or as an illustrated presentation to others. If writing a report you can meet key skill requirements by synthesising relevant information and organising it clearly and coherently using specialist vocabulary.

Information technology

CD-ROMs and web sites can be a rich source of numerical data and descriptive information about elements and their compounds. You can plot values from data bases to discover patterns in the data.

Section four
Organic Chemistry

Contents

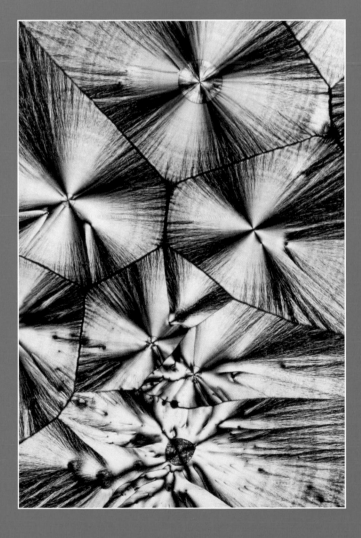

4.1 Organic chemistry in action

We and all other living things on Earth are carbon-based life-forms. Science fiction writers have speculated about life based on other elements but, as far as we know, it is the amazing ability of carbon to form chains, branched chains and rings which makes life possible. We depend on carbon compounds for our food, we burn carbon compounds as fuels and many of the modern materials we use in our homes, wear as clothing and rely on for transport are synthetic carbon compounds.

Figure 4.1.1 ▲
Liquid crystals photographed through a microscope with polarised light. Liquid crystals are used in the displays of computers, calculators and digital watches

Organic synthesis

Many features of modern life depend on the skills of chemists and their ability to synthesise new and complex molecules. New colours for the fashion industry are organic molecules. So are large numbers of compounds made every day for testing as possible drugs to cure disease. Lightweight and thin computer screens depend on liquid crystals, which are carbon compounds that chemists have tailor made so that they can affect light and respond to an electric field.

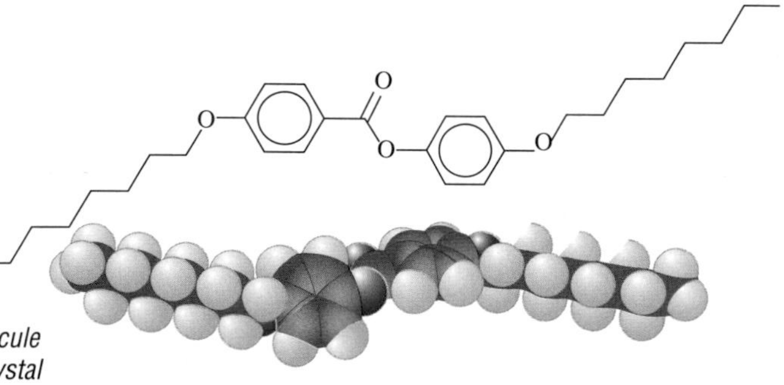

Figure 4.1.2 ▶
Structure of a molecule which is a liquid crystal

The job of a synthetic organic chemist is to start with a proposed structure and devise a practical method of making it from simpler molecules. The scale of events, and some indication of the difficulties involved in the synthesis of a complex molecule are illustrated by the famous synthesis of chlorophyll completed in 1959 at Harvard University by a team of seventeen chemists led by Robert Woodward. Chlorophyll is the green pigment in plants and it was only in 1940 that another famous chemist, Hans Fischer, had proposed a likely structure for the molecule. When Woodward and his team started their work they were not even sure that the proposed structure was correct.

Figure 4.1.3 ▶
Structure of chlorophyll

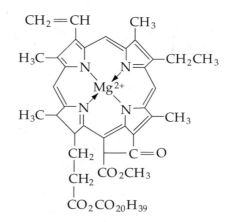

Woodward and his team started work in 1956. From the start, the project was planned in great detail. The chemists read all papers reporting previous studies of chlorophyll very carefully to ensure that no clues to a successful synthesis were missed. They drew on their understanding of the mechanisms of organic reactions to predict the outcomes of reactions and to suggest routes to necessary intermediates. The synthesis would have been impossible without the newer methods of separation, purification and identification. The variety of spectroscopy techniques were crucial to success. They published their paper describing the successful synthesis in 1960, opening up a new stage of research to understand better how chlorophyll contributes to the processes of photosynthesis.

Organic analysis

Chemists use a variety of methods for both qualitative analysis and quantitative analysis, to identify organic compounds and work out their composition and structure. Traditionally, chemical tests helped to identify functional groups in organic molecules. In a modern laboratory, organic analysis is based on a range of automated and instrumental techniques including combustion analysis, mass spectrometry, infra-red spectroscopy, ultraviolet spectroscopy and nuclear magnetic resonance spectroscopy.

Biochemistry

Organic chemistry is closely linked to biochemistry, which is the study of chemical changes in living organisms. Many of the natural compounds in living cells are large molecules such as carbohydrates, fats, proteins and nucleic acids.

Some biochemical reactions are involved in the hydrolysis and oxidation of food. Others build up small molecules into the complex molecules which cells need as they grow. All these reactions depend on the amazing catalytic effect of enzymes which are themselves proteins.

The techniques of chemists seem violent and crude when compared to the processes in living cells. Boiling alkali, liquid bromine and metallic sodium may be suitable as laboratory reagents but they destroy life. So for many years chemists have envied the powers of enzymes. Thanks to the study of molecular shapes and organic reaction mechanisms, chemists now have a much better understanding of the ways in which enzymes work.

In recent years chemists have learnt to design catalysts based on transition metal ions which share some of the properties of enzymes. These new catalysts work under relatively mild conditions and are much more specific than older catalysts. Like enzymes, these catalysts have a definite three dimensional shape and an active site where the chemical changes take place.

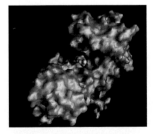

Figure 4.1.4 ▲
Computer graphic of an enzyme with a molecule of glucose bound to its active site. The enzyme is a protein which catalyses the conversion of glucose to glucose-6-phosphate

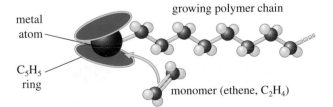

Figure 4.1.5 ▲
The polymerisation of ethene using a metallocene catalyst. Molecules of ethene add to the growing chain one at a time at the active site

Labels in figure: growing polymer chain; metal atom; C_5H_5 ring; monomer (ethene, C_2H_4)

Organic chemists must know the formula and structure of a compound before they can explain its chemistry or devise a method of synthesis. Thanks to modern methods of analysis, chemists now have a powerful array of techniques for exploring molecular structures.

CD-ROM

Formulae of organic molecules

Empirical formulae

Finding the simplest formula of a compound is the first step towards understanding its chemistry. The usual way to find the empirical formula of an organic compound containing carbon, hydrogen and possibly oxygen, is to burn a measured sample and then determine the masses of carbon dioxide and water which form. This makes it possible to calculate the masses of the elements carbon and hydrogen in the sample and hence the empirical formula.

The empirical formula of a compound shows the simplest ratios of the numbers of each atom in the compound. More interesting is the molecular formula which shows the number of atoms of each element in a molecule of the compound. All that is needed to find the molecular formula from the empirical formula is the molar mass of the compound. This can be found from the mass spectrum of the compound (see pages 226–229).

Molecular formulae

The molecular formula is always a simple multiple of the empirical formula. Benzene, for example has the empirical formula CH and the molecular formula C_6H_6.

Figure 4.2.1 ▶
The empirical, molecular, structural and displayed formulae of cyclohexene

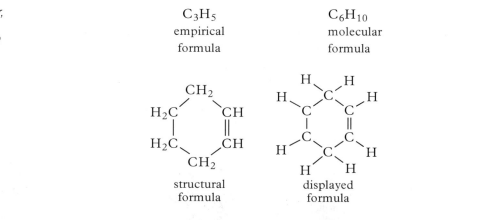

Test yourself **D**

1 The empirical formula of an ingredient of antifreeze is CH_3O and its relative molecular mass is 62. What is the molecular formula of the compound?

2 Determine the empirical and molecular formulae of an amino acid with the composition 32.12% carbon, 6.71% hydrogen, 42.58% oxygen and 18.59% nitrogen. The relative molecular mass of the compound is 75.

3 Complete combustion of 0.667 g of a crystalline organic acid produced 0.651 g carbon dioxide and 0.134 g water. The molar mass of the acid is 90. What are the empirical and molecular formulae of the acid? Suggest a structure and name for the acid.

Structural formulae

The structural formula of a molecule shows how the atoms link together. Given the molecular formula of an organic compound, and knowing something about its reactions, it is often possible to suggest possible structures, assuming that atoms form the number of covalent bonds shown in Figure 4.2.2. Chemists use a variety of spectroscopic methods to work out the true structures (see section 4.20).

◀ *Figure 4.2.2*

Element	Number of bonds
carbon, C	4
hydrogen, H	1
oxygen, O	2
nitrogen, N	3
sulfur, S	2
halogens, F, Cl, Br or I	1

Sometimes it is enough to show structural formulae in a condensed form such as $CH_3CH_2CH_2CH=CHCH_3$ for hex-2-ene.

Often it is clearer to write the full structural formula showing all the atoms and all the bonds. This type of formula is also called a displayed formula.

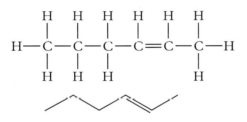

◀ *Figure 4.2.3*
Displayed and skeletal formulae for hex-2-ene. Drawn like this the formulae do not show the true shape in three dimensions

Chemists have devised a useful shorthand for showing the structure of more complex molecules. These skeletal formulae need careful study because they only represent the hydrocarbon part of the molecule with lines for the bonds between carbon atoms while leaving out the symbols for the carbon and hydrogen atoms.

Often it is important to understand the three-dimensional shape of the molecules. Chemists use various types of models including computer models to study how molecules interact in three dimensions.

◀ *Figure 4.2.4*
Ball-and-stick and space filling models of a hydrocarbon

Test yourself

4 Work out the empirical, molecular, structural, displayed and skeletal formulae of the hydrocarbon in Figure 4.2.4. What is the name of the hydrocarbon?

Hydrocarbons

Hydrocarbons are compounds of carbon and hydrogen only. They are of great practical importance because they make up most of the fuels we burn and are the main raw materials for the chemical industry.

Figure 4.2.5 ▶
Classification of hydrocarbons

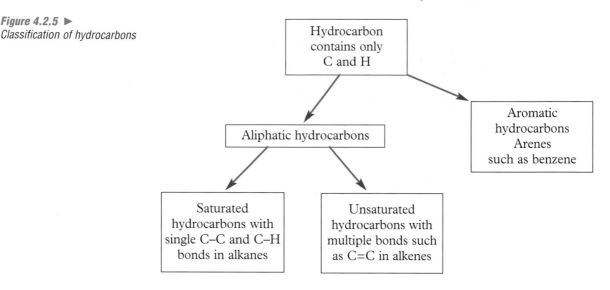

The systematic names of organic compounds are based on the name of the parent hydrocarbon which has the same number and arrangement of carbon atoms.

Functional groups

A functional group is the atom or atoms which gives a series of organic compounds its characteristic properties. Chemists often think of an organic molecule as a hydrocarbon skeleton with one or more functional groups in place of hydrogen atoms.

The functional group of a molecule is responsible for most of its reactions. Carbon–carbon and carbon–hydrogen bonds are both strong and non-polar so they are unreactive towards most common reagents.

Figure 4.2.6 ▶
The structure of the steroid, cortisone, labelled to show both the reactive functional groups and the hydrocarbon skeleton. The carbon skeleton does not change in most of the reactions of cortisone

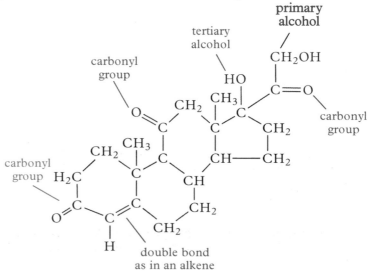

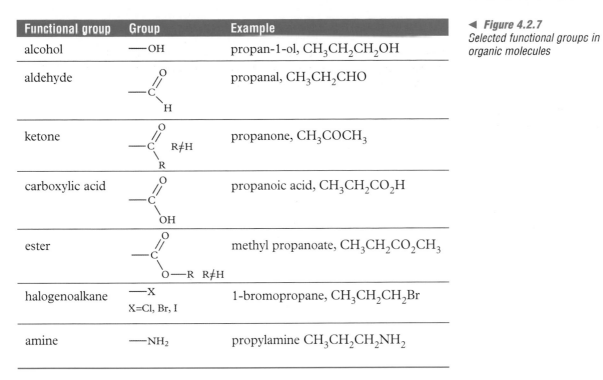

Functional group	Group	Example
alcohol	—OH	propan-1-ol, $CH_3CH_2CH_2OH$
aldehyde	(C=O with H)	propanal, CH_3CH_2CHO
ketone	(C=O with R, R≠H)	propanone, CH_3COCH_3
carboxylic acid	(C=O with OH)	propanoic acid, $CH_3CH_2CO_2H$
ester	(C=O with O—R, R≠H)	methyl propanoate, $CH_3CH_2CO_2CH_3$
halogenoalkane	—X X=Cl, Br, I	1-bromopropane, $CH_3CH_2CH_2Br$
amine	—NH_2	propylamine $CH_3CH_2CH_2NH_2$

◄ *Figure 4.2.7*
Selected functional groups in organic molecules

Organic Chemistry

Section four

Test yourself

5 Name each of these compounds:

a)

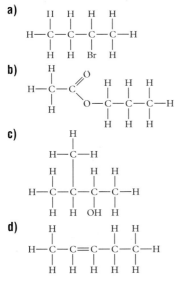

b)

c)

d)

6 Anaerobic respiration in muscle cells breaks down glucose to simpler compounds including these two molecules. Identify the functional groups in these molecules:

a) $CH_2OH–CHOH–CHO$

b) $CH_3–CO–CO_2H$

7 Pheromones are messenger molecules produced by insects to attract mates or to give an alarm signal. Identify the functional groups in this pheromone produced by queen bees.

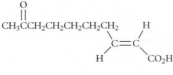

Definition

An **homologous series** consists of closely related organic compounds. The compounds in an homologous series have the same functional group and can be described by a general formula. The formula of one member of the series differs from the next by CH_2.

Structural isomerism

Isomers are compound with the same molecular formula but different structures. The number and variety of isomers adds to the multitude of carbon compounds.

There are three main types of structural isomerism.

■ Chain isomers in which the chain of carbon atoms is arranged in different ways.

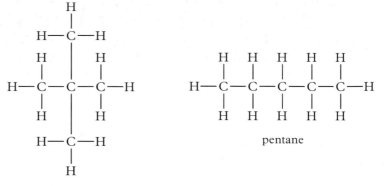

2,2-dimethylpropane   pentane

■ Position isomers where the functional group is in different positions.

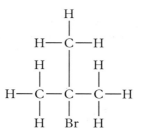   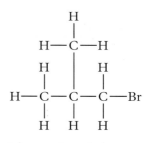

2-bromo-2-methylpropane   1-bromo-2-methylpropane

■ Functional group isomerism in which the characteristic reactive groups are different.

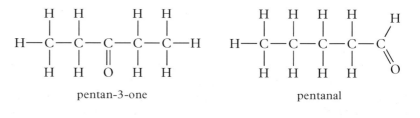

pentan-3-one   pentanal

Test yourself

8 Draw the structures and name the chain isomers of 2,2-dimethylbutane.

9 Draw the structures and name the two position isomers of 1-bromopentane.

10 Draw the structures and name two functional group isomers with the molecular formula C_4H_8O.

11 There are three isomers with the formula C_5H_{12}. Their boiling points are: 10 °C, 28 °C and 36 °C. Draw the structures of the three compounds and match the structures with the boiling points.

4.3 Stereochemistry

It was the study of isomerism which led chemists to start thinking about molecules in three dimensions. This was an essential step in the development of our understanding of the chemical reactions in living thing. Stereochemistry is the study of molecular shapes and the effect of shape on chemical properties. Smell and taste seem to depend on molecular shape. In one form the hydrocarbon limonene, for example, smells of oranges while in another form with a different shape it smells of lemons. Molecular shape can also subtly alter the physiological effects of drugs.

CD-ROM

Stereoisomers are distinct compounds. They have the same molecular formulae and structural formulae but have different three-dimensional shapes. There are two kinds of stereoisomerism: geometrical isomerism and optical isomerism.

Geometrical isomerism

Alkenes and other compounds with C=C double bonds may have geometrical isomers because there is no rotation about the double bond. The isomers are labelled *cis* and *trans*. In the *cis* isomer, similar functional groups are on the same side of the double bond. In the *trans* isomer similar functional groups are on opposite sides of the double bond.

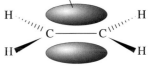

Figure 4.3.1 ▲
The pi (π) bond in an alkene prevents rotation about the double bond

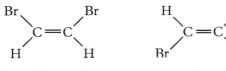

cis-1,2-dibromoethene
m.p. −53 °C, b.p. 110 °C

trans-1,2-dibromoethene
m.p. −9 °C, b.p. 108 °C

Figure 4.3.2 ▲
Geometric isomers of 1,2-dibromoethene. They are distinct compounds with different melting points, boiling points and densities

The female silk moth secretes a pheromone called bombycol. This messenger molecule strongly attracts male moths of the same species. Analysis shows that two double bonds help to determine the shape of the compound. Chemists are interested in pheromones because they offer an alternative to pesticides for controlling damaging insects. By baiting insect traps with sex attractants it is possible to capture large numbers of insects before they mate.

Note

'Trans' is a prefix in many words meaning 'across', such as transport, transplant and transmute. Hence the use of the term to label the trans form of a pair of geometric isomers.

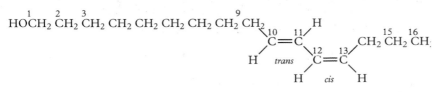

Figure 4.3.3b ▲
The sex attractant bombycol which is hexadeca-trans,10-cis,12-dien-1-ol

Figure 4.3.3a ▲
Silk moths (male right and female lower) with eggs on a cocoon

147

1 Which of these compounds have geometric isomers: propene, but-1-ene, but-2-ene, 1,1-dichloroethene, 1,2-dichloroethene?

Vision also depends on geometric isomerism. The light sensitive cells in the retina of an eye contain pigments which absorb light. Part of each pigment is a molecule called retinal (Figure 4.3.4). The light is absorbed by the electrons in one of the double bonds. The energy from the light breaks the π-bond so that the molecule can change shape from the *cis* form to the *trans* form. In a way that is not yet understood, this change of shape is the start of a sequence of molecular events which sets off a nerve impulse to the brain.

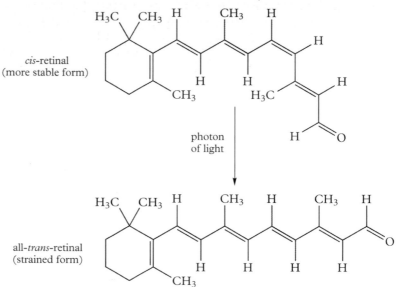

Figure 4.3.4 ▶
The effect of light on retinal

Optical isomerism

Mirror image molecules

Every molecule has a mirror image. Generally the mirror image of a molecule can be turned about to show that it is identical with the original molecule. Sometimes, however, it turns out that a molecule and its mirror image are subtly different. The molecule and its mirror image cannot be superimposed.

A molecule is chiral if, like one of your hands, it cannot be superimposed on its mirror image. The name comes from the Greek for 'hand'.

Figure 4.3.5 ▶

2 Identify the chiral objects in Figure 4.3.5.

3 With the help of molecular models decide which of these molecules are chiral: H_2O, NH_3, CH_4, CH_2Cl_2, CH_2ClBr, $CH_3CHClBr$, $CH_3CHOHCO_2H$.

Chirality and molecular shape

Chiral molecules are asymmetric. This means that they have mirror image forms which are not identical. The commonest chiral compounds are organic molecules in which there is a carbon atom attached to four different atoms or groups.

Organic Chemistry

Section four

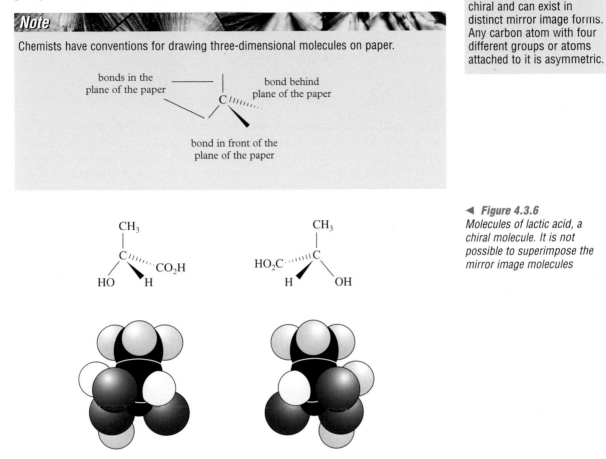

> **Note**
>
> Chemists have conventions for drawing three-dimensional molecules on paper.
>
> bonds in the plane of the paper ———
> bond behind plane of the paper
> bond in front of the plane of the paper

> **Definition**
>
> **Asymmetric molecules** are molecules with no centres, axes or planes of symmetry. Asymmetric molecules are chiral and can exist in distinct mirror image forms. Any carbon atom with four different groups or atoms attached to it is asymmetric.

◀ **Figure 4.3.6**
Molecules of lactic acid, a chiral molecule. It is not possible to superimpose the mirror image molecules

The two forms of a compound such as lactic acid behave identically in ordinary chemical reactions and their main physical properties are also the same. They differ in their effect on polarised light. This is the only way of telling the two forms apart. So chemists call them optical isomers.

Optical isomerism and polarised light

A light beam becomes polarised after passing through a sheet of Polaroid – the material used to make sunglasses. In polarised light all the waves are vibrating in the same plane.

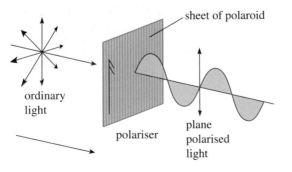

◀ **Figure 4.3.7**
Ordinary light and polarised light

Figure 4.3.8 ▶
Effect of passing a light beam through two sheets of Polaroid. After the first sheet the light is polarised. The beam passes through the second sheet if it is aligned the same way as the first. No light gets through if the second sheet is rotated through 90°

Light is plane polarised after passing through a sheet of Polaroid.

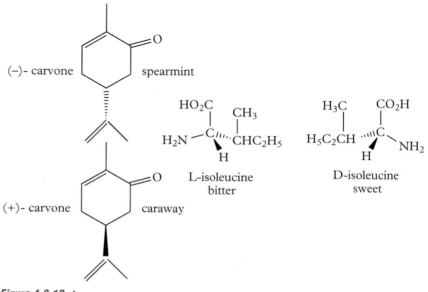

When polarised light passes through a solution of just one form of a chiral compound it rotates the plane of polarisation. One isomer rotates the plane of polarised light clockwise (the + isomer). The other isomer rotates the plane of polarised light in the opposite direction (the − isomer). For accurate results, chemists measure the rotations with monochromatic light in an instrument called a polarimeter.

Figure 4.3.9 ▶
Effect of passing a light beam through a solution of a chiral compound. After the first sheet the light is polarised. On passing through the solution the plane of polarisation rotates a small amount. Now the second sheet has to be twisted to let the polarised light beam to pass through

tube containing solution of sample rotates the plane-polarised light

second polaroid needs to be rotated to allow the plane-polarised light through

angle of rotation

first polaroid produces plane-polarised light

plane-polarised light has been rotated to the left

Definitions

Optical isomers are mirror image forms of a chiral compound. One mirror image form rotates the plane of polarised light in one direction. The other form has the opposite effect.

Chemists use the term **enantiomer** to describe these mirror image molecules which are optical isomers. They derived the word from the Greek word for 'opposite'.

A **racemic mixture** is a mixture of equal amounts of the two mirror-image forms of a chiral compound. The mixture does not rotate polarised light because the two optical isomers have equal and opposite effects so they cancel each other out.

Chirality and living things

Human senses are sensitive to molecular shape. The optical isomers of some molecules have different tastes and smells.

(−)- carvone   spearmint

L-isoleucine
bitter

D-isoleucine
sweet

(+)- carvone   caraway

Figure 4.3.10 ▲
Optical isomers with differing tastes and smells

What is true of the sensitive cells in the nose and on the tongue, is also true of most of the rest of the human body. Living cells are full of messenger and carrier molecules which interact selectively with the active sites and receptors in other molecules such as enzymes. These messenger and carrier molecules are all chiral and the body works with only one of the mirror image forms. This is particularly true of amino acids and proteins (see section 4.14).

The chemists who synthesise and test new drugs have to pay close attention to chirality. Dextropropoxyphene, for example, is a painkiller. The molecule has two asymmetric carbon atoms. Its mirror image is no use for treating pain but is a useful ingredient of cough mixtures.

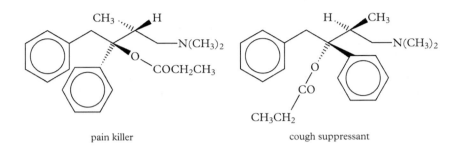

pain killer   cough suppressant

Figure 4.3.11 ▲
Dextropropoxyphene and its mirror image

Test yourself

4   Identify the chiral centres in: isoleucine, carvone and dextropropoxyphene.

5   Explain why the amino acid glycine, $CH_2NH_2CO_2H$, does not have optical isomers.

6   Which of these alcohols are chiral: propan-1-ol, propan-2-ol, butan-2-ol?

4.4 Organic reaction mechanisms

A mechanism describes how a reaction takes place showing step by step the bonds which break and the new bonds which form. Thanks to this knowledge chemists can plan more ambitious syntheses, such as the synthesis of chlorophyll (see page 140). In industry they can devise new methods of manufacture which are more efficient and produce less waste. Also biochemists can make sense of the remarkable properties of the enzymes which control and catalyse the reactions in living cells.

Figure 4.4.1 ▲
By measuring the properties of lasers as they pass through chemical reactions, chemists can determine rates of reaction

Figure 4.4.2 ▶
Use of labelling to investigate bond breaking during ester hydrolysis

Investigating reaction mechanisms

The techniques which chemists use to study reactions are becoming more and more sophisticated. With the help of laser beams and spectroscopy it is now possible to follow extremely fast reactions and to watch molecules breaking apart and rearranging themselves in fractions of a second.

Isotopic labelling

Isotopic labelling uses isotopes as markers to trace what happens to particular atoms during a chemical change. Chemists replace atoms of the normal abundant isotope of an element in a molecule with a different isotope.

Isotopes have the same chemical properties so it is possible to use them to follow what happens during a change without altering the normal course of the process. Radioactive isotopes can be tracked by detecting their radiation. The fate of non-radioactive isotopes can be followed by analysing samples with a mass spectrometer.

Isotopic labelling has been used to investigate the mechanism of ester hydrolysis. Using water labelled with oxygen-18 instead of the normal isotope oxygen-16, it was possible to show which bond breaks in the ester. Oxygen-18 is not radioactive. Mass spectrometry showed that the heavier oxygen atoms from the water ended up in the acid and not in the alcohol.

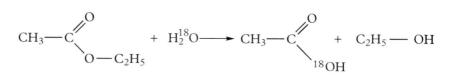

Test yourself

1 a) How does an oxygen-18 atom differ from an oxygen-16 atom?

   b) Why do oxygen-18 and oxygen-16 atoms have the same chemical properties?

2 a) According to Figure 4.4.2, which bond in the ester breaks during the reaction?

   b) Where would the oxygen-18 atoms have appeared if the mechanism involved breaking the other C–O bond in the ester?

Trapping intermediates

Some reaction mechanisms involve not just one step but a series of steps. Intermediates in reactions are the atoms, molecules or ions which do not appear in the balanced equation but are formed during one step of a reaction and are then used up in the next step. Chemists often use spectroscopy to detect intermediates which only exist for a short time during a reaction.

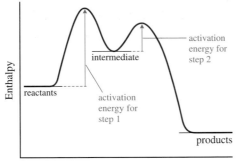

◀ *Figure 4.4.3*
Reaction profile for a reaction with an intermediate. Note the energy dip as the intermediate forms. The dip is not deep enough for the intermediate to exist as a stable product

An ingenious experiment to investigate the addition of bromine to ethene showed that this must be a two-step reaction with a reactive intermediate. Dr Francis and his colleagues mixed ethene and bromine in the presence of a solution of sodium chloride. Ethene reacts rapidly with bromine but it does not react with chloride ions. In this experiment, however, the product was an oily liquid consisting of roughly equal amounts of 1,2-dibromoethane and 1-chloro-2-bromoethane. Based on this evidence chemists suggested a two-step mechanism for the addition reaction with a positively charged intermediate (see page 156).

Studying reaction rates

From measurements of reaction rates, chemists can deduce the rate equation (see section 2.2) for a reaction. The form of the rate equation can often provide valuable clues about the likely mechanism of the reaction. It is generally the molecules or ions involved in the rate-determining step of a reaction which appear in the rate equation for the reaction.

This is illustrated by the reaction of aqueous hydroxide ions with tertiary halogenoalkanes such as 2-bromo-2-methylpropane. The hydroxide ion concentration does not appear in the rate equation which has the form: rate = $k[C_4H_9Br]$. This suggests that the rate-determining step only involves the halogenoalkane molecules (see section 2.4).

Test yourself

6 Draw the displayed formula of 2-bromo-2-methylpropane and use it to explain the term tertiary halogenoalkane.

7 a) Write an equation for the reaction of 2-bromo-2-methylpropane with hydroxide ions in aqueous solution and show that this is an example of a substitution reaction.

 b) Describe a test to show that the reaction produces bromide ions.

 c) Why does the rate equation suggest that the mechanism for this reaction must involve at least two steps?

Test yourself

3 Sketch a curve to show how the reaction profile for a reaction with just a transition state between reactants and products differs from Figure 4.4.3.

4 Write an equation for the expected reaction of ethene with bromine and name the product.

5 a) Suggest a method which Dr Francis might have used to separate and identify the two products of the reaction of ethene with bromine in the presence of sodium chloride solution.

 b) How do the results of the experiment show that the reaction of ethene with bromine does not happen in one step?

 c) Why is it reasonable to suggest that the intermediate in the addition reaction is a positive ion?

Organic Chemistry

Section four

Homolytic bond breaking

A normal covalent bond consists of a shared pair of electrons. There are two ways that a bond can break. In one way the covalent bond breaks leaving one electron with each of the atoms that were linked by the bond. This is homolytic bond breaking. This type of bond breaking produces free radicals which are reactive particles with unpaired electrons.

Figure 4.4.4 ▶
Homolytic bond breaking. A curly arrow with only half an arrow head indicates the movement of a single electron

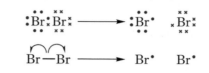

Free radicals are intermediates in reactions taking place:

- ■ in the gas phase at high temperature or in ultraviolet light
- ■ in a non-polar solvent, either when irradiated by ultraviolet light or with an initiator.

Examples of free radical processes include the reaction of alkanes with chlorine or bromine and the addition polymerisation of alkenes (see page 203) as well as the cracking of alkanes and the formation and destruction of the ozone layer.

Free radical reactions

Free radical reactions are often chain reactions. One example is the free radical substitution reaction of chlorine with alkanes such as methane. Alkanes react with chlorine (or bromine) on heating or when exposed to ultrviolet light.
    Typically the reaction has three stages:

1   **initiation** – the step which produces free radicals,
    $Cl–Cl \longrightarrow Cl^\bullet + Cl^\bullet$
2   **propagation** – steps which produce products and more free radicals,
    $CH_4 + Cl^\bullet \longrightarrow CH_3^\bullet + HCl$
    $CH_3^\bullet + Cl_2 \longrightarrow CH_3Cl + Cl^\bullet$
3   **termination** – steps which remove free radicals by turning them into molecules.
    $CH_3^\bullet + CH_3^\bullet \longrightarrow CH_3CH_3$
    $CH_3^\bullet + Cl^\bullet \longrightarrow CH_3Cl$

Note

The symbol for a free radical generally represents the unpaired electron by a dot. Other paired electrons in the outer shells are often not shown. A free chlorine atom has one unpaired electron in its outer shell and is a free radical.

Test yourself

8   Why is a high temperature (or ultraviolet light) needed to form free radicals?

9   Draw structures and use curly arrows to show what happens when an ethane molecule splits into two methyl radicals.

Figure 4.4.5 ▶
The reaction of methane with chlorine as a chain reaction. The chorine atoms and methyl radicals are the carriers which sustain the process until eliminated by a termination step

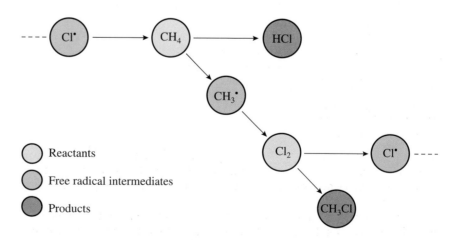

Test yourself

10 What is the practical importance of the reaction of alkanes with halogens?

11 a) What are the two main products of the reaction of methane with chlorine?

   b) Account for the formation of some ethane during the reaction.

   c) Explain why the reaction may also produce other products such as dichloromethane, trichloromethane and tetrachloromethane.

12 Why is it only necessary for a small proportion of the chlorine molecules to be split into free radicals by the initiation step?

Heterolytic bond breaking

Covalent bonds can also break in a way which produces ionic intermediates. In heterolytic bond breaking a covalent bond breaks so that one of the atoms joined by the bond takes both of the shared pair of the electrons while the other is left with none. This type of bond breaking is favoured when reactions take place in polar solvents such as water.

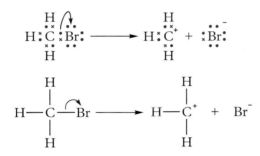

The bond which breaks to produce ionic intermediates is often polar to start with. So heterolytic bond breaking is characteristic of compounds with electronegative atoms such as chlorine, oxygen and nitrogen. Bonds may also break this way if they are easily polarised.

The reagents which attack polar bonds and bring about heterolytic fission are electrophiles and nucleophiles.

Electrophiles are reactive ions and molecules which attack parts of molecules which are rich in electrons. They are 'electron-loving' reagents. Electrophiles form a new bond by accepting a pair of electrons from the molecules they attack during a reaction.

Nucleophiles are molecules or ions with a lone-pair of electrons to donate to form a new covalent bond. They are electron-pair donors (see page 157). Nucleophiles are reagents which attack molecules where there is a partial positive charge, $\delta+$. So they seek out positive charges – they are 'nucleus-loving'.

Electrophilic addition

A two-step mechanism

Electrophilic addition reactions are characteristic of the alkenes. The electrophile attacks the electron-rich region of the double bond between two carbon atoms. Electrophiles which add to alkenes include hydrogen bromide and bromine.

Hydrogen bromide molecules are polar. The hydrogen atom, with its $\delta+$ charge is the electrophilic end of the molecule. The reaction takes place in two steps. The first step forms an intermediate with a positive charge on a carbon atom. It is a carbocation.

Definition

In a free radical **chain reaction** the products of one step of the reaction is a radical which takes part in the next step. The second step again produces a radical which can carry on the chain.

◀ *Figure 4.4.6*
Heterolytic bond breaking. Note the use of double-headed curly arrows to show what happens to pairs of electrons as the bond breaks. One atom takes both the electrons from the shared pair. The tail of the arrow starts at the electron pair. The head of the arrow points to where the electron pair ends up

Test yourself

13 Explain why there is a positive charge on the carbon atoms and a negative charge on the chlorine atoms after the C–Cl bond breaks heterolytically in methyl chloride.

14 Classify these reagents as free radicals, electrophiles or nucleophiles: Cl, H_2O, Br_2, Br^-, Br, H^+, CN^-, CH_3, OH^-, NO_2^+, NH_3.

15 Draw the displayed formulae of these molecules and indicate any polar bonds using $\delta+$ and $\delta-$ signs: C_2H_5Cl, CH_3OH, CH_3COCH_3, C_3H_8, CH_3NH_2.

Definition

A **carbocation** is an intermediate formed during organic reactions in which a carbon atom carries a positive charge. Cations are positively charged ions.

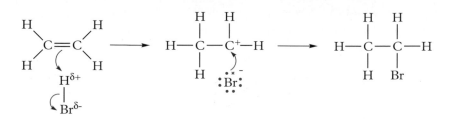

Figure 4.4.7 ▲
Electrophilic addition of hydrogen bromide to ethene. Bond breaking is heterolytic and the intermediates are ions. Curly arrows show the movement of pairs of electrons

Bromine molecules are not polar but they become polarised as they approach the electron-rich double bond. Electrons in the double bond repel electrons in the bromine molecule. The $\delta+$ end of the molecule is then electrophilic.

Test yourself

16 Explain how the mechanism for the addition of bromine to ethene can account for the results of A. W. Fisher's experiments (see page 153), which showed that when bromine reacts with ethene in the presence of sodium chloride:

a) there are two main products one of which is 1-chloro-2-bromoethane

b) no 1,2–dichloroethane forms.

17 Which bonding electrons between the carbon atoms in ethene would you expect to take part in the addition reaction: the electrons in the σ bond or the electrons in the π bond?

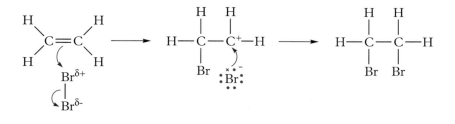

Figure 4.4.8 ▲
Electrophilic addition of bromine to ethene

Addition to unsymmetrical alkenes

The Russian chemist Vladimir Markovnikov studied a great many alkene addition reactions during the 1860s. He found that when a compound HX (such as H–Br or H–OH) is added to an unsymmetrical alkene (such as propene) the rule is that the hydrogen atom adds to the carbon atom (of the carbon–carbon double bond) which already has more hydrogen atoms attached to it.

The mechanism for electrophilic addition helps to account for this rule. On adding HBr to propene there are two possible intermediate carbocations. The secondary carbocation is preferred because it is slightly more stable than the primary carbocation. The secondary ion has two alkyl groups pushing electrons towards the positively charged carbon atom. This helps to stabilise the ion by 'spreading' the charge over the ion.

Definition

The **inductive effect** describes the extent to which electrons are pushed towards or pulled away from a carbon atom by the atoms or group to which it is bonded. Alkyl groups, have a slight tendency to push electrons towards the carbon atom to which they are bonded. As a result any positive charge on the carbon atom is spread more widely over the molecule and this tends to make the molecule more stable. One of the effects of this kind of inductive effect is that a tertiary carbocation is more stable than a secondary carbocation which, in turn, is more stable than a primary carbocation.

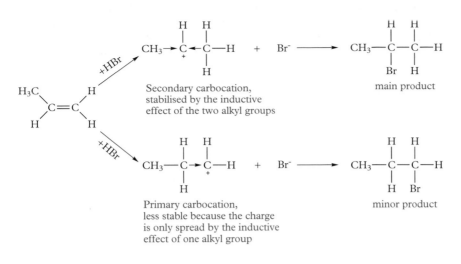

◀ *Figure 4.4.9*
An explanation for Markovnikov's observations

Test yourself

18 Name and give the displayed formula of the main product of the reaction of hydrogen bromide with:

a) but-1-ene

b) *trans* but-2-ene

c) pent-1-ene

19 Arrange these carbocations in order of stability with the most stable ion first: $CH_3CH_2CH^+CH_3$, $CH_3CH_2CH_2CH_2^+$, $(CH_3)_3C^+$.

Nucleophilic substitution

The carbon–halogen bond in a halogenoalkane is polar with the carbon atom at the δ+ end of the dipole. The characteristic reactions of these compounds are substitution reactions brought about by nucleophiles which attack the δ+ end of the carbon–halogen bond.

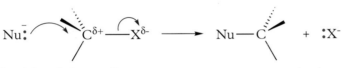

nucleophile   halogenoalkane   leaving group

◀ *Figure 4.4.10*
A generalised description of the substitution reactions of halogenoalkanes

Study of the rate equations suggests that there are two different mechanisms for the nucleophilic substitution reactions of halogenoalkanes.

Hydrolysis of primary halogenoalkanes, such as bromobutane, is overall second order. The rate equation has the form: rate = $k[C_4H_9Br][OH^-]$. The suggested mechanism shows the C—Br bond breaking as the nucleophile, OH^-, forms a new bond with carbon.

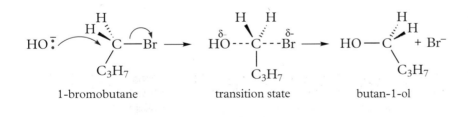

1-bromobutane   transition state   butan-1-ol

◀ *Figure 4.4.11*
The nucleophile is the OH^- ion. The S_N2 mechanism. Substitution–Nucleophilic–2 (the 2 shows that two molecules or ions are involved in the rate-determining step)

Definition

A **leaving group** is an atom or group of atoms which breaks away from a molecule during a reaction. Some groups are better at leaving than others. An iodide ion is a better leaving group than a bromide ion which is in turn a better leaving group than a chloride ion.

Hydrolysis of tertiary halogenoalkanes such as 2-bromo-2-methylpropane, is first order. The rate equation has the form: rate = $k[C_4H_9Br]$. The suggested mechanism shows the C–Br bond breaking first to form a carbocation intermediate. Then the nucleophile, OH^-, forms a new bond with carbon.

Figure 4.4.12 ▶
The S_N1 mechanism. Substitution–Nucleophilic–1 (the 1 shows that one molecule or ion is involved in the rate determining step)

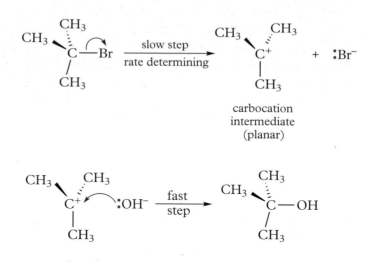

20 Show the steps in the reaction of cyanide ions with 1-bromobutane by the S_N2 mechanism.

21 Suggest reasons why the S_N1 mechanism is favoured rather than the S_N2 mechanisms for tertiary halogenoalkanes, especially if they have bulky groups around carbon atom bonded to the halogen atom. (Hint: when comparing the two mechanisms consider the ease with which the nucleophile can attack the δ+ carbon atom and the stability of possible intermediates).

22 In general the relative rates of hydrolysis of halogenoalkanes are:
RI > RBr > RCl.
Does this trend correlate better with the polarity of the bonds or the strength of the bonds?

4.5 Aromatic hydrocarbons

Today, biotechnology companies are scrambling to patent parts of the human genome in the hope of developing profitable medical treatments. Over a hundred years ago there was a similar rush to patent new dyes based on chemicals from coal tar following the discovery of exciting new colours such as William Perkin's mauve. At the heart of these colourful compounds was a remarkably stable ring of six carbon atoms – the benzene ring.

Arenes

CD-ROM

Benzene is one of the arene family of hydrocarbons. Traditionally chemists called the arenes 'aromatic' ever since the German chemist Friedrich Kekulé was struck by the fragrant smell of oils such as benzene. In the modern name the '**ar–**' comes from **ar**omatic and the ending '**–ene**' means that these hydrocarbons are unsaturated compounds like the alk**ene**s.

The structure of benzene

Kekulé played a crucial part in the history of the chemistry of benzene thanks to a dream which helped him to arrive at a possible structure for a compound with the empirical formula CH and the molecular formula C_6H_6. Kekulé had been working at the problem for some time but one day, in 1865, while dozing in front of the fire he dreamed of a snake biting its own tail and this inspired him to think of a ring structure for the compound.

Figure 4.5.1 ▲
Fabric dyed with Perkin's mauve: the first of the coal tar dyes

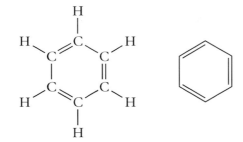

Figure 4.5.2 ▲
Representations of the Kekulé structure for benzene

The problem with this structure is that it suggests that there should be two isomers of 1,2-dichlorobenzene.

In practice it has never been possible to separate isomers of di-substituted benzenes. To get round this problem, Kekulé suggested that benzene molecules might somehow rapidly alternate between the two possible structures but this is not the modern explanation.

The Kekulé structure shows a molecule with alternating double and single bonds. Chemists now know that double bonds are shorter than single bonds (see Figure 4.5.4). They have also discovered that the carbon atoms in a benzene molecule are at the corners of a regular hexagon. All the bonds are the same length; they are shorter than single bonds but longer than double bonds.

An inexperienced chemist looking at the Kekulé structure might expect benzene to behave chemically like a very reactive alkene and to take part in addition reactions with bromine, hydrogen bromide and other reagents. Benzene does not do this. The compound is much less reactive than alkenes and its characteristic reactions are substitution reactions.

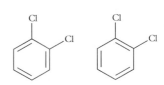

Figure 4.5.3 ▲
Isomers of 1,2-dichlorobenzene suggested by the Kekulé structure. Isomers with this formula have never been separated. There is only one form of the compound

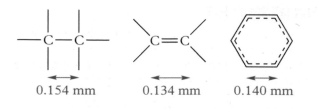

0.154 mm   0.134 mm   0.140 mm

Test yourself   D

1 Given that the empirical formula for benzene is CH, what further information is needed to show that its molecular formula is C_6H_6? What methods can chemists use to obtain this information?

2 Draw a possible structure for C_6H_6 which is not a ring. Why does the structure not fit with what Kekulé knew about benzene?

3 An arene consists of 91.3% carbon.

   a) What is the empirical formula of the arene?

   b) What is the molecular formula of the arene if its molar mass is 92 g mol⁻¹?

   c) Draw the structure of the arene.

The stability of benzene

A study of energy changes shows that benzene is more stable than expected for a compound with the Kekulé formula. The argument is based on a comparison of the enthalpy changes for hydrogenating benzene and cyclohexene.

Cyclohexene is a cyclic hydrocarbon with one double bond. Like other alkenes it adds hydrogen in the presence of a catalyst to form cyclohexane.

Figure 4.5.5 ▶
Enthalpy changes for hydrogenating cyclohexene

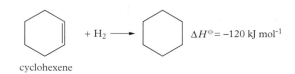

$+ H_2 \longrightarrow$ $\Delta H^\ominus = -120$ kJ mol⁻¹

cyclohexene

Benzene is actually more stable than might be expected as Figure 4.5.6 shows.

Test yourself

4 a) Redraw Figure 4.5.6 as an energy level diagram.

   b) According to Figure 4.5.6 (or your energy level diagram), how much more stable is real benzene than might be expected for Kekulé benzene?

   c) Predict the enthalpy change for the complete hydrogenation of cyclohexa-1,3-diene.

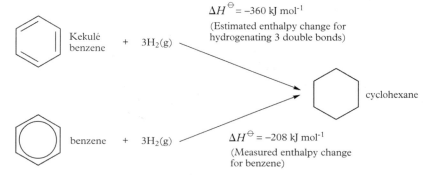

$\Delta H^\ominus = -360$ kJ mol⁻¹
(Estimated enthalpy change for hydrogenating 3 double bonds)

Kekulé benzene   + $3H_2(g)$

cyclohexane

benzene   + $3H_2(g)$

$\Delta H^\ominus = -208$ kJ mol⁻¹
(Measured enthalpy change for benzene)

Figure 4.5.6 ▲
Comparing the enthalpy change for hydrogenating real benzene with the calculated enthalpy change for hydrogenating a ring compound with three normal double bonds

Organic Chemistry

Delocalisation in benzene

Figure 4.5.7 shows benzene with normal covalent bonds (sigma, σ, bonds) between the carbon atoms and hydrogen atoms. Each carbon atom uses three of its electrons and, by sharing with neighbouring atoms, forms three σ bonds. This then leaves each atom with one electron in an atomic p-orbital.

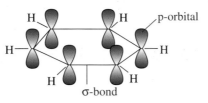

◀ *Figure 4.5.7*
Sigma bonds in benzene with one electron per atom remaining in p-orbitals

Instead of these six electrons pairing up to form three π orbitals, they are shared evenly between all six atoms giving rise to circular clouds of negative charge above and below the ring of carbon atoms. This is an example of electron delocalisation which takes place in any molecule where the conventional structure shows alternating double and single bonds.

Molecules with delocalised electrons are more stable than might otherwise be expected. In benzene this accounts for the compound being 152 kJ mol^{-1} more stable than expected for the Kekulé structure.

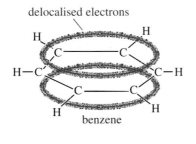

◀ *Figure 4.5.8*
Representations of the bonding in benzene. The circle in a benzene ring represents six delocalised electrons. This way of showing the structure explains the shape and stability of benzene. The Kekulé structure can be more helpful when describing the mechanism of the reactions of benzene

Test yourself

5 How does the model of benzene molecules with delocalised electrons account for these facts:

a) the benzene ring is a regular hexagon

b) there are no isomers of 1,2-dichlorobenzene

c) benzene is less reactive than cycloalkenes.

Physical properties

Arenes are non-polar. The forces between the molecules are weak intermolecular forces. The boiling points of arenes depend on the size of the molecules. The bigger the molecules the higher the boiling points. Benzene and methylbenzene are liquids at room temperature. Naphthalene is a solid. Arenes, like other hydrocarbons, do not mix with water but they do mix freely with non-polar solvents.

Definition

Delocalised electrons are bonding electrons which are not fixed between two atoms in a bond but are shared between three or more atoms. Delocalisation also affects ions where an atom with a lone pair of electrons and a negative charge is separated by a single bond from a double bond (see pages 167 and 177).

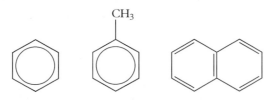

Figure 4.5.9 ▲
Three arenes: benzene, methylbenzene and naphthalene

Section four

Test yourself

6 Account for the existence of weak attractive forces between benzene molecules which are uncharged and non-polar.

7 Explain why benzene does not mix with water given that the forces between benzene molecules are weak.

8 Name a solvent with which you would expect benzene to mix freely.

Chemical properties

The important reactions of benzene and other arenes are very different from alkenes because of the delocalised electrons. Instead of electrophilic addition reactions the useful changes are electrophilic substitution reactions.

Oxidation

Arenes burn in air. Unlike alkanes and alkenes they burn with a very smoky flame because of the high ratio of carbon to hydrogen in the molecules.

Partial oxidation is possible for arenes with hydrocarbon side chains. Heating such an arene with with acidic potassium manganate(VII) oxidises the hydrocarbon side chain to a carboxylic acid group.

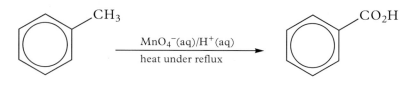

methylbenzene   benzoic acid

Figure 4.5.11 ▲
Oxidation of the hydrocarbon side chain of methylbenzene

Figure 4.5.10 ▲
A sample of methylbenzene burning in air showing the yellow flame and very smoky fumes

Electrophilic substitution reactions

Electrophilic substitution is the characteristic reaction of arenes such as benzene. The electrophile attacks the electron rich benzene ring with its six delocalised electrons. When describing the mechanism of this type of reaction it is easier to keep track of what is happening to the electrons by using the Kekulé structure for benzene.

Halogenation

Electrophilic substitution of bromine in benzene is rapid on warming in the presence of iron. Iron reacts with some bromine to form iron(III) bromide which is the catalyst. Iron(III) bromide is an electron pair acceptor which produces the electrophile, Br^+.

Figure 4.5.12 ▶
Formation of the electrophile

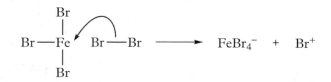

The first step of electrophilic substitution in benzene is very similar to the first step of electrophilic addition to alkenes (see page 156) but a more reactive electrophile is needed to start the reaction – hence the need for a catalyst.

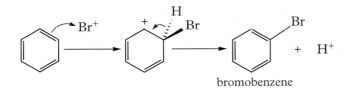

bromobenzene

The intermediate formed in the first step of the reaction could complete the reaction by adding a bromide ion (as happens when bromine adds to alkenes) but in fact what happens is that the loss of H^+ leads to substitution. Loss of H^+ is preferred because it keeps the benzene ring stabilised by six delocalised electrons.

Definition

Chemists sometimes use the term **halogen carrier** to describe the catalysts for the reaction of benzene with chlorine or bromine. Iron(III) halides and aluminium halides are halogen carriers. So is iron itself because it first reacts with the halogen to form the corresponding iron(III) halide.

Test yourself

9 Boiling benzene reacts with chlorine gas in the presence of iron(III) chloride or aluminium chloride.

a) Sketch the apparatus which could be used to carry out this reaction on a small scale (though not in a teaching laboratory where benzene is banned).

b) Write an equation for the reaction.

c) What is the electrophile in this reaction and how is it formed?

Nitration

Nitration of benzene and other arenes is important because it produces a range of products including the powerful explosive TNT (trinitrotoluene, now called 1-methyl-2,3,5-trinitrobenzene). Nitro compounds are also intermediates in the synthesis of chemicals used to make polyurethanes. Nitro groups are easily reduced to amine groups (see page 193).

◀ *Figure 4.5.14*
Explosions are important in mining, tunnelling and road building. Nitrated organic compounds such as TNT are useful explosives

Organic Chemistry

Section four

The reagent which nitrates benzene is a mixture of concentrated nitric and sulfuric acids. The purpose of the concentrated sulfuric acid is to produce the electrophile which is the nitryl cation, NO_2^+ (also called the nitronium ion).

$$HO-NO_2 + H_2SO_4 \rightarrow H_2O^+-NO_2 + HSO_4^-$$

$$\downarrow$$

$$H_2O + NO_2^+$$

Figure 4.5.15 ►
The concentrated sulfuric acid gives a proton to the –OH group in nitric acid which then loses a molecule of water forming the nitryl cation

Test yourself

10 Which is the acid and the base in the first step of the formation of nitryl cations?

Heating the nitrating mixture with benzene below 60 °C leads to the formation of nitrobenzene as the main product. At higher temperatures more di- and tri- nitro benzene form.

Figure 4.5.16 ►
Mechanism for the nitration of benzene. In the first step the nitryl cation accepts a electron pair from the benzene ring. Loss of H^+ in the second step then restores delocalisation over the six carbon atoms of the ring

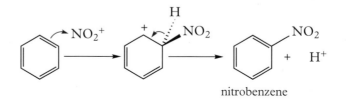

nitrobenzene

Friedel–Crafts reaction

The Friedel–Crafts reaction is an important method for forming C—C bonds. It is used to build up the carbon skeleton by adding a side chain to an arene. This reaction was discovered and developed jointly by the French organic chemist Charles Friedel (1832 – 1899) and the American, James Crafts (1839 – 1917). The reaction is used both on a laboratory and on an industrial scale.

In a Friedel–Crafts reaction a halogenoalkane or an acyl chloride undergoes an electrophilic substitution reaction with an arene. The reaction takes place in the presence of a catalyst such as aluminium chloride. One example is the substitution of a group with three carbon atoms in place of a hydrogen atom in a benzene ring using 2-chloropropane.

Figure 4.5.17 ►
Friedel–Crafts alkylation with a chloroalkane. Benzene reacting with 2-chloropropane

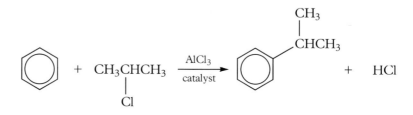

The reaction with an acyl chloride produces a ketone.

Figure 4.5.18 ►
Friedel–Crafts acylation with an acyl chloride. Benzene reacting with ethanoyl chloride

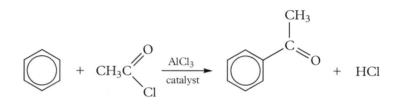

The purpose of the aluminium chloride is to produce the electrophile which attacks the benzene ring.

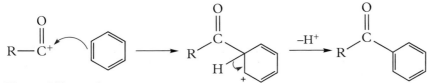

Formation
of the electrophile

◀ *Figure 4.5.19*
The mechanism of
Friedel–Crafts alkylation. The
electrophile in this example is a
carbocation formed by reaction
of the halogenoalkane with the
aluminium chloride

Electrophilic attack
and substitution

A Friedel-Crafts reaction to make ethyl benzene is the first step in the
production of the addition polymer poly(phenylethene), more often called
polystyrene (see page 204). In this case the carbocation is formed from ethene
in the presence of hydrogen chloride and aluminium chloride.

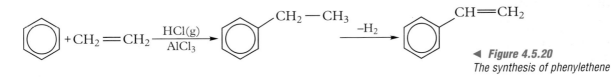

◀ *Figure 4.5.20*
The synthesis of phenylethene

Test yourself

11 Why must the reaction mixture be completely dry during a Friedel–Crafts reaction?

12 a) Show that aluminium chloride acts as a Lewis acid when used as a catalyst in a
Friedel–Crafts reaction (see page 100).

  b) Show how reaction with H^+ from the last step of the reaction regenerates the
catalyst from $AlCl_4^-$.

13 Draw the structures of the products of a Friedel–Crafts reaction of benzene with

  a) propanoyl chloride

  b) 2-iodo-2-methylpropane.

14 Suggest a reason for using an iodoalkane instead of a chloroalkane in a
Friedel–Crafts reaction.

4.6 Phenols

Phenol is an example of a compound with a functional group directly attached to a benzene ring. In phenol the functional group is –OH. It turns out that the –OH group affects the behaviour of the benzene ring while the benzene ring modifies the properties of the –OH group. As a result phenol and other related compounds have distinctive and useful properties.

Figure 4.6.1 ▲
Crystals of phenol

Structures and names

In general phenols are compounds with one or more –OH groups directly attached to a benzene ring.

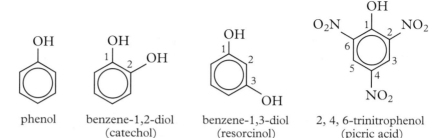

phenol   benzene-1,2-diol   benzene-1,3-diol   2, 4, 6-trinitrophenol
        (catechol)   (resorcinol)   (picric acid)

Figure 4.6.2 ▲
Structure of phenol and other phenols. To name a phenol the numbering of the carbon atoms in the benzene ring starts from an –OH group and runs in the direction which gives the lowest possible numbers for the other groups substituted in the ring

Reactions of the –OH group in phenol

Reaction with sodium

Phenol reacts with sodium in a similar way to ethanol. Bubbles of hydrogen form on adding a small cube of sodium to molten phenol or to a solution of phenol in an inert solvent.

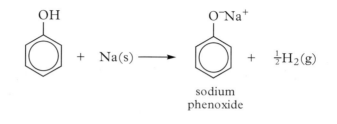

sodium
phenoxide

Figure 4.6.3 ▲
The reaction of phenol with sodium

Phenol as an acid

The benzene ring makes the –OH group more acidic in phenol than it is in alcohols. Phenol is not, however, as acidic as carboxylic acids (see page 177). It does not ionise significantly in water and it does not react with carbonates to

Test yourself   D

1 Explain, in terms of intermolecular forces, why:

  a) phenol is a solid while benzene is a liquid

  b) phenol, unlike benzene, is soluble in water but does not mix with water as freely as ethanol.

2 Phenol is manufactured from benzene, propene and oxygen. This is a three step process which produces equimolar amounts of phenol and propanone. The yield is about 86%. Calculate the mass of benzene required to make 8 tonnes of phenol and the mass of propanone formed as the other product.

3 What would you expect to observe on heating phenol until it burns?

produce carbon dioxide. Phenol is acidic enough to form soluble salts with alkalis. So phenol dissolves easily in a solution of sodium hydroxide.

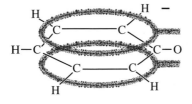

◀ *Figure 4.6.4*
Phenol reacting with alkali

The phenoxide ion is stabilised because the negative charge can be spread by delocalisation over the whole molecule. This cannot happen in alcohols.

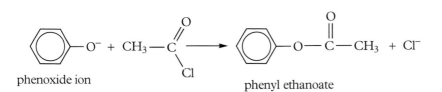

◀ *Figure 4.6.5*
The phenoxide ion showing the delocalisation of the negative charge

Esterification

Phenols do not react directly with carboxylic acids to form esters. The phenoxide ion, however, will react with an acid chloride to form an ester.

◀ *Figure 4.6.6*
The reaction of ethanoyl chloride with a solution of phenol in sodium hydroxide to form an ester (phenyl ethanoate)

Reactions of the benzene ring in phenol   *CD-ROM*

The –OH group makes the benzene ring more reactive. A lone pair on the –OH group interacts with the delocalised electrons in the benzene rings releasing electrons into the ring and making electrophilic attack easier. As a result, electrophilic substitution takes place under much milder conditions with phenol than with benzene.

Adding aqueous bromine to a solution of phenol, for example, produces an immediate white precipitate of 2,4,6-tribromophenol as the bromine colour fades.

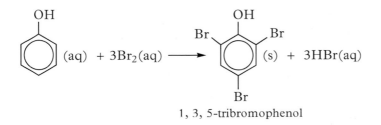

Figure 4.6.7 ▲
Reaction of phenol with bromine. The reaction is rapid at room temperature and there is no need for a catalyst

Note

The names still used by chemists for compounds with a benzene ring can be confusing. The phenyl group is C_6H_5- which is present in many compounds where one of the hydrogen atoms in benzene has been replaced by another atom or group. The use of phenyl in this way dates back to the first studies of benzene when 'phene' was suggested as an alternative name for the compound based on a Greek word for 'giving light'. This was because benzene had been discovered in the tar formed on heating coal to produce gas for lighting.

4 Identify two ways in which the reactions of the –OH group are similar in phenol and ethanol and two ways in which they differ.

5 What would you expect to observe on adding enough dilute hydrochloric acid to a solution of phenol in sodium hydroxide to make the mixture acidic?

6 Explain the meaning of the terms 'acyl group' and 'acylation' using the reaction of phenol with ethanoyl chloride as the example.

7 Suggest a reason why phenol, unlike alcohols, does not react directly with carboxylic acids to form esters.

8 Dilute nitric acid nitrates phenol rapidly at room temperature to form 2-nitrophenol and 4-nitrophenol. Draw the structures of the two products. How do these conditions for nitrating phenol compare with the conditions for nitrating benzene?

A test for phenols

Phenol forms a complex ion with a purple colour when mixed with a neutral solution of iron(III) chloride. Other compounds with an –OH group attached directly to a benzene ring behave in the same way.

Uses of phenols

The UK chemical industry produces about 100 000 tonnes of phenol each year. The main uses are to make:

- thermosetting plastics such as Bakelite
- polycarbonate plastics used as a tough substitute for glass
- epoxy resins for adhesives
- intermediates for the manufacture of nylon (see page 206).

Figure 4.6.8 ▶
Bakelite is made from phenol and methanal and it was the first commercially successful plastic

Phenol itself is a powerful disinfectant which has been used to kill germs ever since it was isolated from coal tar in the 19th century. For a time phenol was famous as the chemical which made surgery safe. In the mid-19th century many doctors were in despair because infection in hospitals was so common that most of their patients died of gangrene after operations to repair severe wounds. Then in 1865 a young surgeon called Joseph Lister, working in Glasgow, read of Louis Pasteur's researches into the germ theory of disease and realised that he could use a disinfectant chemical to prevent infection. He developed a technique for spraying a solution of phenol over the open wounds during operations and as a result many more of his patients survived. He soon realised, however, that keeping the germs away through cleanliness was better than trying to kill them with chemicals.

Phenol is an unpleasant chemical which burns the skin so it is not suitable as an antiseptic. Since Lister's time chemists have discovered substituted phenols which are suitable as antiseptics. Among them is 2,4,6–trichlorophenol (the active ingredient of TCP).

> **Definitions**
>
> **Disinfectants** are chemicals which destroy micro-organisms. Phenol is a disinfectant, and so is chlorine bleach. Unlike antiseptics, disinfectants cannot be used on skin and other living tissues.
>
> **Antiseptics** are chemicals which kill micro-organisms but, unlike disinfectants, they can be used safely on the skin.

Test yourself

9 Which of these compounds would you expect to give a violet colouration when mixed with a solution of iron(III) chloride?

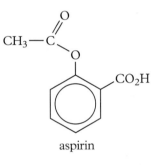

aspirin

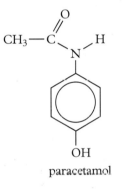

paracetamol

10 Draw the structure of TCP.

4.7 Carbonyl compounds

The two main classes of carbonyl compounds are the aldehydes and the ketones. The reactive carbonyl group plays an important part in the chemistry of living things, in laboratory chemistry and in industry. As expected for a double bond, the characteristic reactions are addition reactions. The C $=$ O bond is polar because oxygen is highly electronegative. As a result the mechanism of addition to carbonyl compounds is not the same as the mechanism for addition to alkenes.

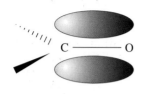

Figure 4.7.1 ▲
Representations of the carbonyl group in aldehydes and ketones. The bond is polar because the oxygen atom is much more electronegative than carbon

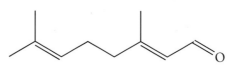

citral (lemon): a perfume chemical

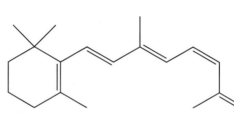

retinal: a component of the visual pigment in the eye

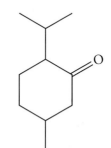

menthone (mint)
a flavouring

Figure 4.7.2 ▲
Skeletal formulae of naturally occurring compounds which contain a carbonyl group in their molecules

Aldehydes

CD-ROM

Names and structures

Aldehydes are carbonyl compounds in which the carbonyl group is attached to two hydrogen atoms or to a hydrocarbon group and a hydrogen atom. So the carbonyl group is at the end of a carbon chain. The names are based on the alkane with the same carbon skeleton with the ending changed from **–ane** to **–anal**.

Note

Always write the aldehyde group as –CHO. Writing –COH is unconventional and can easily lead to confusion with alcohols.

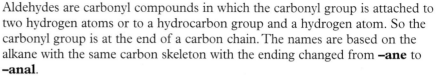

methanal

ethanal

Figure 4.7.3 ▶
Structures and names of aldehydes. The –CHO group is the functional group which gives aldehydes their characteristic reactions

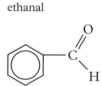

propanal

benzenecarbaldehyde
(benzaldehyle)

Physical properties

Methanal is a gas at room temperature. Ethanal boils at 21 °C so it may be a liquid or gas at room temperature depending on the conditions. Other common aldehydes are all liquids. The simpler aldehydes such as methanal and ethanal are freely soluble in water.

Formation and uses

Oxidation of primary alcohols by heating with a mixture of dilute sulfuric acid and potassium dichromate(VI) produces aldehydes under conditions which allow the aldehyde to distil off as it forms. Unlike ketones, aldehydes can easily be oxidised further to carboxylic acids by continued heating with an excess of the reagent.

Biologists use a solution of methanal to preserve specimens. Since the beginning of the 20th century it has been the main ingredient of the fluids used by embalmers. Methanal is also an important industrial chemical because it is a raw material for the manufacture for a range of thermosetting plastics.

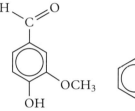
vanillin
vanilla flavour

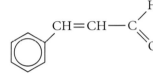
3-phenylpropenol
(cinnamaldehyde)
cinnamon flavour

benzenecarbaldehyde
(benzaldehyde)
almond flavour

Figure 4.7.4 ▲
Three aldehydes used to flavour food

Ketones   *CD-ROM*

Names and structures

In ketones, the carbonyl group is attached to two hydrocarbon groups. Chemists name ketones after the alkane with the same carbon skeleton by changing the ending **–ane** to **–anone**. Where necessary a number in the name shows the position of the carbonyl group.

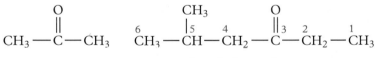
propanone   5-methylhexan-3-one

Figure 4.7.5 ▲
Structures and names of ketones

Physical properties

All the common ketones are liquids with boiling points similar to those of the corresponding aldehydes. The simplest ketone, propanone, mixes freely with water.

Occurrence and uses

Oxidation of secondary alcohols with hot, acidified potassium dichromate(VI) produces ketones which, unlike aldehydes, are not easily oxidised further.

Organic Chemistry

Section four

Test yourself

1 Write the structure of 2-methylbutanal.

2 **a)** Show that the boiling points of aldehydes are higher than the alkanes with similar relative molecular masses (see page 245) but lower than the corresponding alcohols.

   b) Account for the values of the boiling points of aldehydes relative to those of alkanes and alcohols in terms of intermolecular forces.

3 Why does an aldehyde such as ethanal mix freely with water while benzaldehyde is much less soluble?

4 **a)** Write an equation for the oxidation of propan-1-ol to propanal. (Represent the oxygen from the oxidising agent as [O].)

   b) Explain why the product in **a)** must be allowed to escape from the reaction mixture as it forms.

   c) Sketch a diagram of the apparatus you would use to make propanal in this way.

Test yourself

5 Write the structure of 4,4-dimethylpentan-2-one.

6 Show that propanone and propanal are functional group isomers.

7 Write an equation for the oxidation of butan-2-ol to butanone. (Represent the oxygen from the oxidising agent as [O].)

8 What are the reagents and conditions for converting benzene to the ketone, phenylethanone? (See page 164)

9 Refer to Figures 4.7.2 and 4.7.4 and work out the molecular formulae of:

a) vanillin

b) citral.

10 Identify the carbonyl group in each of the compounds illustrated in Figure 4.7.2. In each case state whether the carbonyl group is like that of an aldehyde or a ketone.

Propanone is widely used as a solvent. It has a low boiling point and evaporates quickly making it suitable for cleaning and drying parts of precision equipment. Propanone is also the starting point for producing the monomer of the glass-like addition polymer in display signs, plastic baths, and the cover of car lights.

Reactions of aldehydes and ketones

Reduction

Sodium tetrahydridoborate(III), $NaBH_4$, reduces aldehydes to primary alcohols, and ketones to secondary alcohols.

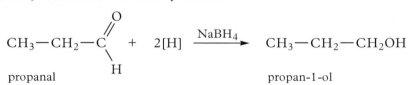

Figure 4.7.6 ▲
Reduction of propanal and propanone. The 2[H] comes from the reducing agent. This is a shorthand way of balancing a complex equation involving reduction

Oxidation

Oxidising agents easily convert aldehydes to carboxylic acids. It is much harder to oxidise ketones. Oxidation of ketones is only possible with powerful oxidising agents which break up the molecules. Chemical tests to distinguish aldehydes and ketones are based on the difference in the ease of oxidation (see pages 174–175).

Figure 4.7.7 ▶
Oxidation of propanal to propanoic acid

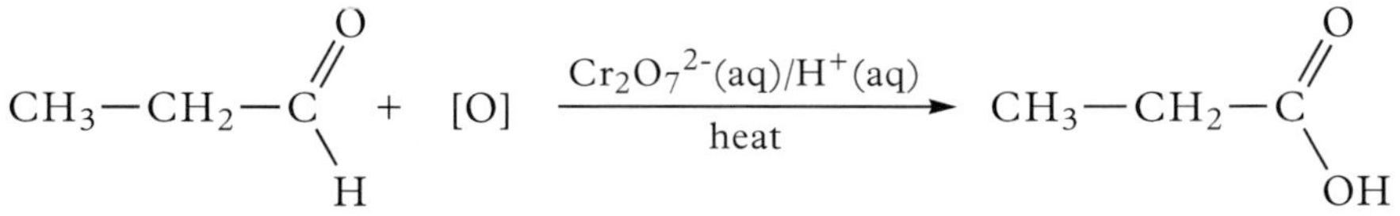

$$CH_3-CH_2-C{\Large\overset{O}{\underset{H}{\diagdown}}} \quad + \quad [O] \quad \xrightarrow[\text{heat}]{Cr_2O_7{}^{2-}(aq)/H^+(aq)} \quad CH_3-CH_2-C{\Large\overset{O}{\underset{OH}{\diagdown}}}$$

Acidified potassium dichromate(VI) is orange and contains $Cr_2O_7{}^{2-}$ ions. After oxidising an aldehyde to a carboxylic acid the reagent turns green giving a solution of green Cr^{3+} ions (see Figure 3.5.7 on page 105).

Addition of hydrogen cyanide

Hydrogen cyanide rapidly adds to carbonyl compounds at room temperature. Hydrogen cyanide is a highly toxic gas which forms in the reaction mixture by adding potassium cyanide and dilute sulfuric acid. The potassium cyanide must be in excess to ensure that there are free cyanide ions ready to start the reaction.

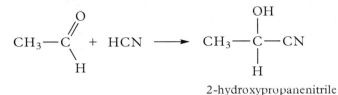

2-hydroxypropanenitrile

This reaction is a useful step in synthetic routes to valuable compounds because the –CN group can be hydrolysed to a carboxylic acid or reduced to an amine (see pages 190–191). The reaction extends the carbon skeleton of the molecule by one carbon atom.

Nucleophilic addition

The reduction of a carbonyl compound with $NaBH_4$ and the reaction with hydrogen cyanide are both examples of nucleophilic addition reactions. The carbon atom in a carbonyl group is open to attack by nucleophiles because the electronegative oxygen draws electrons away from it.

The incoming nucleophile uses its lone pair to form a new bond with the carbon atom. This displaces one pair of electrons in the double bond onto oxygen. Oxygen has thus gained one electron from carbon and now has a negative charge.

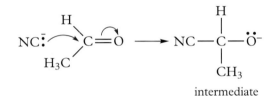

intermediate

Figure 4.7.9 ▲
First step of nucleophilic addition of hydrogen cyanide to ethanal

To complete the reaction, the negatively charged oxygen acts as a base and gains a proton.

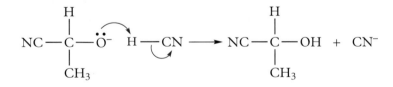

Figure 4.7.10 ▲
Second step of nucleophilic addition of hydrogen cyanide to ethanal. Note that taking a proton from HCN produces another cyanide ion

Tests for aldehydes and ketones

Recognising carbonyl compounds

Chemists can identify aldehydes and ketones with the help of instrumental techniques such as mass spectrometry and infra-red spectroscopy (see pages 226 – 233). Traditionally, chemists characterised these compounds by measuring their boiling points and measuring the melting points of crystalline derivatives (page 246).

Test yourself

11 Name the products of reducing butanal and butanone. Which is a primary alcohol and which a secondary alcohol?

12 Show that reduction of an aldehyde or ketone with $NaBH_4$ has the effect of adding hydrogen to the double bond.

13 Show that the cyanide ion acts as a base in the formation of hydrogen cyanide from potassium cyanide.

14 a) Show that the hydroxynitrile formed from ethanal and HCN is chiral.

 b) Suggest an explanation for the fact that the solution of a hydroxynitrile formed from ethanal and HCN does not rotate the plane of polarised light.

15 Write the equation for the formation of 2-hydroxy-2-methylpropanenitrile from a ketone.

16 Identify examples of these types of reaction from the chemistry of carbonyl compounds: oxidation, reduction, addition.

17 Write equations to show the mechanisms of the reduction of ethanal by $NaBH_4$. The nucleophile is the hydride ion, H^-. The reaction takes place in aqueous solution so water molecules are available for the second step of the process.

Figure 4.7.11 ▲
*A bright orange 2,4-
dinitrophenylhydrazone
derivative*

The reagent 2,4-dinitrophenylhydrazine reacts with carbonyl compounds forming 2,4-dinitrophenylhydrazone derivatives which are solid at room temperature and bright yellow or orange. The solid derivative has no practical use but it can be filtered off, recrystallised and identified by measuring its melting point. This makes it possible to identify the carbonyl compound.

> ### Definition
> Chemists can use **derivative**s to identify unknown organic compounds. Converting a compound to a crystalline derivative produces a product which can be purified by recrystallisation and then identified by measuring its melting point.

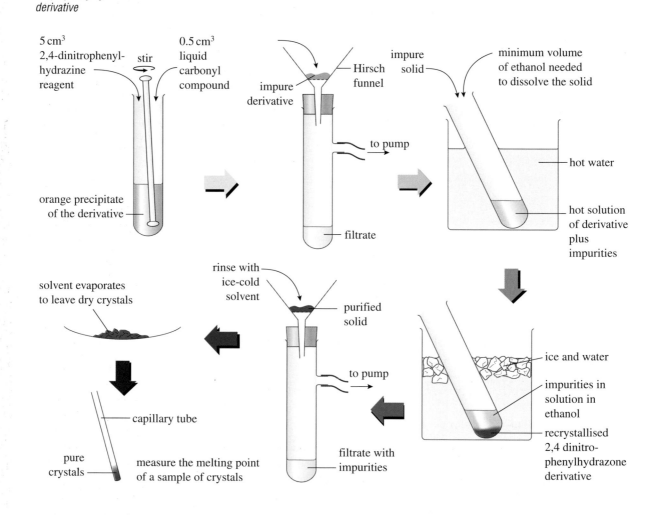

Figure 4.7.12 ▲
*Stages in the formation,
purification and identification
of a derivative of a carbonyl
compound*

Distinguishing aldehydes and ketones

Fehling's solution and Tollen's reagent (ammoniacal silver nitrate) are mild oxidising agents which are used to distinguish aldehydes from ketones. Aldehydes are easily oxidised but ketones are not.

Fehling's reagent does not keep so it is made when required by mixing two solutions. One solution is copper(II) sulfate in water. The other solution is a solution of 2,3-dihydroxybutanedioate (tartrate) ions in strong alkali. The 2,3-dihydroxybutanedioate salt forms a complex with copper(II) ions so that they do not precipitate as copper(II) hydroxide with the alkali.

Figure 4.7.13 ▲
Fehling's reagent before (left) and after (right) warming with an aldehyde. The deep blue solution turns greenish and then loses its blue colour as an orange–red precipitate of copper(I) oxide appears. There is no change with a ketone

Tollen's reagent consists of an alkaline solution of diamminesilver(I) ions, $[Ag(NH_3)_2]^+$. Aldehydes reduce the silver ions to metallic silver. Ketones do not react.

Figure 4.7.14 ▲
Warming Tollen's reagent with an aldehyde produces a precipitate of silver which coats clean glass with a shiny layer of silver so that it acts like a mirror. There is no reaction with ketones

The tri-iodomethane reaction

A compound containing a CH_3CO- group produces a yellow precipitate of tri-iodomethane (iodoform) when warmed with a mixture of iodine and sodium hydroxide. The test shows the presence of a methyl group next to a carbonyl group in an organic molecule.

The reaction takes place in two steps.

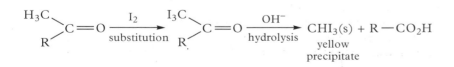

Figure 4.7.15 ▲
Equation for the tri-iodomethane reaction

Test yourself   **D**

18 Identify the compound which boils at 80 °C, does not react with Fehling's solution and forms a 2,2-dinitrophenylhydrazone derivative, which melts at 115 °C.

19 Hydrolysis of C_4H_9Cl with hot, aqueous sodium hydroxide produces $C_4H_{10}O$ which can be oxidised by an acidic solution of potassium dichromate(VI) to C_4H_8O. The compound C_4H_8O does not react with Fehling's solution but it does give a yellow precipitate with 2,4-dinitrophenylhydrazine. Give the names and structures of these compounds.

20 a) Explain why alcohols with this group CH_3CHOH- will also give a positive result with the tri-iodomethane reaction.

   b) Explain why a mixture of iodine with sodium hydroxide solution is chemically equivalent to a mixture of potassium iodide and sodium chlorate(I) solutions.

21 Name and write out the displayed formulae of the two isomers, of $C_5H_{10}O$, which form a yellow precipitate when they react with iodine in the presence of alkali.

22 Which is the only aldehyde to undergo the tri-iodomethane reaction?

Note

An alternative reagent for the iodoform reaction is a mixture of potassium iodide and sodium chlorate(I).

4.8 Carboxylic acids

Many organic acids are instantly recognisable by their odours. Ethanoic acid, for example, gives vinegar its taste and smell. Butanoic acid is responsible for the foul smell of rancid butter while the body odour of goats is a blend of the three unbranched organic acids with 6, 8 and 10 carbon atoms. Organic acids play a vital part in the biochemistry of life because of the great variety of their reactions. The acids are also important in laboratory and industrial chemistry and helped to give rise to a range of new materials: especially synthetic fibres and plastics.

Structures and names

Carboxylic acids are compounds with the formula $R-CO_2H$ where R represents an alkyl group, aryl group or a hydrogen atom. The carboxylic acid group $-CO_2H$ is the functional group which gives the acids their characteristic properties.

Figure 4.8.1 ▲
The traditional names for organic acids were based on their natural origins. The original name for methanoic acid was formic acid because it was first obtained from red ants and the Latin name for 'ant' is formica. The acid adds bite to the sting of an ant

Figure 4.8.2 ▲
Many vegetables contain ethanedioic acid, better known as oxalic acid. The level of the acid in rhubarb leaves is high enough for it to be dangerous to eat the leaves. The acid kills by lowering the concentration of calcium ions in blood to a dangerously low level

The carboxylic acid group can be regarded as a carbonyl group, $C=O$, attached to an $-OH$ group but is better seen as a single functional group with distinctive properties.

Chemists name carboxylic acids by changing the ending of the corresponding alkane to **-oic** acid. So ethane becomes ethan**oic** acid.

Note

Many writers show the carboxylic acid group as $-COOH$ but for a group with this structure

$-CO_2H$ is more accurate and avoids confusion with $-C-O-O-H$.

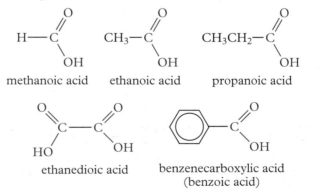

methanoic acid   ethanoic acid   propanoic acid

ethanedioic acid   benzenecarboxylic acid (benzoic acid)

Figure 4.8.3 ▶
Names and structures of carboxylic acids

Physical properties

Even the simplest acids such as methanoic acid and ethanoic acid are liquids at room temperature because of hydrogen bonding between the carboxylic acid groups. Also because of hydrogen bonding, these acids mix freely with water. Hydrogen bonding leads to carboxylic acids pairing up to form dimers in non-polar solvents.

Carboxylic acids with more than eight carbon atoms are solids. Benzoic acid is a solid at room temperature. It is only very slightly soluble in cold water but more soluble in hot water.

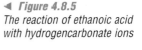

Figure 4.8.4 ▲
Ball-and-stick model of an organic acid

Test yourself

1 Write out the structural formulae and give the systematic names of the three carboxylic acids which were traditionally derived from the Latin word *caper* meaning goat: caproic acid (6C), capryllic acid (8C) and capric acid (10C).

2 Give the molecular formula, displayed formula and name of the acid shown in Figure 4.8.4.

3 **a)** Draw a diagram to show hydrogen bonding between ethanoic acid molecules and water molecules.

   b) In a non-polar solvent, ethanoic acid molecules dimerise through hydrogen bonding. Draw a diagram to show how this happens.

Chemical reactions

Reactions as acids

Carboxylic acids are weak acids (see page 37). Carboxylic acids are sufficiently acidic to produce carbon dioxide when added to a solution of sodium carbonate or sodium hydrogencarbonate. This reaction distinguishes carboxylic acids from weaker acids such as phenols.

◀ **Figure 4.8.5**
The reaction of ethanoic acid with hydrogencarbonate ions

$$CH_3-C\overset{O}{\underset{OH}{\big\backslash}}(aq) + HCO_3^-(aq) \longrightarrow CH_3-C\overset{O}{\underset{O^-}{\big\backslash}}(aq) + H_2O(l) + CO_2(g)$$

The –OH group in the carboxylic acid group is more acidic than the –OH group in alcohols

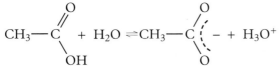

Figure 4.8.6 ▲
Ionisation of ethanoic acid

Delocalisation of electrons stabilises the carboxylate ion. This favours ionisation of the acid and helps to shift the equilibrium to the right.

Note

As a general rule, delocalisation is more effective in a negative ion which can delocalise the charge over more oxygen atoms. So ethanoic acid turns out to be more acidic than phenol (see pages 166–167).

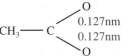

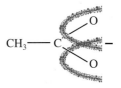

Figure 4.8.7 ▲
Bond lengths in the ethanoate ion. Part of the evidence for this description of the bonding in this ion is the fact that both the carbon–oxygen bond lengths are the same. They are shorter than a normal C–O bond but longer than a C=O bond

Organic Chemistry

Section four

Formation of acyl chlorides

Phosphorus pentachloride, PCl_5, reacts with carboxylic acids at room temperature. The reaction replaces the –OH group with a chlorine atom.

Figure 4.8.8 ▶
The reaction of PCl_5 with ethanoic acid

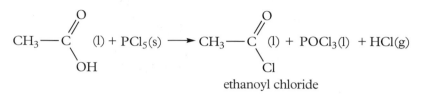

ethanoyl chloride

Formation of esters

Carboxylic acids react with alcohols to form esters (see page 183). The two organic compounds are mixed and heated under reflux in the presence of a small amount of a strong acid catalyst such as concentrated sulfuric acid.

Figure 4.8.9 ▶
Formation of an ester from ethanoic acid and propan-1-ol

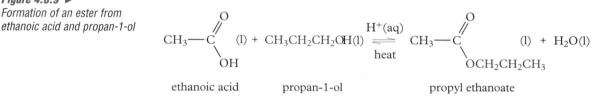

ethanoic acid   propan-1-ol   propyl ethanoate

This reaction is reversible. The conditions for reaction have to be arranged to increase the yield of the ester. One possibility is to use an excess of either the acid or the alcohol, depending on which is the more available or cheaper. Using more concentrated sulfuric acid than needed for its catalytic effect can help because the acid reacts with water. In some esterification reactions it is possible to distil off either the ester or the water as they form.

Reduction

Reduction of a carboxylic acid to an alcohol is possible with the powerful reducing agent $LiAlH_4$. The reagent is suspended in dry ether. Addition of water after reduction destroys any excess reducing agent.

Figure 4.8.10 ▶
Reduction with $LiAlH_4$

> ### Note
>
> Lithium tetrahydridoaluminate(III), $LiAlH_4$, is a powerful reducing agent used to reduce aldehydes, ketones, esters and carboxylic acids to alcohols. Many chemists still use the older name, lithium aluminium hydride. In its reactions the tetrahydridoaluminate(III) ion can be regarded as a source of hydride ions, H^-. The reagent is rapidly hydrolysed by water so has to be used in an anhydrous solvent such as dry ether. Where possible sodium tetrahydridoborate(III) is preferred because it is easier and safer to use (see page 172).

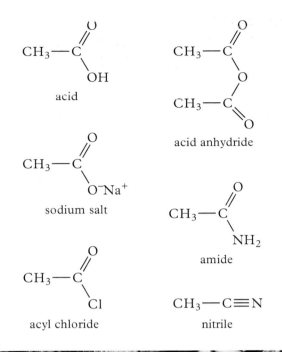

acid

acid anhydride

sodium salt

amide

acyl chloride

nitrile

Test yourself

4 Benzoic acid is only very slightly soluble in water but it dissolves freely in aqueous sodium hydroxide. Use Le Chatelier's principle to explain why this happens. Start by writing an equation for the equilibrium between solid benzoic acid and its aqueous ions in solution.

5 Write an equation for the reaction of propanoic acid with sodium carbonate.

6 Suggest an explanation, in terms of the inductive effect (see page 156) for the fact that:

a) methanoic acid is a stronger acid than ethanoic acid

b) chloroethanoic acid, CH_2ClCO_2H, is stronger than ethanoic acid.

7 How can the reaction of carboxylic acids (and alcohols) with PCl_5 be used as a test for the presence of an –OH group?

8 Use Le Chatelier's principle to explain the methods used to increase the yield of an ester formed from an acid and an alcohol.

Organic Chemistry

Section four

4.9 Acylation

Chemists value acid chlorides and acid anhydrides as reactive compounds for synthesis both on a small laboratory scale and on a large scale in industry. These reagents often provide the easiest way to make valuable products such as esters. The synthesis of the pain-killer aspirin is an example of acylation.

Acylating agents

Acylation is a reaction which substitutes an acyl group for a hydrogen atom. The H atom may be part of an –OH group, an –NH$_2$ group or a benzene ring. Acylating agents are either acyl chlorides or acid anhydrides.

Figure 4.9.1 ▶
Examples of acyl groups. Note that an acyl group consists of all parts of a carboxylic acid except the –OH group

ethanoyl group

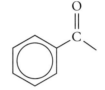

benzenecarbonyl (benzoyl) group

Figure 4.9.2 ▶
Examples of acyl chlorides. Acyl chlorides form when carboxylic acids react with PCl$_5$ (see page 178)

ethanoyl chloride

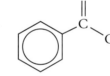

benzenecarbonyl (benzoyl) chloride

Figure 4.9.3 ▶
The functional group in an acid anhydride is formed by eliminating a molecule of water from two carboxylic acid groups. Sometimes this happens simply on heating the acid but generally anhydrides are made in other ways

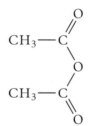

ethanoic anhydride

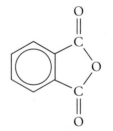

benzene-1,2-dicarboxylic anhydride

Acyl chlorides

Acyl chlorides are very reactive which makes them powerful acylating agents. Ethanoyl chloride, for example, is a colourless liquid which fumes as it reacts with moisture in the air. The reaction between ethanoyl chloride and water is violent at room temperature.

Ethanoyl chloride and other acyl chlorides, also react rapidly at room temperature with alcohols to form esters.

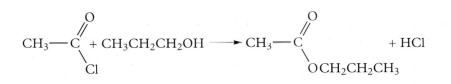

Acid chlorides such as ethanoyl chloride also reacts with ammonia and amines to form amides (see also page 188).

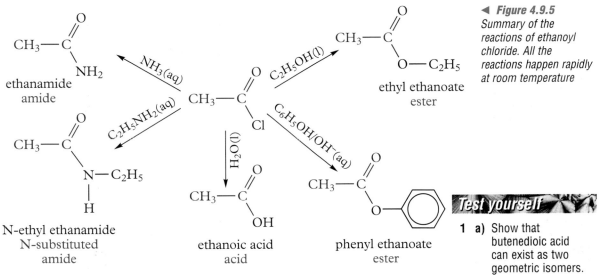

◀ **Figure 4.9.5**
Summary of the reactions of ethanoyl chloride. All the reactions happen rapidly at room temperature

Acid anhydrides

CD-ROM

Acid anhydrides react in a similar way as acyl chlorides with water, alcohols and amines. They are, however, less reactive than acyl chlorides which makes them safer and more convenient to handle. As a result they are often preferred for laboratory and industrial syntheses. Another advantage is that they do not form hydrogen chloride as they react. The reactions of acid anhydrides generally require heating in a flask fitted with a reflux condenser.

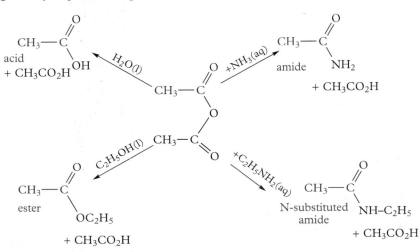

Figure 4.9.6 ▲
Reactions of ethanoic anhydride

Organic Chemistry · Section four

Test yourself

1 a) Show that butenedioic acid can exist as two geometric isomers.

   b) One isomer forms an anhydride relatively easily on heating above its melting point. Which isomer would you expect it to be?

2 Write an equation for the reaction between ethanoyl chloride and water. Show that this is an example of hydrolysis.

3 What are the conditions for the reaction of phenol with an acid chloride? (See page 167)

4 Write an equation for the formation of ethanamide from ethanoyl chloride. Show that two moles of ammonia are required for reaction with one mole of the acyl chloride.

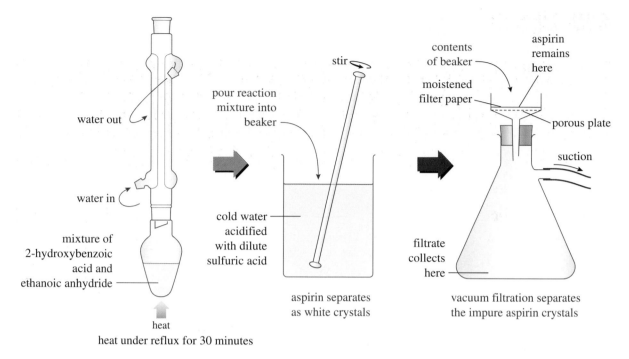

water out

water in

mixture of
2-hydroxybenzoic
acid and
ethanoic anhydride

heat

heat under reflux for 30 minutes

stir

pour reaction
mixture into
beaker

cold water
acidified
with dilute
sulfuric acid

aspirin separates
as white crystals

contents
of beaker

moistened
filter paper

filtrate
collects
here

aspirin
remains
here

porous plate

suction

vacuum filtration separates
the impure aspirin crystals

Figure 4.9.7 ▲
Stages in the laboratory acylation converting 2-hydroxybenzoic acid to aspirin. The aspirin can be purified by recrystallisation from hot water (see Figure 4.7.13 on page 174). **CD-ROM**

Note

Addition–elimination reactions are also known as condensation reactions because the overall effect is to split off a small molecule (such as water) when two larger molecules join together (see page 206).

Test yourself ✕ D

5 Write a balanced equation for the reaction of ethanoic anhydride with propan-1-ol.

6 Why is it an advantage to use an acylating agent which does not produce hydrogen chloride as a by-product of the reaction?

7 Calculate the percentage yield if 10 g of 2-hydroxybenzoic acid reacts with excess ethanoic anhydride to give 8 g of aspirin.

Addition–elimination reactions of acid chlorides

The reactions of acyl chlorides with the nucleophiles water, alcohols, ammonia and amines are addition–elimination reactions.

During an addition–elimination reaction two molecules first add together and then immediately split off a small molecule such as water or hydrogen chloride.

Addition–elimination reactions take place in three stages:

■ nucleophilic addition
■ gain and loss of hydrogen ions (protons)
■ elimination of water.

Test yourself ✕

8 Why is the carbon atom, attached to the chlorine atom in ethanoyl chloride highly susceptible to attack by nucleophiles?

9 Write out the mechanism for the reaction of ammonia with ethanoyl chloride.

Figure 4.9.8 ▶
Mechanism of the reaction of an alcohol (ethanol) with an acyl halide (ethanoyl chloride) to form an ester (ethyl ethanoate)

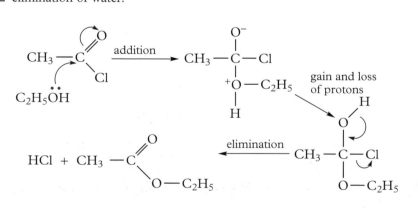

4.10 Esters

Many of the sweet smelling compounds found in perfumes and fruit flavours are esters. Some drugs used in medicine are esters, including aspirin, paracetamol and the local anaesthetics novocaine and benzocaine. The insecticides malathion and pyrethrin are also esters. Compounds with more than one ester link include fats and oils as well as polyester fibres. Other esters are important as solvents and plasticisers.

Names and structures

CD-ROM

The general formula for an ester is RCO_2R', where R and R' are alkyl or aryl groups.

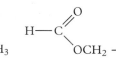
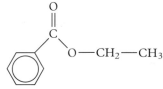

methyl ethanoate

ethyl methanoate

ethyl benzenecarboxylate
(ethyl benzoate)

Figure 4.10.2 ▲
Names and structures of esters

Physical properties

Common esters such as ethyl ethanoate are volatile liquids and only slightly soluble in water.

Figure 4.10.1 ▲
Natural fruit flavours are complex mixtures. Some simpler esters on their own have odours which resemble fruit flavours. Examples are: propyl ethanoate (pear), ethyl butanoate (pineapple), octyl ethanoate (orange), 2-methylpropyl ethanoate (apple)

Test yourself
D

1 Give the name and displayed formulae of the esters formed when:

   a) butanoic acid reacts with propan-1-ol

   b) ethanoyl chloride reacts with methanol

   c) ethanoic anhydride reacts with butan-1-ol.

2 Why is the boiling point of ethyl ethanoate similar to that of ethanol and lower than that of ethanoic acid?

3 In Figure 4.10.3 on page 184:

   a) Identify what is happening at each of the stages A, B, C and E.

   b) What is the purpose of the concentrated sulfuric acid?

   c) What are the visible signs of reaction during stage C and what practical precautions are necessary during this stage?

   d) The calcium chloride in step D removes unchanged ethanol from the impure product. What type of reaction would you expect between the alcohol and the $CaCl_2$?

   e) A volatile by-product distils off the boiling range 35–40 °C before the ester in stage E. Suggest a structure for this by-product which has the molecular formula $C_4H_{10}O$.

   f) Calculate the percentage yield if the actual yield is 50 g from 40 g ethanol and 52 g ethanoic acid.

Definition

Essential oils are the oils which chemists extract from plants as a source of chemicals for perfumes, food flavourings and other uses. They often include esters.

Figure 4.10.3 ▶
Stages in the laboratory preparation of ethyl ethanoate

CD-ROM

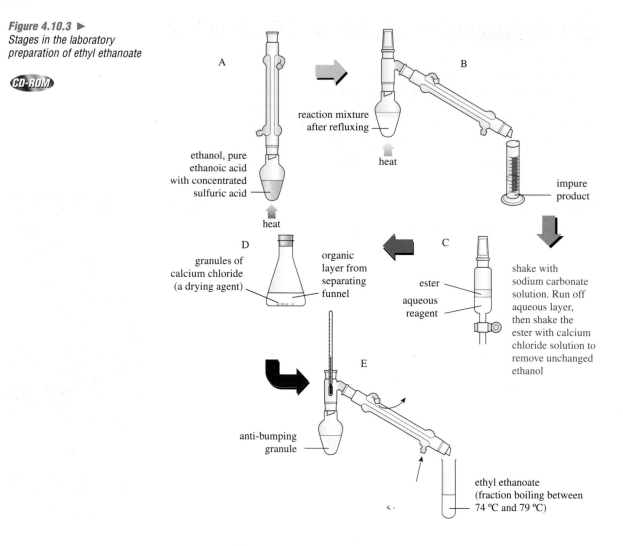

A

ethanol, pure ethanoic acid with concentrated sulfuric acid

heat

reaction mixture after refluxing

heat

B

impure product

C

ester

aqueous reagent

shake with sodium carbonate solution. Run off aqueous layer, then shake the ester with calcium chloride solution to remove unchanged ethanol

D

granules of calcium chloride (a drying agent)

organic layer from separating funnel

E

anti-bumping granule

ethyl ethanoate (fraction boiling between 74 °C and 79 °C)

Hydrolysis of esters

Hydrolysis splits an ester into an acid and an alcohol.

Figure 4.10.4 ▶
Hydrolysis of an ester

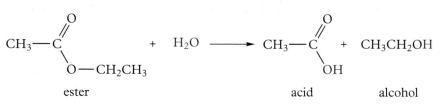

$$CH_3 - C\!\!\begin{array}{c}O\\\\O-CH_2CH_3\end{array} \quad + \quad H_2O \longrightarrow CH_3 - C\!\!\begin{array}{c}O\\\\OH\end{array} \quad + \quad CH_3CH_2OH$$

ester   acid   alcohol

Acids or bases can catalyse the hydrolysis. Hydrolysis catalysed by acid is the reverse of the reaction used to synthesise esters from carboxylic acids (see page 178).

Base catalysis is generally more efficient because it is not reversible. This is because the acid formed loses its proton and turns into a negative ion which does not react with the alcohol.

Figure 4.10.5 ▶
Base catalysed hydrolysis of ethyl ethanoate

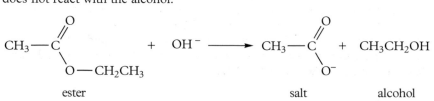

$$CH_3 - C\!\!\begin{array}{c}O\\\\O-CH_2CH_3\end{array} \quad + \quad OH^- \longrightarrow CH_3 - C\!\!\begin{array}{c}O\\\\O^-\end{array} \quad + \quad CH_3CH_2OH$$

ester   salt   alcohol

4 Identify the products of heating:

a) propyl butanoate with hydrochloric acid

b) ethyl methanoate with aqueous sodium hydroxide.

5 Under acid conditions the reaction of ethyl ethanoate with water is reversible.

a) What conditions favour the hydrolysis of the ester?

b) How do these conditions compare with those for the synthesis of the ester?

Fats and oils

CD-ROM

Fats and vegetable oils belong to the family of compounds call lipids. Chemically fats and oils are esters of long-chain carboxylic acids and the alcohol propan-1,2,3-triol better known as glycerol.

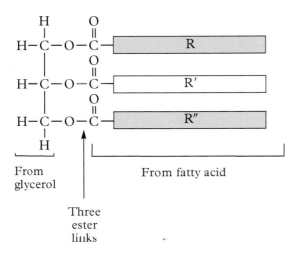

Definition

Lipids are a broad class of biological compounds which are soluble in organic solvents such as ethanol, but insoluble in water. Lipids are very varied and include fatty acids, fats, vegetable oils, phospholipids and steroids. Lipids release more energy per gram when oxidised than carbohydrates. This makes lipids important as a concentrated energy store in living organisms.

◄ *Figure 4.10.6*
General structure of a triglyceride. In natural fats and vegetable oils the hydrocarbon chains may be the same or they may be different

The carboxylic acids in fats are usually referred to as fatty acids. Saturated fatty acids do not have double bonds in the hydrocarbon chain but there are double bonds in the molecules of unsaturated fatty acids.

Fatty acid	Chemical name	Formula	Type
palmitic	hexadecanoic	$CH_3(CH_2)_{14}CO_2H$	saturated
stearic	octadecanoic	$CH_3(CH_2)_{16}CO_2H$	saturated
oleic	*cis*-octadec-9-enoic	$CH_3(CH_2)_7CH=CH(CH_2)_7CO_2H$	unsaturated
linoleic	*cis, cis*-octadec-9,12-dienoic	$CH_3(CH_2)_4CH=CHCH_2CH=CH(CH_2)_7CO_2H$	unsaturated

Figure 4.10.7 ▲
Examples of fatty acids

Fats are solid at around room temperature (below 20 °C) because they contain a high proportion of saturated fatty acids. Solid triglycerides are generally found in animals. In lard, for example, the main fatty acids are palmitic acid (28%), stearic acid (8%) and only 56% oleic acid.

The triglycerides with unsaturated fatty acids have a less regular structure than saturated fats. The molecules do not pack together so easily to make solids so they have lower melting points and have to be cooler before they solidify. Triglycerides of this kind occur in plants and are liquids at around room temperature. In a vegetable oil such as olive oil the main fatty acids are oleic acid (80%) and linoleic acid (10%).

Figure 4.10.8 ▲
Green olives harvested for their oil, California, USA

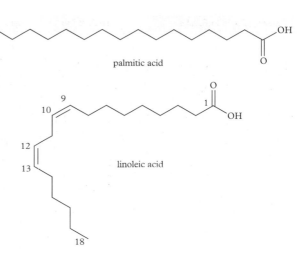

palmitic acid

linoleic acid

Figure 4.10.9 ▲
Skeletal formulae to compare the shape of palmitic acid and linoleic acid

Chemicals from fats and oils

CD-ROM

Hardening

Hydrogenation is used industrially to add hydrogen to double bonds in oils. This produces saturated fats which are solid at room temperature. So the process is sometimes called 'hardening'. The catalyst is finely divided nickel. Non-dairy spreads are a blend of vegetable oils with a high enough proportion of partly-hardened fats to make the product a spreadable solid.

Hydrolysis

Hydrolysis of triglycerides produces soaps and glycerol. Soaps are the sodium or potassium salts of fatty acids.

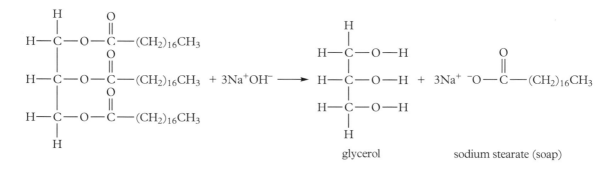

glycerol   sodium stearate (soap)

Figure 4.10.10 ▲
Saponification – the hydrolysis of a fat or vegetable oil (triglyceride) with alkali to make soap

Soaps are surfactants which help to remove greasy dirt because they have an ionic (water-loving) head and a long hydrocarbon tail (water-hating). Soaps help to separate greasy dirt from surfaces. Then they keep dirt dispersed in water so that it rinses away.

Most toilet soaps are made from a mixture of animal fat and coconut palm oil. Soaps from animal fat are less soluble and longer lasting. Soaps from palm oils are more soluble so that they lather quickly but wash away more quickly. A bar of soap also contains a dye and perfume together with an antioxidant to stop the soap and air combining to make irritant chemicals.

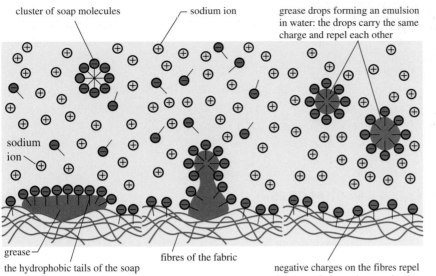

cluster of soap molecules — sodium ion — grease drops forming an emulsion in water: the drops carry the same charge and repel each other

sodium ion

grease —

the hydrophobic tails of the soap molecules escape from the water by dissolving in the grease

fibres of the fabric

negative charges on the fibres repel the charged grease drops and stop them returning to the fabric

Definition

Surfactants are surface-active agents. They are chemicals with molecules which seek out the boundary surface between two liquids or between a liquid and a gas. One of the important effects of surfactants is that they change the surface tension of water so that it wets surfaces more effectively. A surfactant molecule has an ionic or polar group attached to a hydrocarbon chain. The polar group is water-loving (hydrophilic). The hydrocarbon chain is water-hating (hydrophobic).

Figure 4.10.11 ▲
Soap molecules help to wash away grease which normally does not mix with water. The hydrocarbon chains of the soap molecules dissolve in the grease. The ionic heads of the molecules then surround the grease droplet making it polar and negatively charged

Test yourself

6 Draw the skeletal formula of *cis,cis,cis*-octadec-9,12,15-trienoic acid.

7 Write an equation to show the hydrogenation of oleic acid. What it the name of the product?

8 Why is hydrolysis with alkali preferred to acid hydrolysis when making soaps from fats and oils?

Organic Chemistry

Section four

4.11 Amides

Amides are organic nitrogen compounds derived from carboxylic acids. In an amide, the $-NH_2$ group replaces the $-OH$ group of the acid. The chemistry of amides is important because it is amide groups which link up the small molecules in both proteins and in synthetic polymers such as nylon and Kevlar.

Names and structures

CD-ROM

Figure 4.11.1 ▶
Names and structures of amides. Note that N-ethyl propanamide has an ethyl group substituted for one of the hydrogen atoms of the $-NH_2$ group

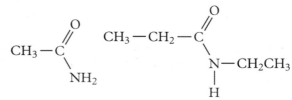

ethanamide   N-ethyl propanamide

Test yourself

1 Give the name and displayed formula of the amide formed when ethanoyl chloride or ethanoic anhydride reacts:

   a) with ammonia

   b) with CH_3NH_2.

Physical properties

Amides are white crystalline solids at room temperature, apart from methanamide which is a liquid. The simpler amides are soluble in water.

Reactions which form amides

Amides form rapidly at room temperature when acyl chlorides or acid anhydrides react with ammonia or with amines (see page 181).

Hydrolysis

Heating with dilute acid or dilute alkali hydrolyses amides. Under acid conditions the products are the corresponding carboxylic acid and an ammonium salt. Under alkaline conditions the products are a salt of the carboxylic acid and ammonia.

Figure 4.11.2 ▶
Hydrolysis of ethanamide

$$CH_3-C\overset{O}{\underset{NH_2}{}} + H_3O^+(aq) \longrightarrow CH_3-C\overset{O}{\underset{OH}{}} + NH_4^+(aq)$$

Dehydration

Heating an amide with phosphorus(v) oxide dehydrates the compound and produces a nitrile.

Figure 4.11.3 ▶
Dehydration of propanamide. Note that the nitrile formed has three carbon atoms and so is called propanenitrile

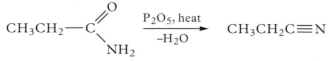

propanenitrile

The Hofmann degradation

The Hofmann degradation is a reaction which turns amides into primary amines while removing a carbon atom from the molecule. The reaction can be useful in synthetic pathways because it shortens the carbon chain. Hence the term 'degradation'. The reaction was discovered by August von Hofmann (1818 – 1892) who was a pioneering German organic chemist and head of the Royal College of Chemistry in London for 20 years from 1845. Amides react in this way when treated with bromine and concentrated sodium or potassium hydroxide solution.

Note
Chemists often name types of reaction after the chemist who first discovered the reaction or developed it as a practical technique.

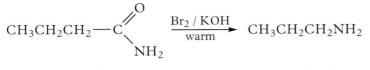

butanamide   propylamine

◀ *Figure 4.11.4*
Equation for the so-called Hofmann degradation of an amide to a primary amine with one less carbon atom

Organic Chemistry

Test yourself

2 Account for the differences in the products of hydrolysis of an amide under acidic and alkaline conditions.

3 Identify the products of hydrolysing:

   a) butanamide with dilute hydrochloric acid

   b) N-methyl propanamide with aqueous potassium hydroxide.

4 Give the names and formulae of the organic products of heating butanamide:

   a) with phosphorus(V) oxide

   b) with bromine in aqueous sodium hydroxide.

5 State the conditions needed to convert:

   a) $CH_3CH_2CH_2CONH_2 \longrightarrow CH_3CH_2CH_2NH_2$

   b) $CH_3CONH_2 \longrightarrow CH_3CN$

6 Identify compound P and Q. Also state the reagents and conditions for the third step.

$$CH_3CH_2CO_2H \xrightarrow{PCl_5} P \xrightarrow{NH_3} Q \longrightarrow CH_3CH_2NH_2$$

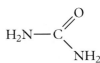

Figure 4.11.5 ▲
Structure of urea. Urea is a diamide of carbonic acid so it has the alternative name carbamide

Test yourself   D

7 Write an equation for the manufacture of urea.

8 **a)** Why do plants need sources of soluble, inorganic nitrogen in the soil?

   b) Suggest an advantage of using the organic compound urea as a fertiliser instead of the inorganic fertiliser ammonium nitrate.

   c) Compare the percentage by mass of nitrogen in urea with the percentage of nitrogen in NH_4NO_3.

Urea: a naturally occurring diamide

Urea is a white crystalline solid. It is an end product of the metabolism of proteins and excreted as a waste product in urine.

Urea is manufactured on a large scale by heating ammonia and carbon dioxide under pressure. The main use of urea is as a fertiliser. Urea has the useful property of slowly hydrolysing in water to release ammonia.

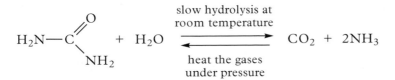

Figure 4.11.6 ▲
Hydrolysis and manufacture of urea

4.12 Nitriles

The nitrile functional group occurs in the molecules of the polymer which makes up the tough plastic ABS which is widely used for parts of telephones, suitcases and business machines. Otherwise nitriles are useful intermediates in organic synthesis, especially when there is a need to add a carbon atom to the skeleton of the molecule.

Names and structures

CD-ROM

Nitriles are organic compounds with the functional group –CN. Their general formula is R–CN where R is an alkyl or an aryl group.

Figure 4.12.1 ▶
Names and structures of nitriles. Note that the carbon atom in the –CN group is included in deciding the name of the related alkane

$$CH_3-CH_2-C\equiv N \qquad CH_3-CH_2-CH_2-C\equiv N$$

propanenitrile   butanenitrile

Physical properties

The simpler nitriles are colourless liquids which are soluble in water.

Formation of nitriles

Figure 4.12.2 ▶
Two ways of making nitriles

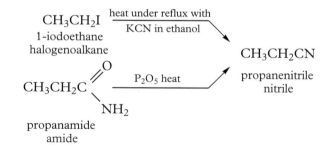

CH_3CH_2I
1-iodoethane
halogenoalkane

heat under reflux with
KCN in ethanol

CH_3CH_2CN
propanenitrile
nitrile

$CH_3CH_2C \overset{O}{\underset{NH_2}{\big\langle}}$
propanamide
amide

P_2O_5 heat

Hydrolysis

Hydrolysis converts a nitrile to a carboxylic acid. Acids or alkalis can catalyse this reaction.

Figure 4.12.3 ▶
Hydrolysis of a nitrile under acid and alkaline conditions

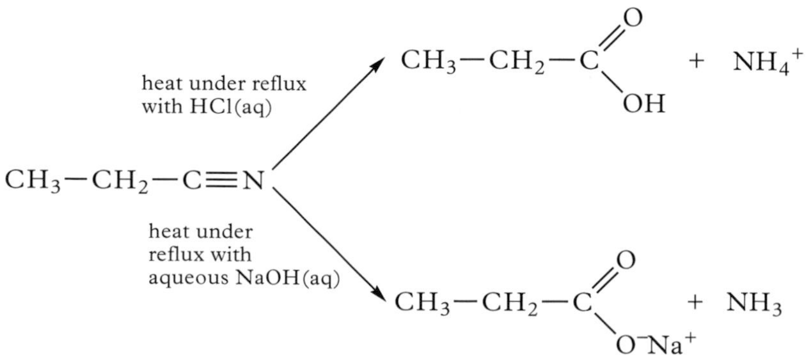

heat under reflux
with HCl(aq)

$$CH_3-CH_2-C\overset{O}{\underset{OH}{\big\langle}} \quad + \quad NH_4^+$$

$$CH_3-CH_2-C\equiv N$$

heat under
reflux with
aqueous NaOH(aq)

$$CH_3-CH_2-C\overset{O}{\underset{O^-Na^+}{\big\langle}} \quad + \quad NH_3$$

Reduction

Reduction with lithium tetrahydridoaluminate(III) converts nitriles to amines (see page 193).

$$CH_3CH_2-C{\equiv}N \ + \ 4[H] \ \xrightarrow[\text{in dry ether}]{LiAlH_4} \ CH_3-CH_2-CH_2-NH_2$$
propylamine

◀ **Figure 4.12.4**
Reduction of a nitrile

Test yourself

1 Draw the displayed formula of butanenitrile.

2 Name this nitrile: CH_3CN.

3 **a)** Draw the structure of acrylonitrile given that its systematic name is propenenitrile.

   b) Draw a short length of the addition polymer formed by propenenitrile – a polymer used to make clothing fibres (see page 203).

4 Classify the reactions used to make:

   a) nitriles from halogenoalkanes

   b) nitriles from amides.

5 Give the name and structure of the compound formed when butanenitrile is:

   a) heated under reflux with aqueous sodium hydroxide

   b) treated with $LiAlH_4$ in dry ether and followed by water or acid.

Organic Chemistry

Section four

4.13 Amines

Amines can be very smelly. Ethylamine, for example, has a fishy smell. The great significance of the amine functional group, however, is its importance in biochemistry because of the central role of amino acids (see page 197) and proteins in metabolism. As a result the amine functional group also appears in many medical drugs. In the chemical industry, aromatic (aryl) amines have commercial value because they are the basis of the manufacture of a wide range of colourful dyes.

Names and structures

CD-ROM

Amines are nitrogen compounds in which one or more of the hydrogen atoms in ammonia, NH_3, is replaced by an alkyl group or an aryl group. The number of alkyl groups determines whether the compound is a primary amine, a secondary amine or a tertiary amine.

The names of amines are based on the nature and number of alkyl groups attached to the nitrogen atom. Chemists use two systems. For the simplest compounds they treat the compounds as a combination of one or more alkyl groups and an amine group. Chemists also use the prefix 'amino' in compounds such as amino acids in which there is a group such as $-CO_2H$ which appears at the end of the name.

Figure 4.13.1 ▶
Names and structures of primary, secondary and tertiary amines. Phenylamine is a primary amine with an $-NH_2$ group attached to a benzene ring

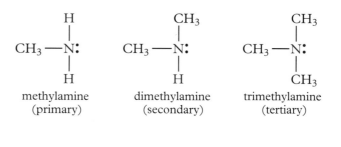

methylamine (primary)   dimethylamine (secondary)   trimethylamine (tertiary)

Note

The terms primary, secondary and tertiary do not have the same meaning for amines as they do for alcohols and halogenoalkanes.

Test yourself

1 Write out the structures of:

a) diethylamine

b) hexane-1,6-diamine

c) 1,2-diaminopentane which contributes to the smell of rotting flesh and has the trivial name cadaverine

d) 1-phenyl-2-aminopropane, an amphetamine which is an addictive stimulant.

2 Suggest a reason why an amine has a higher boiling point than the alkane with the same number of carbon atoms.

Physical properties

The simplest amines such as methylamine and ethylamine are gases at room temperature. Most other common amines are liquids including phenylamine.

Alkyl amines with short hydrocarbon chains are freely soluble in water. Phenylamine is only slightly soluble.

Formation of amines

Reactions to make alkyl amines

Two common methods for making primary amines are:

■ the reduction of a nitrile (see page 191)
■ the substitution reaction of a halogenoalkane with ammonia.

Ammonia is a nucleophile. Warming a halogenoalkane with a solution of ammonia in ethanol produces an amine.

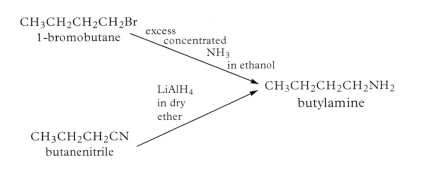

◀ *Figure 4.13.2*
Two reactions forming butylamine

Test yourself

3 a) Suggest a mechanism for the nucleophilic substitution reaction of ammonia with 1-bromopropane (see page 157).

b) Suggest a reason for using a solution of ammonia in ethanol rather than in water when making an amine from a halogenoalkane.

Synthesis of aryl amines

The usual laboratory method for introducing an amine into an aromatic (aryl) compound is a two step route: first nitration to make a nitro compound and then reduction.

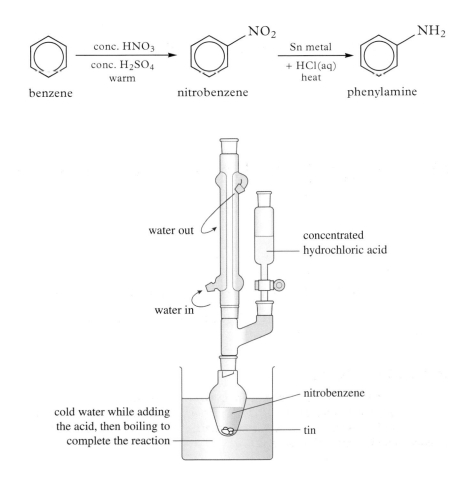

benzene   conc. HNO$_3$ / conc. H$_2$SO$_4$ / warm   nitrobenzene   Sn metal + HCl(aq) / heat   phenylamine

◀ *Figure 4.13.3*
Formation of phenylamine from benzene

water out

concentrated hydrochloric acid

water in

cold water while adding the acid, then boiling to complete the reaction

nitrobenzene

tin

◀ *Figure 4.13.4*
Apparatus for reducing nitrobenzene to phenylamine

Organic Chemistry

Section four

Amine reactions

 CD-ROM

Primary amines, like ammonia can acts as bases, form complex ions with metal ions (see section 3.3) and react as nucleophiles.

Figure 4.13.5 ▶
Methylamine and ammonia compared

Amines as bases

The lone pair on a nitrogen atom is a proton acceptor.

Figure 4.13.6 ▶
Ethylamine acting as a base

$$CH_3CH_2NH_2(aq) + H_2O(l) \rightleftharpoons CH_3CH_2NH_3^+(aq) + OH^-(aq)$$

 **Note**

The chemical industry makes phenylamine from phenol and ammonia with an alumina catalyst.

Primary alkyl amines are stronger bases than ammonia. Chemists explain this in terms of the inductive effect (see page 156). The shift of electrons from an alkyl group increases the electron density on the nitrogen atom so that it can hold a proton more strongly.

Phenylamine is a weaker base than ammonia. A lone pair on the nitrogen atom interacts with the delocalised electrons of the benzene ring. This extended delocalisation stabilises the free amine, makes the lone-pair less available and so reduces the tendency of the molecule to form a dative covalent bond with a proton.

Figure 4.13.7 ▶
Delocalisation of electrons in phenylamine

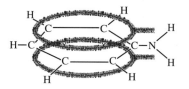

 **Test yourself**

4 Arrange these compounds in order of base strength: ammonia, butylamine, 4-methylphenylamine.

5 Write an equation to show the formation of a salt when ethylamine reacts with hydrogen chloride gas. Explain why the two gases react to form a solid product.

 **Definition**

Quaternary ammonium cations form when all the hydrogen atoms in an ammonium ion are replaced by alkyl groups.

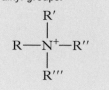

Amines as nucleophiles

Reaction with halogenoalkanes

Alkyl amines are more reactive than ammonia as nucleophiles. They react with halogenoalkanes. As a result when ammonia reacts with a halogenoalkane the result is usually a mixture of primary, secondary and tertiary amines. If there is sufficient halogenoalkane the final product is a quaternary ammonium salt.

The cationic surfactants used in fabric and hair conditioners are quaternary ammonium salts such as dodecyltrimethylammonium bromide, $CH_3(CH_2)_{11}N^+(CH_3)_3Br^-$.

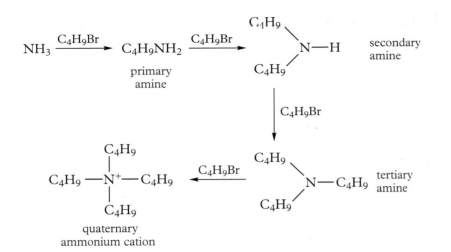

Acylation

Acid anhydrides and acyl chlorides both acylate primary amines. The amine acts as a nucleophile in the first step of this addition–elimination reaction (see page 182).

Diazonium salts and azo dyes

CD-ROM

Phenylamine and other arylamines are important intermediates in the manufacture of azo dyes. This is because they react with nitrous acid, HNO_2, below 10 °C to produce diazonium salts.

Nitrous acid is unstable. It is a weak acid. A solution of nitrous acid forms on adding dilute hydrochloric acid to sodium nitrite, $NaNO_2$. A solution of nitrous acid is blue. It starts to decompose at room temperature giving off nitrogen monoxide which turns brown as it meets the air as it turns into nitrogen dioxide.

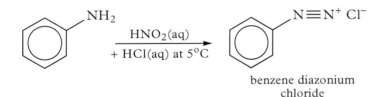

Figure 4.13.9 ▲
Formation of benzene diazonium chloride

Diazonium salts are unstable. The diazonium salts of aryl amines are stabilised by delocalisation so they are useful reagents if kept cool. The diazonium salts of alkyl amines are much more unstable and decompose immediately giving off nitrogen gas.

Being unstable, diazonium salts are made as needed and kept cold. Above 10 °C, benzene diazonium chloride decomposes to phenol and nitrogen.

The commercial importance of diazonium salts is based on their coupling reactions to form azo dyes. Azo dyes are the products of coupling reactions between diazonium salts and coupling agents such as phenols or amines. Benzene diazonium chloride, for example, couples with phenylamine to make a yellow azo compound.

Most of the dyes are red, orange or yellow. The acid-base indicator methyl orange is an azo dye.

6 Write an equation for the reaction of 1-aminobutane (butylamine) with 1-bromopropane.

7 Why is an excess of ammonia in ethanol used when making a primary amine from a halogenoalkane?

8 a) Compare an ammonium ion with a quaternary ammonium ion.

   b) Why is quaternary methylammonium chloride a solid at room temperature?

9 Write equations to describe the mechanism of the reaction of 1-aminopropane with ethanoyl chloride.

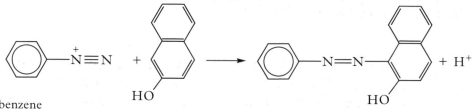

benzene
diazonium
chloride

naphthalene-2-ol

an azo dye

Figure 4.13.10 ▲
*Equation for a coupling reaction
to make an azo dye*

Figure 4.13.11 ▶
This shirt is dyed with azo dyes

Test yourself

10 Write an ionic equation for the formation of nitrous acid in solution.

11 a) When nitrous acid decomposes it disproportionates to the +5 and the +2 states. Write a balanced equation for this decomposition reaction.

b) Write a second equation to show why the gas given off when HNO_2 decomposes turns brown in air.

12 Write out the structures of the amine and the phenolic compound which can be used to make this azo dye:

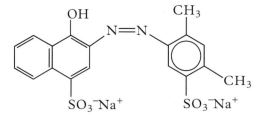

4.14 Amino acids and proteins

Proteins make up 15 per cent of the human body. There are many different protein molecules, each able to do a special job. Skin, muscle and hair consist of fibrous proteins. Similar proteins make up the connective tissues which hold the body together by tying muscle to bone, and bone to bone. Other proteins coil up into a globular shape and dissolve in body fluids where they act as enzymes, hormones and oxygen carriers.

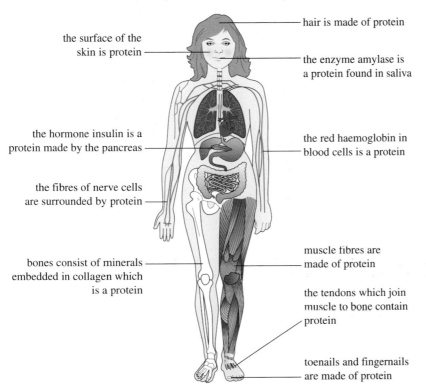

◀ *Figure 4.14.1*
Proteins in the human body

the surface of the skin is protein

hair is made of protein

the enzyme amylase is a protein found in saliva

the hormone insulin is a protein made by the pancreas

the red haemoglobin in blood cells is a protein

the fibres of nerve cells are surrounded by protein

bones consist of minerals embedded in collagen which is a protein

muscle fibres are made of protein

the tendons which join muscle to bone contain protein

toenails and fingernails are made of protein

Amino acids

Amino acids are the compounds which join together in long chains to make proteins. They are carbon compounds with two functional groups: an amino group and a carboxylic acid group. There are about 20 different amino acids which link together to make proteins.

◀ *Figure 4.14.2*
Examples of amino acids in proteins. Note that in all of them the amino group is attached to the carbon atom next to the carboxylic acid group. Each acid has a three-letter abbreviation

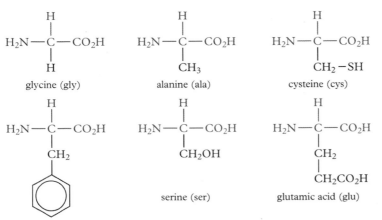

glycine (gly)

alanine (ala)

cysteine (cys)

phenylalanine (phe)

serine (ser)

glutamic acid (glu)

Definition

Chemists sometimes use the term **alpha–amino acid** to describe the amino acids in proteins. In these compounds the amino group is linked to the carbon atom next to the carboxylic acid group: the α–carbon atom.

Figure 4.14.3 ▶
Mirror image forms of the amino acid alanine

Amino acid structures

Each of the amino acids in proteins has a central carbon atom attached to four other groups so, except for glycine, these are chiral molecules which can exist in mirror image forms (see page 148).

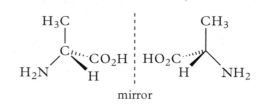

mirror

Test yourself

1 What is the systematic name for:

 a) glycine

 b) alanine?

2 Explain why the amino acid glycine is not chiral.

3 How could you distinguish between solutions of the two mirror image forms of an amino acid by experiment?

Acid–base properties

The $-NH_2$ group is basic while the $-CO_2H$ group is acidic. In aqueous solution, amino acids form ions as the amino groups accept protons and the acid groups give away protons. The ions formed are unusual in that they have both positive and negative charges. Chemists call them zwitterions from a German word meaning 'mongrel'.

Figure 4.14.4 ▶
Glycine forming a zwitterion

Definition

Zwitterions are ions with both a positive and a negative charge.

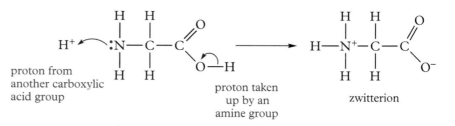

proton from another carboxylic acid group

proton taken up by an amine group

zwitterion

Amino acids are very soluble in water and they crystallise as colourless solids from solution. In crystals and in aqueous solution, amino acids exist largely as zwitterions.

Test yourself

4 Explain why alanine is a solid while propanoic acid and 1-aminopropane are liquids.

5 **a)** Write equations to show the reactions of alanine with:

 (i) dilute hydrochloric acid

 (ii) aqueous sodium hydroxide.

 b) How do the products of these two reactions differ from the zwitterion of glycine dissolved in water?

Peptides

Peptides are compounds made up of chains of amino acids. The simplest example is a dipeptide with just two amino acids linked. For chemists this is an example of an amide bond (see page 188). The tradition in biochemistry is to call it a 'peptide bond'. The formation of a peptide bond is an example of a condensation reaction.

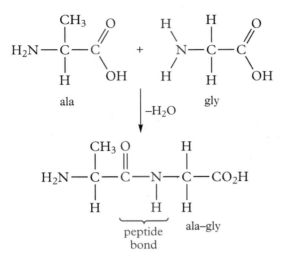

◄ *Figure 4.14.5*
Formation of a peptide bond between two amino acids

Digestive enzymes in the stomach and small intestine catalyse the hydrolysis of peptide bonds, splitting polypeptide chains into the amino acids. Chemists achieve the same effect by heating the polypeptide chain (or protein) with hydrochloric acid under reflux.

Definition

During a **condensation reaction** molecules join together by splitting off a small molecule such as water.

Test yourself

6 Draw the structures of the two dipeptides which alanine and phenylalanine can form.

7 Show that splitting a dipeptide into two amino acids is an example of hydrolysis.

8 **a)** Identify the functional groups in the sweetener Aspartame and show how it differs from a dipeptide.

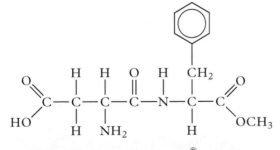

aspartame (Nutrasweet®)

b) Suggest a reason for the fact that Aspartame cannot be used for for sweetening foods that will be cooked.

c) Suggest a reason why soft drinks sweetened with Aspartame carry a warning for people with the genetic disorder which means that they must not consume any phenylalanine.

Note

Polypeptides are long-chain peptides. There is no precise dividing line between a peptide and a polypeptide.

Some chemists make a distinction between polypeptides and the longer amino acid chains in proteins. They restrict the definition of polypeptides to chains with 10 – 50 or so amino acids.

Section four   Organic Chemistry

Proteins

A protein molecule consists of one or more polypeptide chains. Biochemists describe the structure of proteins at a series of levels:

- The **primary structure** is the sequence of amino acids in the polymer chains.
- The **secondary structure** is the ways in which the chains are arranged and held in place by hydrogen bonding within and between chains – this includes the coiling of chains into helices in proteins such as keratin and the formation of layers of parallel chains as in the pleated sheets of protein in silk.
- The **tertiary structure** describes the three dimensional folding of protein chains which gives some proteins, such as enzymes, a definite three-dimensional shape held in place by hydrogen bonding, disulfide bonds and interactions between amino acid side chains with surrounding water molecules.
- The **quaternary structure** describes the linking of two or more amino acid-chains as in a haemoglobin molecule which consists of four chains.

Figure 4.14.6 ▶
The alpha helix, an example of the secondary structure of a protein. Note the importance of hydrogen bonding between amino acids in the same molecule to hold the shape of the helix

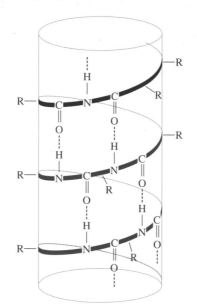

Figure 4.14.7 ▼
The four levels of protein structure

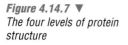

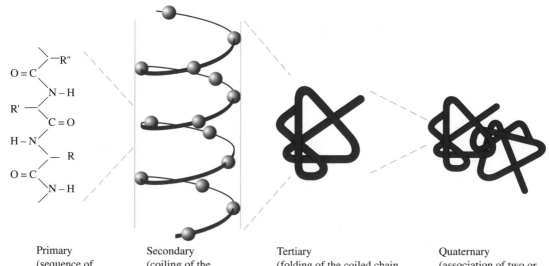

| Primary (sequence of amino acids) | Secondary (coiling of the amino acid chain) | Tertiary (folding of the coiled chain to create an active site) | Quaternary (association of two or more coiled chains) |

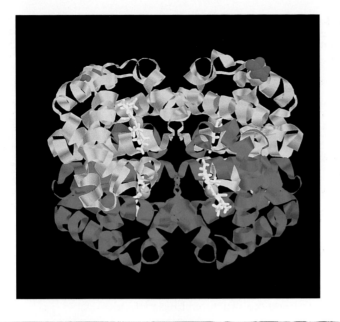

Definition

Cross linking is the formation of chemical bonds between the polymer chains. Cross-links in protein molecules take various forms including:

■ hydrogen bonding between amino acid side-chains

■ disulfide bridges formed by covalent bonding between cystine side-chains.

Test yourself

9 This structure shows a short section of a protein molecule.

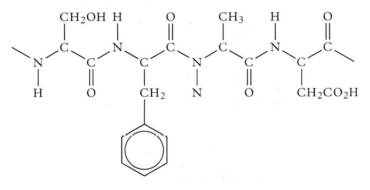

a) Identify the amino acids which have joined to make this chain (see Figure 4.14.2)

b) Which of these amino acids have: **(i)** side chains which are non-polar **(ii)** polar side chains **(iii)** side chains which can ionise?

10 Biochemists talk about enzymes being 'denatured' by strong acids, strong bases or by a rise in temperature. Suggest a reason for the loss of the catalytic activity of enzymes when they are treated in any of these ways.

4.15 Polymers

Polymers are long chain molecules. Natural polymers include proteins, carbohydrates such as starch and cellulose, and nucleic acids. Synthetic polymers include polyesters, polyamides and the many polymers formed by the addition polymerisation of compounds with carbon–carbon double bonds. The properties of polymers are very varied. Polymeric materials include plastics, elastomers and fibres.

Polymer chemistry

Polymer chemistry is the study of the synthesis, structure and properties of polymers. It is a branch of chemistry, which developed during the 20th century starting with Leo Baekland's discovery of Bakelite in 1905 with little help from theory. This first thermosetting polymer was discovered before chemists understood the structure of big molecules.

Figure 4.15.1 ▶
Computer graphic of low density polythene showing the branches which stop the chains packing closely together. The colour code shows carbon atoms light green and mauve. Hydrogen atoms are pink and blue

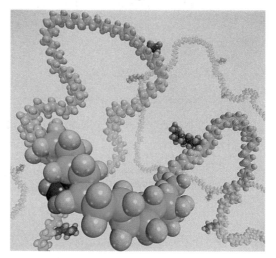

The German chemist, Hermann Staudinger published his theory that natural rubber, cellulose and related substances consist of long chain molecules, in 1922. He had to fight hard to persuade other chemists to accept his theory which is now taken for granted.

One of the people who was convinced by the new theory was the American industrial chemist Wallace Carothers. He introduced the terms addition polymerisation and condensation polymerisation in an article published in 1931. His research team at Du Pont invented the synthetic rubber, Neoprene. They also discovered the first completely synthetic polymer nylon which went into production in 1939.

The 1930s were probably the most important years in the development of the plastics industry. It was the time when the addition polymers polythene, pvc, polystyrene and Perspex were all developed commercially. The high-pressure process for making polythene was discovered and developed in the mid-1930s as an unexpected offshoot of the study of high-pressure gas reactions by Eric Fawcett and Reginald Gibson working for the UK chemical company ICI.

It was in the early 1950s that Karl Ziegler in Germany discovered how to polymerise ethene at low temperatures using a new kind of catalyst. At the same time Guilio Natta in Italy discovered the benefits of producing addition polymers with a regular three-dimensional shape.

Definition

Polymerisation is a process in which many small molecules (**monomers**) join up in long chains by addition polymerisation or condensation polymerisation.

Research and development in the second half of the 20th century led to the development of many new specialised polymer materials including the polyamide Kevlar and biodegradable polymers.

Addition polymerisation

Addition polymerisation is a process for making polymers from compounds containing double bonds. Many molecules of the monomer add together to form a long chain polymer. Ethene, for example, polymerises to form poly(ethene).

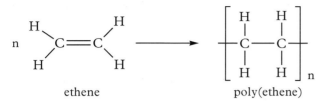

ethene   poly(ethene)

Figure 4.15.3 ▲
Formation of polythene from ethene. This shows the repeat unit in the polymer chain

One technique for making addition polymers involves free radical chain reactions (see page 154).

The initiator can be an organic peroxide such as benzoyl peroxide which acts as a source of free radicals, R•. Organic peroxides can act as a source of free radicals to initiate addition polymerisation of unsaturated compounds.

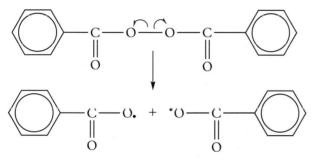

Figure 4.15.4 ▲
A benzoyl peroxide molecule splitting to form two free radicals. The O—O bond is relatively weak

This leads to the propagation steps which are repeated many times producing a very long chain molecule.

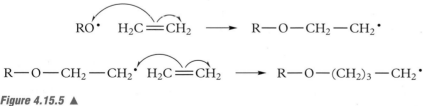

Figure 4.15.5 ▲
Chain propagation

Termination takes place when two free radicals combine to make a molecule.

$$2R-O-(CH_2)_n-CH_2^{\bullet} \longrightarrow R-O-(CH_2)_n-CH_2-CH_2-(CH_2)_n-O-R$$

Figure 4.15.6 ▲
Chain termination

Figure 4.15.2 ▲
False colour scanning electron micrograph of the waterproof but breathable fabric Gore-tex. The pink outer layers are nylon. The yellow and white layers consist of teflon (ptfe). Magnification × 100

Test yourself

1 Draw a short length of the polymer chain formed by these monomers. Name the monomers and the polymers and match them with two of Figures 4.15.7 – 4.15.10.

a)

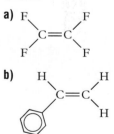

b)

2 Identify these polymers and give the displayed formulae of the monomers. Name the monomers and the polymers and match them with two of Figures 4.15.8 – 4.15.11.

a)

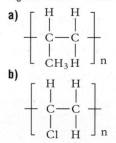

b)

Organic Chemistry

Section four

Figure 4.15.7 ▲
The domes consist of interconnecting steel hexagons 'glazed' with panels of ETFE. ETFE is a plastic made by polymerising a mixture of two monomers, ethene and tetrafluorethene. The panels consist of three layers of the polymer film and are individually inflated to give an insulting material which is lightweight and lets through just the right spectrum of light for plants

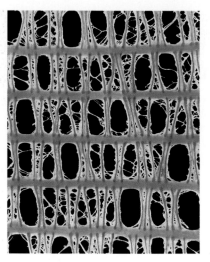

Figure 4.15.8 ▲
Poly(propene) which is often called polypropylene. A coloured scanning electron micrograph of netting made from poly(propene). This strong polymer is used to make pipes, wrapping films, carpet fibres and ropes. Magnification $\times$ 13

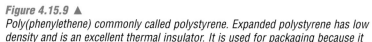

Figure 4.15.9 ▲
Poly(phenylethene) commonly called polystyrene. Expanded polystyrene has low density and is an excellent thermal insulator. It is used for packaging because it absorbs shocks

THERE ARE BAGS OF REASONS
FOR THE USE OF CHLORINE IN PACKAGING.
HERE'S ONE OF THEM.

CHLORINE
Life depends on it

Figure 4.15.10 ▲
Poly(chloroethene) commonly called polyvinylchloride (pvc). Unplasticised upvc is a rigid polymer suitable for guttering and window frames. Plasticised pvc is flexible and used for packaging, flooring and cable insulation as well as blood bags

◄ *Figure 4.15.11*
Poly(ethene) commonly called polythene. Extruding recycled poly(ethene) to make plastic sheeting for the building industry. A high-pressure, high-temperature process in the presence of a peroxide initiator produces low-density poly(ethene) with branched chains. A low-pressure, low-temperature process with a Ziegler catalyst produces high-density poly(ethene) in which the polymer chains pack closer because they have no side branches)

Definitions

Plastic materials are materials which can be moulded by gentle pressure. Examples are wet clay and Plasticine. Once moulded, a plastic material keeps its new shape; unlike an elastomer which tends to spring back to its original shape.

Plastics are materials made of long-chain molecules which at some stage can be easily moulded into shape. Thermoplastics become plastic materials when they are hot and harden on cooling. Some thermoplastics are rigid and brittle at room temperature such as polystyrene and upvc. Other thermoplastics are flexible such as polythene.

Plasticisers are additives mixed with a polymer material to make it more flexible. A plasticiser is usually a liquid with a high boiling point, such as a large ester molecule. The molecules of the plasticiser get between the polymer chains and reduce the intermolecular forces between the chains so that they can slide past each other.

An alternative method for speeding up the polymerisation of alkenes such as ethene and propene is to use a Ziegler–Natta catalyst. These catalysts make it possible to produce addition polymers at relatively low temperatures and pressures. They are combinations of titanium(IV) chloride and aluminium alkyls (such as triethyl aluminium) in a hydrocarbon solvent.

There is little chain branching in the polymers so the poly(ethene) chains formed by this method can pack closely together forming the high-density form of poly(ethene).

Poly(propene) made with a Ziegler–Natta catalyst has a regular arrangement of the methyl side chains. All the methyl side chains are on the same side of the carbon chain. The molecules coil into a regular helical shape and pack together to form a highly crystalline polymer which is very strong. This is the form of the polymer which is useful for hardwearing fibres, tough mouldings in motor vehicles and containers which can hold boiling water.

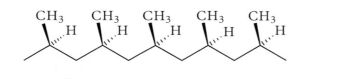

◄ *Figure 4.15.12*
Structure of isotactic poly(propene) with all the methyl groups on the same side of the carbon chain

Test yourself

3 Explain the terms homolytic fission and free-radical using as your example the use of benzoyl peroxide as an initiator.

4 The paste supplied with D.I.Y. wood fillers often contains phenylethene. The paste is supplied with a small tube of hardener.

   a) Why is the hardener supplied in a separate tube and only mixed with the paste shortly before use?

   b) Suggest a chemical for the hardener.

   c) What factors will determine how fast the wood filler sets?

Definitions

In **isotactic** poly(propene) all the methyl side chains are on the same side of the carbon chain.

A **syndiotactic** polymer the arrangement of side chains is also regular but the side chains alternate from one side of the chain to the other.

In **atactic** poly(propene) the side groups along the polymer chain are randomly orientated. Atactic poly(propene) is an amorphous, rubbery polymer. One of its uses is to make sealants and flexible roofing materials.

CD-ROM

Figure 4.15.13 ▶
Part of a chain of atactic poly(propene)

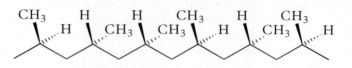

Test yourself

5 Draw a short length of a molecule of syndiotactic poly(propene).

Condensation polymers

Condensation polymers are produced by a series of condensation reactions splitting off water between the functional groups of the monomers. Examples of condensation polymers are polyamides and polyesters. Where each monomer has two functional groups this type of polymerisation produces chains. Cross-linking is possible if one of the polymers has three functional groups.

Polyesters

Polyesters are polymers formed by condensation polymerisation between acids with two carboxylic acid groups and alcohols with two or more –OH groups. The units in the polymer chains are linked by a series of ester bonds.

The monomers for making polyester fibres are benzene-1,4-dicarboxylic acid and ethan-1,2-diol.

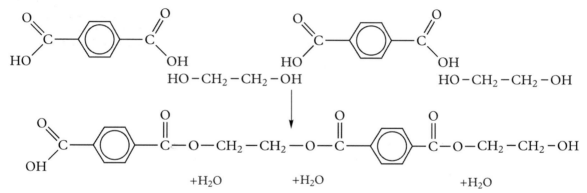

Figure 4.15.14 ▲
Condensation polymerisation to produce polyester

The traditional name for these compounds are terephthalic acid and ethylene glycol. Hence the commercial name Terylene for one brand of polyester. The alternative name for the polymer is polyethylene terephthalate which gives rise to the name PET when the same polymer is used as a plastic to make bottles for carbonated drinks.

Polyamides

Polyamides are polymers in which the monomer molecules are linked by amide bonds. The first important polyamides were formed by condensation polymerisation between diamines and dicarboxylic acids.

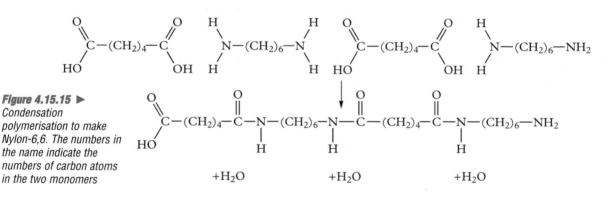

Figure 4.15.15 ▶
Condensation polymerisation to make Nylon-6,6. The numbers in the name indicate the numbers of carbon atoms in the two monomers

Chemists describe polymeric aryl amides as aramids. Aramids, such as Kevlar, are extremely strong, flexible, fire resistant polymers with a low density. Bullet proof vests are made of Kevlar and punctures are less likely with bicycle tyres which include the polymer. Kevlar ropes are much stronger than the same weight of steel ropes. The polymer can also replace steel in motor vehicle tyres.

The rigid, linear polymer chains in Kevlar line up parallel to each other held together by hydrogen bonding.

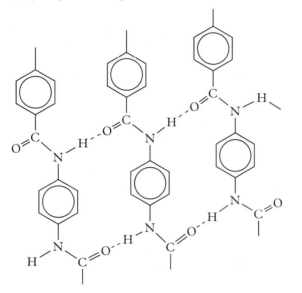

Figure 4.15.16 ▲
False colour scanning electron micrograph of a Velcro hook, with loops made from nylon. Magnification × 20

Figure 4.15.17 ▲
The structure of Kevlar

Test yourself

6 This compound can form a polymer:

CO_2H

... NH_2

a) Identify the functional groups in the monomer.

b) What other product forms during polymerisation?

c) Draw a short length of the polymer chain.

7 Identify one similarity and two differences between the structures of nylon and the structure of a protein (see pages 199–200).

8 a) Show that Kevlar is a polymer of benzene-1,4-diamine and benzene-1,4-dicarboxylic acid.

b) Suggest a reasons why the polymer is made from monomers with functional groups in the 1,4 position and not from the possible isomers of these monomers.

9 Identify the types of intermolecular forces which act between the polymer chains in:

a) poly(ethene)

b) nylon.

Polymer properties

The properties of polymers are very varied. Polymeric materials include plastics, elastomers and fibres. As polymer science has developed, chemists and materials scientists have learnt how to develop new materials with particular properties.

Some of the ways of modifying polymeric materials include:

- altering the average length of the polymer chains
- changing the structure of the monomer perhaps by adding side groups which increase intermolecular forces
- selecting a monomer which produces a polymer which is biodegradable
- varying the degree of cross-linking between chains
- co-polymerisation
- changing the alignment of the polymer chains, for example by spinning the polymer into fibres and then stretching the fibres
- adding fillers, pigments and plasticisers
- making composites.

Figure 4.15.18 ▶
A scanning electromicrograph of a glass fibre composite showing rod-shaped glass fibres embedded in a polyester matrix. Magnification × 330

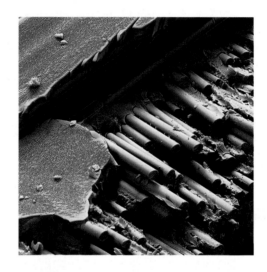

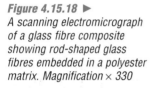

Biodegradable materials break down in the environment due to the action of micro-organisms.

Photodegradable plastics break down when exposed to sunlight.

Test yourself

10 A bottle made of poly(phenylethene) can be used to store dilute potassium hydroxide but holes gradually appear in a polyester lab coat which has soaked up splashes of the same reagent. Account for the difference in the behaviour of the two polymers.

11 Suggest a reason why many polyesters and polyamides are biodegradable while poly(alkenes) are not.

12 Which type of plastic makes up most of the plastic waste from households: condensation or addition polymers? What are the implications of this when it comes to disposing of the waste?

13 What difficulties may arise during waste management because of the mixture of polymers in plastic waste from households?

Polymers after use

Plastics make up only about 8% of household waste in Western Europe. Even so plastics are seen as a problem now that it is becoming more and more expensive to dump waste in landfill sites. The choices for dealing with plastic waste include recycling and energy recovery.

Mechanical recycling

Some plastics are easier to recycle than others. Automatic systems for separating bottles are now available and as a result the percentage of plastic recovered from waste has increased markedly. The bottles used for carbonated drinks are particularly valuable because the polyester (PET) can be re-melted and spun into fibres. Recycling of pvc produces sewage pipes and flooring tiles.

Chemical recycling

Several processes are being developed to recover useful chemicals from waste plastics. One technique involves cracking the waste into small hydrocarbon molecules. Another technique depolymerises the plastic to recover the monomer.

Energy recovery

It is hard to prove the environmental and economic benefits of recycling plastics. The energy needed to transport plastic waste to a processing plant and then to sort and clean the large volume of material may well be greater than making fresh plastic from oil.

The energy from burning plastic waste can be nearly as high as the energy from the same mass of oil. Only about 4% of the oil used in Europe is converted to plastic. So one option is to regard the manufacture of plastics as a means of making greater use of oil on its way to being burnt as a fuel.

There is often strong opposition from the public when plans to build incineration plants are announced. Modern plants for burning waste have to meet tough environmental standards and much of their cost arises from the investment in equipment to generate electricity and control pollution. Despite the higher standards, however, many people remain suspicious of the emissions from incinerators and worry that they are a health risk. The anxiety is that if the waste includes chlorine compounds such as pvc, an incinerator will give off corrosive or toxic chemicals such as hydrogen chloride or dioxins.

Figure 4.15.19 ▲
The managing director of a Dutch recycling company standing on a pile of bottle tops and holding a roll of plastic made from recycled caps

Figure 4.15.20 ▲
Siphoning an oil from a plant which breaks down plastic waste by heating it to over 400 °C. The product can be used in place of crude oil as a feedstock for oil refineries and the petrochemical industry

Organic Chemistry

Section four

4.16 Synthesis

Synthesis is at the heart of all the research which develops new drugs. Synthesis provides the fashion industry with new colours and provides us all with the chemicals to clean and ornament ourselves and our homes. Synthesis is also important to the understanding of organic molecules. It is not until someone has synthesised a molecule that chemists are confident they have determined its structure precisely.

Organic routes

Organic chemists synthesise new molecules using their knowledge of functional groups, reaction mechanisms and molecular shapes, as well as the factors which control the direction and extent of chemical change.

A synthetic pathway leads from the reactants to the required product in one step or several steps. Organic chemists often start by examining the 'target molecule', then they work back through a series of steps to find suitable starting chemicals which are cheap enough and available. In recent years chemists have developed computer programmes to help with the process of working back from the target molecule to a range of possible starting points in a systematic way.

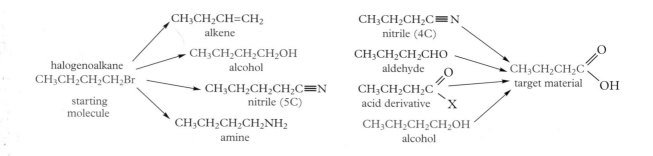

Figure 4.16.1 ▲
Tracking back from the target molecule to find a two-step synthesis of butanoic acid from 1-bromobutane

Chemists seek a synthetic route which has the least number of steps and produces a high yield of the product (see page 219). The larger the scale of production the more important it is to keep the yield high to avoid producing large quantities of wasteful by-products.

Figure 4.16.2 ▶
Combinatorial chemistry automates the synthesis of large numbers of new compounds. Vast numbers of new compounds can be made and tested in a short time

Changing the functional groups

All the reactions in organic chemistry convert one compound to another, but there are some reactions which are particularly useful for developing a synthetic route, such as:

- addition of hydrogen halides to alkenes
- substitution reactions to replace halogen atoms with other functional groups such as –OH or –NH_2
- substitution of a chlorine atom for the –OH group in an alcohol or carboxylic acid
- elimination of a hydrogen halide from a halogenoalkane to introduce a double bond
- oxidation of primary alcohols to aldehydes and then carboxylic acids,
- reduction of carbonyl compounds to alcohols.

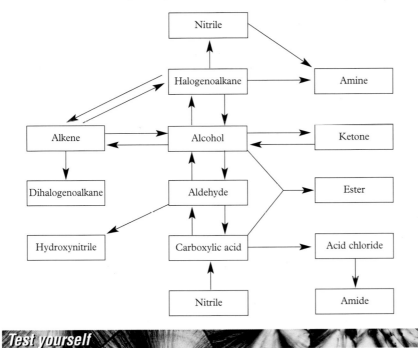

◀ **Figure 4.16.3**
A flow diagram for summarising methods for converting one functional group to another

Organic Chemistry

Section four

Test yourself

1 Give the structural formula of the main organic product of these reactions. Classify each reaction as addition, substitution or elimination. Also classify the reagent as a free-radical, nucleophile, electrophile or base.

   a) $CH_3CHBrCH_3 \xrightarrow{\text{KOH in ethanol}}$

   b) $CH_2{=}CH_2 \xrightarrow{\text{HBr}}$

   c) $CH_3CH_2CH_2Br \xrightarrow{\text{KOH(aq)}}$

   d) ⬡ $\xrightarrow{\text{conc. HNO}_3+ \atop \text{conc. H}_2\text{SO}_4}$

2 Make a copy of the flow chart in Figure 4.16.3. Beside each arrow write the reagents and conditions for the conversion.

3 Suggest syntheses for converting:

   a) ethene to ethanoic acid

   b) butan-1-ol to butan-2-ol

   c) ethanol to ethyl ethanoate (using ethanol as the only carbon compound).

Adding to the carbon skeleton

Fundamental to any synthesis is the need to create the correct skeleton of carbon atoms. Chemists today have a large repertoire of reactions for achieving the required number and arrangement of carbon atoms in a molecule. One of the best known ways of increasing the number of carbon atoms is to add one carbon atom to the carbon chain by forming a nitrile from a halogenoalkane. Nitriles are not useful in themselves but they can be converted to carboxylic acids and amines (see pages 190–191).

In 1924, two chemists, Levine and Taylor, published a paper describing the methods they used to synthesise pure samples of straight chain saturated fatty acids (see page 185). They wanted to measure their melting points accurately. They began with octadecanoic acid (stearic acid) and added one carbon atom at a time to make all the acids up to docosanoic acid, $CH_3(CH_2)_{20}CO_2H$. They then used a natural source of docosanoic acid to continue to hexacosanoic acid. The conversion of one acid to the next took several days.

Figure 4.16.4 ▶
The reaction scheme used by Levine and Taylor to convert one fatty acid to the next member in the series

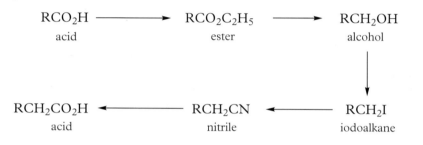

Another general method of building up the carbon skeleton is to use a Grignard reagent (see section 4.17).

The Friedel–Crafts reaction is a valuable technique for adding carbon atoms to aromatic (aryl) compounds (see page 164).

Reducing the carbon skeleton

One method for shortening a carbon chain is to heat an amide with bromine and aqueous sodium hydroxide. This is the degradation reaction named after Hofmann (see page 189).

Vigorous oxidation with potassium manganate(VII) can cut down the number of carbon atoms in an aromatic compound by converting any hydrocarbon side chains to carboxylic acid groups (see page 162).

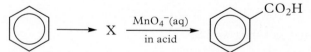

Test yourself

4 Give reagents, conditions and the structures of the organic compounds for converting:

a) ethanol to propylamine in three steps

b) propan-1-ol to butanoic acid in three steps.

5 Complete this flow diagram for a two step synthesis of benzenecarboxylic acid.

6 **a)** Suggest reagents and conditions for each of the reaction steps used by Levine and Taylor.

b) Explain the importance to chemists of accurate tables of melting points of pure compounds.

c) Why were the chemists more concerned to make a very pure product than to obtain a high yield?

7 Give the reagents and conditions for converting butanoic acid to propylamine.

4.17 Grignard reagents

Grignard reagents are reactive organic compounds containing magnesium atoms which are used to form C—C bonds in organic synthesis. They are examples of organometallic compounds. The reagents were discovered by the French organic chemist Victor Grignard (1871 – 1935). At the time he was a graduate student in the laboratories of Professor Barbier at the University of Lyon. Later he succeeded Barbier as the professor at Lyon and won a Nobel prize for his work in 1912.

Grignard showed that halogenoalkanes react with magnesium in dry ether (ethoxyethane).

$$RBr + Mg \rightarrow RMgBr, \text{ where R represents an alkyl group.}$$

The carbon–magnesium bond in a Grignard reagent is polarised with the metal atom at the positive end of the dipole and the carbon atom at the negative end. As a result the carbon atom attached to magnesium is nucleophilic. This means that C—C bonds form when Grignard reagents take part in nucleophilic substitution or addition reactions. This is what makes the reagents so useful in synthesis.

Definition

Organometallic compounds are carbon compounds which contain a metal atom. Grignard reagents are organometallic compounds formed with magnesium.

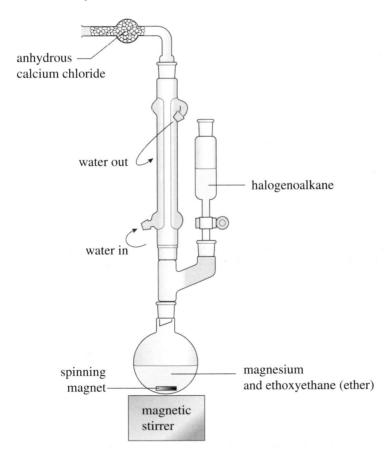

anhydrous calcium chloride

water out

halogenoalkane

water in

spinning magnet

magnesium and ethoxyethane (ether)

magnetic stirrer

Figure 4.17.1 ▲
Apparatus for the preparation of a Grignard reagent

Figure 4.17.2 ▶
A Grignard reagent acting as a
nucleophile when reacting with
a carbonyl compound

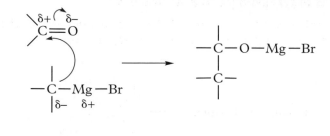

Alkane synthesis by hydrolysis

Hydrolysis converts a Grignard reagent to an alkane. Generally chemists try to stop this happening when working with Grignard reagents by keeping out moisture. There are, however, times when the reaction of a halogenoalkane with magnesium followed by hydrolysis with water can be a convenient method of replacing a halogen atom by a hydrogen atom.

Figure 4.17.3 ▶
Converting 1-bromohexane to
hexane

$$CH_3(CH_2)_4CH_2Br \xrightarrow{\text{Mg}} CH_3(CH_2)_4CH_2MgBr$$

1-bromohexane   hexyl magnesium bromide

$$\downarrow H_2SO_4(aq)$$

$$CH_3(CH_2)_4CH_3$$

hexane

Chemists have long been intrigued by the idea of making compounds which have previously neither been prepared nor isolated from natural sources. Cyclic compounds have had a special fascination. Towards the end of the 19th century many chemists thought that it was unlikely that compounds with rings of 3 or 4 carbon atoms could exist because they would be so highly strained.

In 1883 William Perkin was the first person to announce a synthesis of a derivative of cyclobutane. He made cyclobutane carboxylic acid by a six step process. At that stage he could not find a way to convert it to cyclobutane itself. A satisfactory way of converting the acid to cyclobutane was not discovered until 1949. The conversion depended on making a Grignard reagent to replace a bromine atom by a hydrogen atom.

Figure 4.17.4 ▶
A three-step process to convert
the silver salt of cyclobutane
carboxylic acid to cyclobutane

$$\begin{array}{c} CH_2 - CH_2 - CO_2^-Ag^+ \\ | \qquad | \\ CH_2 - CH_2 \end{array} \xrightarrow[\text{heat}]{\text{Br}_2} \begin{array}{c} CH_2 - CH_2 - Br \\ | \qquad | \\ CH_2 - CH_2 \end{array}$$

$$\downarrow \text{Mg in ether}$$

$$\begin{array}{c} CH_2 - CH_2 \\ | \qquad | \\ CH_2 - CH_2 \end{array} \xleftarrow[\text{hydrolysis}]{\text{H}_2\text{SO}_4(aq)} \begin{array}{c} CH_2 - CH_2 - MgBr \\ | \qquad | \\ CH_2 - CH_2 \end{array}$$

cyclobutane   Grignard reagent

Test yourself

1 Suggest reasons why chemists thought that that a cyclobutane ring would be too unstable to exist.

Alcohols from aldehydes and ketones

The nucleophilic addition reactions of Grignard reagents with aldehydes and ketones produce primary, secondary and tertiary alcohols.

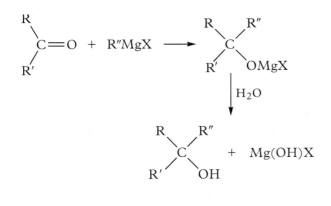

◀ **Figure 4.17.5**
X stands for a halogen atom. R and R' represent alkyl groups or hydrogen. The first step of the reaction forms an intermediate which must be decomposed to the required product with dilute acid. Note the use of $Mg(OH)X$ to represent the magnesium salts formed in the second step for simplicity when balancing equations. These compounds are ionic and it is more likely that they are mixtures of various compounds such as $Mg(OH)_2$ and MgX_2

Organic Chemistry

Section four

Test yourself

2 a) Give the name and structure of the alcohol formed when a Grignard reagent made from 1-bromopropane reacts with methanal and the product is hydrolysed by dilute acid.

b) Is the product a primary, secondary or tertiary alcohol?

3 a) Give the name and structure of the alcohol formed when 2-bromopropane is first converted to a Grignard reagent then mixed with ethanal and then the product is hydrolysed with dilute acid.

b) Is the product a primary, secondary or tertiary alcohol?

4 Identify a halogenoalkane and a ketone which could be used to make the tertiary alcohol 3-methylhexan-3-ol using a Grignard reagent as an intermediate. Write out a reaction scheme for the synthesis.

Carboxylic acids from carbon dioxide

Grignard discovered that a solution of a Grignard reagent will absorb carbon dioxide. The reagent reacts with the gas. Adding dilute hydrochloric acid to the product produces a carboxylic acid.

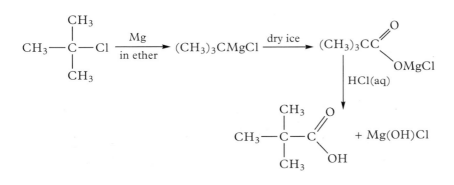

◀ **Figure 4.17.6**
Formation of a carboxylic acid from a Grignard reagent

A convenient procedure is to pour a solution of a Grignard reagent in ether over crushed dry ice (solid carbon dioxide) which sublimes at −78.5 °C. This technique for preparing acids is of more general use than syntheses using cyanide. The cyanide method works well for primary halogenoalkanes but with secondary and tertiary compounds alkenes form as by-products and the yield is low.

Test yourself

5 Suggest reasons for the low yield of nitriles when cyanide ions react with secondary and tertiary halogenoalkanes.

6 Identify the steps needed to convert 1-bromopropane into:

a) butan-1-ol

b) butanoic acid

c) 3-methylhexan-3-ol

d) 2-methyl-pentan-2-ol

e) propane.

4.18 Synthetic techniques

Chemists have developed a range of practical techniques for the synthesis of solid and liquid organic products. These methods allow for the fact that reactions involving covalent molecules are often slow and that it can be hard to avoid the side reactions which produce by-products.

Organic preparations take place in five main stages.

Stage 1: Planning

The starting point is to choose an appropriate reaction or series of reactions (see section 4.16). The next step is to work out suitable reacting quantities from the equation and to decide on the conditions for reaction.

An important part of the planning stage is a risk analysis to make sure that there are no unnecessary hazards and check that the procedure is carried out as safely as possible.

Stage 2: Carrying out the reaction

At this stage the chemist measures out the reactants and mixes them in a suitable apparatus. Most organic reactions are slow at room temperature so it is usually necessary to heat the reactants using a flame, heating mantle or hotplate. One of the commonest techniques is to heat the reaction mixture in a flask fitted with a reflux condenser (see page 182).

Organic reagents do not mix with aqueous reagents and so another common technique is to shaking the immiscible reactants in a stoppered container.

Stage 3: Separating the product from the reaction mixture

Chemists talk of 'working up' the reaction mixture to obtain their crude product. If the product is a solid it can be separated, after cooling by vacuum filtration (see Figure 4.9.7 on page 182).

Liquids can often be separated by simple distillation (see page 184) or steam distillation. Distillation with steam at 100 °C makes it possible to separate compounds which decompose if heated at their boiling points. The technique only works for compounds which do not mix with water. When used to separate products of organic preparations steam distillation leaves behind any reagents and products which are soluble in water.

Test yourself D

1 A possible two-step synthesis of methyl butanoate first converts a carboxylic acid to an acyl chloride and then mixes the acid chloride with an alcohol.

a) Write out a reaction scheme for the synthesis giving reagents and conditions.

b) Calculate the quantities needed to make 10 g of the ester assuming a 55% yield at each stage. Which chemicals might be used in excess?

c) What safety precautions should be taken during the preparation?

◄ *Figure 4.18.1*
Setting up a distillation apparatus to separate chemicals synthesised during research to develop new anti-cancer drugs

Stage 4 Purifying the product

The 'crude' product separated from a reaction mixture is generally heavily contaminated with reactants and by-products.

Purifying solids

The usual technique for purifying solids is recrystallisation (see Figure 4.7.12 on page 174). The procedure for recrystallisation is based on a solvent which dissolves the product when hot but not when cold. The choice of solvent is usually made by trial and error. Use of a Buchner or Hirsch funnel and flask speeds filtering and makes it easier to recover the purified solid from the filter paper.

The procedure is as follows:

- warm the impure solid with the hot solvent
- if the solution is not clear, filter the hot solution though a heated funnel to remove insoluble impurities
- cool the solution so that the product recrystallises leaving the smaller amounts of soluble impurities in solution
- filter to recover the purified product
- wash the solid with small amounts of pure solvent to wash away the solution of impurities
- allow the solvent to evaporate in a stream of air and then in a desiccator.

Purifying liquids

Chemists often purify liquids by shaking with reagents to extract impurities in a separating funnel. This is followed by drying and fractional distillation.

Fractional distillation separates mixtures of liquids with different boiling points. On a laboratory scale, the process takes place in distillation apparatus which has been fitted with a glass fractionating column between the flask and the still-head. Separation is improved if the column is packed with inert glass beads or rings to increase the surface area where rising vapour can mix with condensed liquid running back to the flask. The column is hotter at the bottom and cooler at the top. The thermometer reads the boiling temperature of the compound passing over into the condenser.

Figure 4.18.2 ▶
Apparatus for fractional distillation of a mixture of methanol and water. Methanol boils at a lower temperature so it is the more volatile substance. Boiling any mixture of the liquids produces a vapour which is richer in methanol. The graph to the left is a boiling point-composition diagram inverted to be hotter at the bottom than at the top

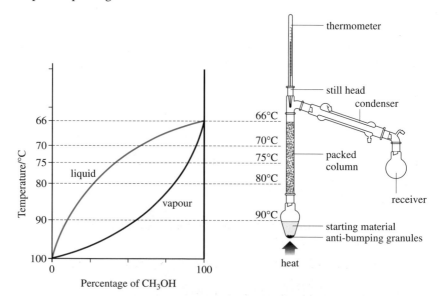

If the flask contains a mixture of two liquids, the boiling liquid in the flask produces a vapour which is richer in the more volatile of the liquids (the one with the lower boiling point).

Most of the vapour condenses in the column and runs back. As it does so it meets more of the rising vapour. Some of the vapour condenses. Some of the liquid evaporates. In this way the mixture repeatedly evaporates and condenses as it rises up the column. It is like carrying out a series of simple distillations. This can be represented as a series of steps in the boiling-point composition diagram. A horizontal line on the diagram shows the composition of the vapour formed when a particular liquid evaporates. A vertical line shows that a vapour condenses to a liquid with the same composition.

Stage 5: Measuring the yield plus checking the identity and purity of the product

Yields

Comparing the actual yield with the yield expected according to the chemical equation is a good measure of the efficiency of the process.

Definitions

The **theoretical yield** is the mass of product expected assuming that the reaction goes according to the chemical equation and the synthesis is 100% efficient.

The **actual yield** is the mass of product obtained.

The **percentage yield** is given by this relationship:

$$\text{percentage yield} = \frac{\text{actual mass of product}}{\text{theoretical yield}} \times 100\%$$

Worked example

What is the theoretical yield of the amino acid glycine when 15.5 g chloroethanoic acid reacts with excess aqueous ammonia? What is the percentage yield if the actual yield of glycine is 7.9 g?

Notes on the method

Start by writing the equation for the reaction. This need not be the full balanced equation so long as the equation includes the limiting reactant and the product.

Since the ammonia is in excess the limiting reactant is the chloroethanoic acid. This means that the ammonia can be ignored during the calculation.

Answer

The equation: $ClCH_2CO_2H \xrightarrow{NH_3(aq)} H_2NCH_2CO_2H$

The molar mass of chloroethanoic acid, $ClCH_2CO_2H = 94.5 \text{ g mol}^{-1}$

The amount of chloroethanoic acid at the start of the synthesis

$$= \frac{15.5 \text{ g}}{94.5 \text{ g mol}^{-1}} = 0.164 \text{ mol}$$

1 mol of the acid produces 1 mol of glycine. The molar mass of glycine, $H_2NCH_2CO_2H = 75 \text{ g mol}^{-1}$

The theoretical yield of glycine = 0.164 mol $\times$ 75 g mol^{-1} = 12.3 g

Percentage yield = $\frac{7.9 \text{ g}}{12.3 \text{ g}} \times 100\% = 64\%$

Definitions

A **volatile liquid** evaporates easily turning to a **vapour**. Vapours are gases formed by evaporation of substances which are usually liquids or solids at room temperature. Chemists talk about 'hydrogen gas' but 'water vapour'. Vapours are easily condensed by cooling or higher pressure because of relatively strong intermolecular forces.

Test yourself **D**

2 Explain why the actual yield from an organic synthesis is always less than the theoretical yield. Give three or four reasons.

3 A two-step synthesis converts 18 g benzene first to 22 g nitrobenzene and then to 12 g phenylamine.

  a) State the reagents and conditions for each step.

  b) Calculate the theoretical and percentage yield for each step.

  c) What is the overall percentage yield?

Identifying the product

■ Qualitative tests

Simple tests for functional groups can help to confirm the identity of the product (see *AS Chemistry* page 246).

■ Measuring melting and boiling points

Pure solids have sharp melting points. Impure solids soften and melt over a range of temperatures. So watching a solid melt can often show whether or not it is pure. Since databases now include the melting points of all known compounds it is possible to check the identity and purity for a product by checking that it melts sharply at the expected temperature.

Boiling points similarly provide a check on the identity of liquids. Chemists can read off the boiling point of a liquid product as it distils over during fractional distillation.

■ Chromatography

Thin-layer chromatography (tlc) is quick, cheap and only needs a very small sample for analysis. The technique is widely used both in research laboratories and in industry. Tlc can be used quickly to check that a chemical reaction is going as expected and making the required product. After attempting to purify a chemical, tlc can show whether or not all the impurities have been removed.

In this type of chromatography the stationary phase is a thin layer of a solid supported on a glass or plastic plate. The mobile phase is a liquid. The rate at which a sample moves up a tlc plate depends on the equilibrium between adsorption on the solid and solution in the solvent. The position of equilibrium varies from one compound to another so the components of a mixture separate.

Figure 4.18.3 ►
Thin layer chromatography. A sample product is spotted at Q. The spots of solution at P, R and S are reference compounds **CD-ROM**

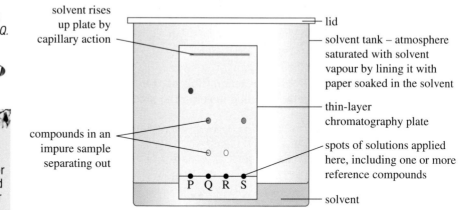

Definitions

All types of **chromatography** have a **stationary phase** (a solid or a liquid held by a solid) and a **mobile phase** (a liquid or a gas). The components of a mixture separate as the mobile phase moves through the stationary phase. Components which tend to mix with the mobile phase move faster. Components which tend to be held by the stationary phase move slower.

Coloured compounds are easy to see on a tlc plate. However, the technique is often used with colourless compounds as well. A quick way of finding the position of colourless organic spots is to stand the plate in a covered beaker with iodine crystals. The iodine vapour stains the spots.

An alternative is to use a tlc plate impregnated with a fluorescent chemical. Under a UV lamp the whole plate glows except in the areas where organic compounds absorb radiation so that they show up as dark spots.

Organic Chemistry

Section four

■ Infra-red spectroscopy

Infra-red spectroscopy is used routinely to check on the identity and purity of products (see pages 231–233). Chemists have built up libraries of spectra which can now be stored electronically and accessed on the Internet. Checking that the spectrum of a product corresponds with the version in the data base shows that the product is as expected and free of impurities.

Shake 20 cm³ of 2-methylpropan-2-ol with 70 cm³ of concentrated hydrochloric acid in a stoppered flask for about 20 minutes until an upper layer of product forms.	Transfer to a separating funnel and run off the aqueous layer. Add 20 cm³ of 0.1 mol dm⁻³ sodium hydrogencarbonate. Stopper and shake, taking care to release gas as the pressure builds up in the funnel. Run off the aqueous layer and repeat the treatment until no more gas forms.	Transfer the product to a small conical flask and add a little anhydrous sodium sulfate. Stopper and leave to stand for a few minutes.	Filter the product into a small distillation flask. Distil the liquid and collect the fraction boiling between 50 °C and 52 °C.

Figure 4.18.4 ▲
A flow diagram describing the preparation of 2-chloro-2-methylpropane

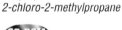

Test yourself

4 **a)** Translate Figure 4.18.4 into labelled diagrams to describe the preparation.

b) Write an equation for the reaction and classify the type of reaction.

c) Explain the purpose of:

(i) adding sodium hydrogencarbonate solution

(ii) adding anhydrous sodium sulfate

(iii) filtering.

d) Suggest chemical tests to show that the product:

(i) is a chloroalkane

(ii) is free of the starting alcohol.

4.19 Drugs for medicines

Chemists seek to discover and synthesise drugs which prevent or cure diseases, or at least alleviate the symptoms. All living things are made up of cells. Drugs act on cells and on the molecules within them.

Drugs and proteins

Most drugs interact with protein molecules. There are two main reasons for this. One reason is that all the chemical reactions in living organisms are controlled by enzymes, which are natural catalysts consisting of proteins. The other reason is that proteins make up the sensitive receptors on the surface of cells. These receptors make cells responsive to the chemical messengers from nerve endings and to hormones.

The structure of proteins (see page 200) means that enzymes and receptors have active sites. The shape of an active site and the functional groups arranged around it give the protein its chemical properties.

Figure 4.19.1 ▶
Computer graphic of an anti-HIV drug (yellow) blocking the active site of the enzyme, reverse transcriptase (green). The drug stops the virus from reproducing.

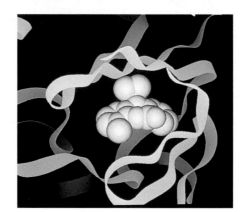

One of the ways in which drugs act is by targeting the active site of a specific protein. Many drugs which kill bacteria or stop their growth work by targetting enzymes. The sulfonamide drugs and penicillin antibiotics work in this way. What is crucial is that the drug affects an enzyme vital to the biochemistry of bacteria but does not damage any enzymes in human beings.

Drugs which target receptor proteins are also very important and they include the active ingredients in medicines to treat pain, heart failure, asthma and Parkinson's disease.

Stereochemistry and drugs

The messenger and carrier molecules in the body all interact selectively with the active sites of enzymes and with protein receptors. These molecules are all chiral (see page 149) but the chemistry of the body works with only one of the mirror image forms. This means that most drugs are chiral too and it is often the case that the two optical isomers behave in different ways in the body. Often one isomer is active while the other isomer is inactive but this is not always the case (see Figure 4.3.11 on page 151).

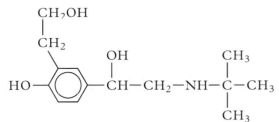

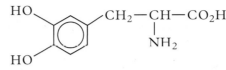

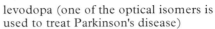

salbutamol
(used to treat asthma)

levodopa (one of the optical isomers is
used to treat Parkinson's disease)

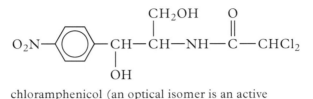

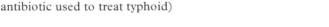

chloramphenicol (an optical isomer is an active
antibiotic used to treat typhoid)

◀ *Figure 4.19.2*
Drugs which are chiral

The pharmaceutical industry was alerted to the importance of chirality in the 1960s. A new drug, thalidomide, appeared on the market in some countries as a safe new sedative and sleeping drug. Soon it emerged that thalidomide could cause serious malformations in babies if taken by women in an early stage of pregnancy.

Thalidomide is chiral. The drug was a racemic mixture of the two isomers (see page 150). Pharmacologists have since discovered that one optical isomer (the + form) is an effective and harmless sedative while the other isomer (the – form) is teratogenic. It turns out that supplying only the + form as the drug would not have been safe either since the body can convert one form to the other.

Licensing authorities now require pharmaceutical companies to distinguish the optical isomers of drugs and test them separately. Increasingly medicines contain just one of the two optical isomers. This means that the dose can be smaller, side effects from the unwanted isomer are eliminated and the total dose can be reduced.

> **Definition**
>
> A **teratogenic** chemical is one which is harmful to babies in the womb.

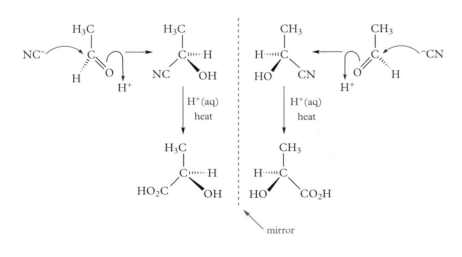

◀ *Figure 4.19.3*
A laboratory synthesis of 2-hydroxypropanoic acid (lactic acid) produces a racemic mixture

> **Test yourself**
>
> 1 Identify the chiral centres in the molecules shown in Figure 4.19.2.
>
> 2 Explain in words, with the help of Figure 4.19.3, why a laboratory synthesis of a chiral compound usually produces a racemic mixture.

The absorption and distribution of drugs in the body

A drug has to be more than effective in curing disease. It also has to have the right properties to be absorbed into the bloodstream, carried round the body's circulatory system and cross the biological barriers on its way to the target enzymes or receptors.

Figure 4.19.4 ▶
A molecular picture of a cell membrane

membrane proteins

phospholipid layer

fatty acid tails to phospholipid molecules

hydrophilic heads to phospholipid molecules

Ideally drugs are taken by mouth in the form of tablets or capsules. This is only possible if the drug can survive in the stomach, which is highly acidic, cross from the gut into the bloodstream and then reach its target without being broken down in the liver or excreted by the kidneys.

Cells membranes consist largely of fat. This means that a drug has to be sufficiently fat-loving (lipophilic) if it is to cross from the gut to the bloodstream. However the fat in the body will extract any drug which is too lipophilic from the bloodstream so that it ceases to be available as an active chemical.

Functional groups which make a molecule hydrophilic (water loving) are: $-OH$, $-NH_2$, $-CO_2H$, $-CO_2^-$.

Functional groups which make a molecule lipophilic (fat loving) include: alkyl and aryl (hydrocarbon) side chains.

A drug which is too water-loving (hydrophilic) may be unable to cross the fatty barriers of cell membranes. It may also be quickly excreted by the kidney into the urine.

Drug design therefore has to be a compromise. One solution is to have a molecule which is a weak acid or weak base with a pK_a in the range 6-8. This means that, at the pH of the blood, the drug in solution is an equilibrium mixture of non-ionised and ionised forms. In its non-ionised form it can cross membranes. In its ionised form it can dissolve in body fluids and bind to an active site.

Penicillin is very effective at killing bacteria. It works by stopping bacteria from synthesising their cell walls. The bacterial cells burst and die. Human cells do not have a cell wall and so they are unaffected by penicillin which is remarkably non-toxic.

The original penicillin–G isolated from a fungus was hydrolysed by acid and hydrophilic. As a result it could not be taken by mouth and was rapidly excreted by the kidneys.

Chemists and pharmacologists have developed a range of semi-synthetic variations of the basic penicillin molecule. These modification can make the molecule stable to acid and less hydrophilic. One example is Pivampicillin.

Test yourself

3 This is the structure of aspirin:

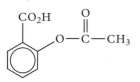

Aspirin is often taken in its more soluble form as a sodium salt.

a) Is aspirin a weak acid or a strong acid?

b) Predict what will happen to soluble aspirin in the acidic conditions of the stomach. Write an equation for the reversible reaction.

c) Which form of aspirin would you expect to be absorbed more quickly from the stomach into the bloodstream?

d) Which form of aspirin would you expect to be excreted more quickly from the kidneys into the urine?

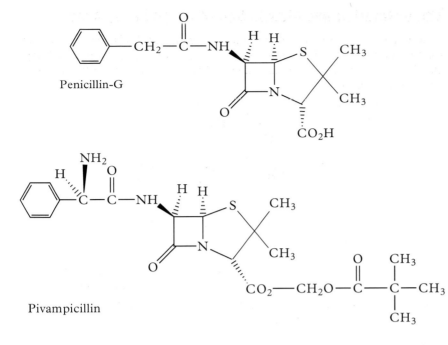

◀ **Figure 4.19.5**
Structures of Penicillin–G and
Pivampicillin

CD-ROM

Penicillin-G

Pivampicillin

Pivampicillin can be taken by mouth because the modifications to its structure mean that it is not broken down by acid. The electronegative nitrogen atom in the amine group makes the drug more stable in acid conditions. Turning the polar carboxylic acid group into an ester makes the molecule less hydrophilic so that it is absorbed from the stomach. Once absorbed into the bloodstream, hydrolysis of the ester link releases the active drug (Ampicillin).

Test yourself

4 a) Identify these functional groups in Penicillin–G: amide, carboxylic acid.

b) Identify these functional groups in Pivampicillin: ester, amide, amine.

c) Draw the structure of Ampicillin. How does this drug differ from Penicillin–G?

Organic Chemistry

Section four

4.20 Instrumental analysis

In a modern laboratory organic analysis is based on a range of automated and instrumental techniques including mass spectrometry, ultraviolet spectroscopy, infra-red spectroscopy, and nuclear magnetic resonance spectroscopy.

CD-ROM

Mass spectrometry

Mass spectrometry is an accurate technique for determining relative molecular masses. Mass spectrometry can also help to determine molecular structures and to identify unknown compounds. The technique is extremely sensitive and requires very small samples which can be as small as one nanogram (10^{-9} g).

The combination of gas-liquid chromatography (glc) with mass spectrometry is of great importance in modern chemical analysis. First glc separates the chemicals in an unknown mixture, such as a sample of polluted water; then mass spectrometry detects and identifies the components.

The instrument

Inside a mass spectrometer there is a very high vacuum so that it is possible to produce and study ionised molecules and fragments of molecules which do not otherwise exist.

Figure 4.20.1 ▶
Diagram of a mass spectrometer showing the stages in the production of a mass spectrum. The instrument is calibrated with a reference compound with a known structure and molecular mass so that the computer can print a scale on the mass spectrum

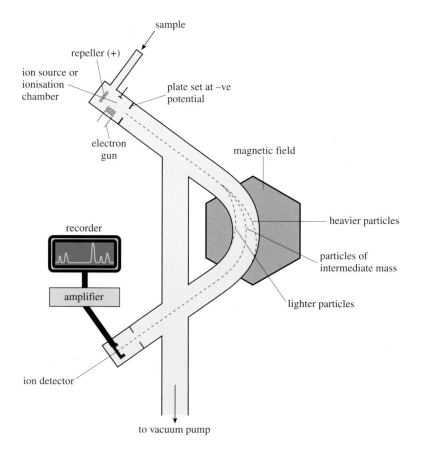

Definition

The **mass to charge ratio** (*m/z*) is the ratio of the relative mass of the ion to its charge where *z* is the number of charges (1, 2 and so on). In most simple examples $z = 1$.

The high energy electrons bombard the molecules of the sample and turn them into ions by knocking out one or more electrons.

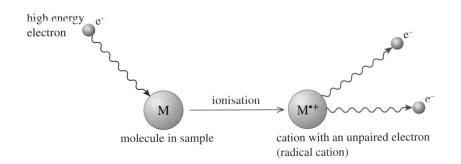

◀ **Figure 4.20.2**
High energy electrons ionising a molecule. Knocking out one electron leaves an unpaired electron behind so the result is a positive ion which is also a free radical. Hence the symbol M•+

Relative molecular masses

When analysing molecular compounds, the peak of the ion with the highest mass is usually the whole molecule ionised. So the mass of this 'parent ion', M, is the relative molecular mass of the compound.

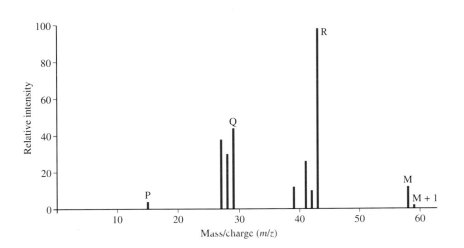

◀ **Figure 4.20.3**
Mass spectrum of butane, C_4H_{10} showing the parent ion, M, with a value 58. This corresponds to the relative molecular mass of 58. The pattern of fragments is characteristic of this compound

Isotopes

The very small peak at 59 in the spectrum of butane (Figure 4.20.3) is the 'M + 1' peak which is present because some molecules of butane include carbon–13 atoms. Carbon–13 makes up 1.1% of natural carbon so in a molecule with 4 carbon atoms the 'M + 1' peak is 4.4% of the molecular ion peak.

The presence of isotopes is much more obvious in spectra of organic compounds containing chlorine or bromine atoms. Chlorine has two isotopes ^{35}C and ^{37}C. Chlorine-35 is three times more abundant than chlorine-37. If a molecule contains one carbon atom its molecular appears as two peaks separated by two mass units. The peak with the lower value of m/z is three times higher than the peak with the higher value.

Bromine consists largely of two isotopes, ^{79}Br and ^{81}Br, which are roughly equally abundant. If a molecule contains one bromine atom the molecular ion shows up as two peaks separated by two mass units with roughly equal intensity.

Test yourself **D**

1 a) Suggest the identity of the fragments labelled P, Q and R in the mass spectrum of butane.

b) Suggest a reason why the peak at m/z = 15 is relatively weak.

c) Use symbols to show how the molecular ion of butane fragments.

Figure 4.20.4 ▶
Mass spectra of two compounds containing halogen atoms

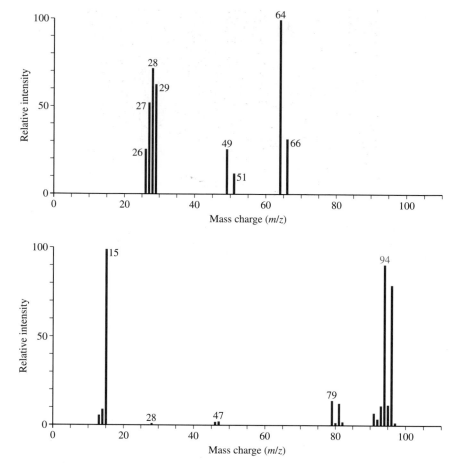

2 Account for these facts about the mass spectrum of dichloroethene:

a) It includes three peaks at *m/z* values of 96, 98 and 100 with intensities in the ratio 9:6:1.

b) It includes two peaks at *m/z* values 61 and 63 with intensities in the ratio of 3:1.

3 One of the mass spectra in Figure 4.20.4 is methyl bromide and the other is chloroethane. Match the spectra to the compounds and identify as many fragments in the spectra as you can.

Fragmentation

Bombarding molecules with high energy electrons not only ionises them but can also split them into fragments. As a result the mass spectrum consists of a 'fragmentation pattern'.

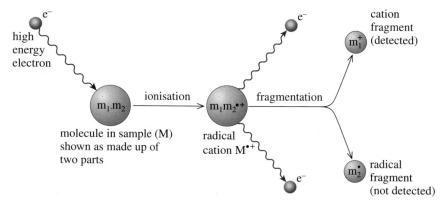

Figure 4.20.5 ▲
Ionisation and fragmentation. Note that only charged species show up in the mass spectrum. Electric and magnetic fields have no effect on neutral fragments

Molecules break up more readily at weak bonds or at bonds which give rise to stable fragments. The highest peaks correspond to positive ions which are relatively more stable such as tertiary carbocations or ions such as RCO^+ (the acylium ion) or the fragment $C_6H_5^+$ from aromatic compounds.

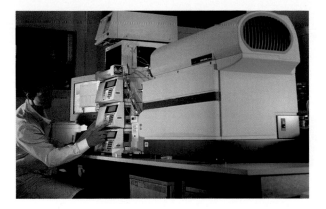

◄ **Figure 4.20.6**
Mass spectrometry is here being used to measure the levels of drugs in the blood as part of a programme of cancer research. This instrument combines liquid chromatography to separate the mixtures and mass spectrometry to identify and measure the components

Chemists study mass spectra with these ideas in mind and as a result can gain insights into the structure of molecules. They identify the fragments from their masses and then piece together likely structures with the help of evidence from other methods of analysis such as infra-red spectroscopy and nmr spectroscopy.

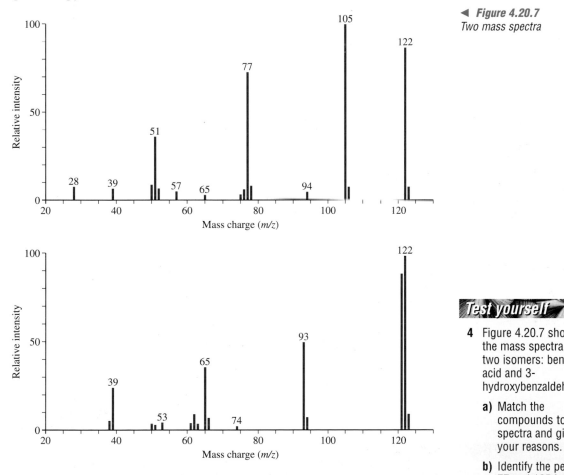

◄ **Figure 4.20.7**
Two mass spectra

Test yourself ✦ **D**

4 Figure 4.20.7 shows the mass spectra of two isomers: benzoic acid and 3-hydroxybenzaldehyde.

a) Match the compounds to the spectra and give your reasons.

b) Identify the peaks at 77 and 105 in the top spectrum.

c) Identify the peaks at 39 and 93 in the bottom spectrum.

Spectroscopy

Spectroscopy is a term which covers a range of practical techniques for studying the composition, structure and bonding of compounds. Spectroscopic techniques are now the essential 'eyes' of chemistry.

Definitions

An **absorption spectrum** is a plot showing how strongly a sample absorbs radiation over a range of frequencies. Absorption spectra from ultraviolet, visible and infra-red spectroscopy give chemists valuable information about the composition and structure of chemicals.

The instruments used are variously called **spectroscopes** (emphasising the uses of the techniques for making observations) or **spectrometers** (emphasising the importance of measurements).

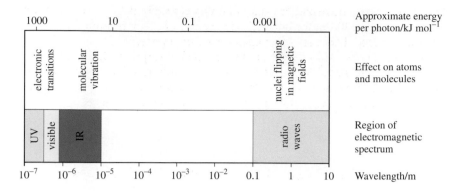

Figure 4.20.8 ▲
The uses of the electromagnetic spectrum in spectroscopy

Figure 4.20.9 ▲
The essential features of a modern single-beam spectrometer

Ultraviolet spectroscopy

Ultraviolet (UV) spectroscopy is particularly useful for studying colourless organic molecules with unsaturated functional groups such as C=C and C=O. The molecules absorb ultraviolet radiation at frequencies which excite shared electrons in double bonds. An ultraviolet spectrometer records the extent to which samples absorb UV radiation across a range of wavelengths.

Figure 4.20.10 ▶
Ultraviolet spectrum of benzene. Note that the plot of a UV spectrum shows the extent to which the radiation is absorbed. So there is a peak wherever the sample absorbs strongly

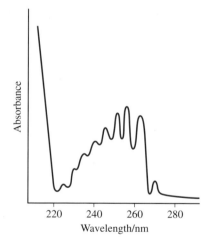

Definition

A **conjugated system** is a system of alternating double and single bonds in a molecule. Organic molecules with extended conjugated systems absorb radiation in the visible part of the spectrum and therefore are coloured.

UV spectra are recorded in solution. The spectra generally appear as broad bands with one or two humps. Sometimes, as with benzene, there is some fine structure superimposed on the broad bands. Chemists record the wavelength of the highest part of each peak and the relative degree of absorption.

The height of the absorption peak at a particular wavelength varies with the concentration of the absorbing substance. This means that UV spectroscopy can be used to estimate amounts of substance in solution for colourless substances just as colorimetry can be used for coloured molecules and ions (see pages 127–128).

In organic molecules with conjugated systems the UV absorption peak moves to longer wavelengths as the number of alternating double and single bonds increases. The maximum absorption by ethene is at $\lambda_{max} = 185$ nm, this shifts to $\lambda_{max} = 220$ nm for buta-1,3-diene and to $\lambda_{max} = 263$ nm for octa-2,4,6-triene. These are three colourless compounds.

The longer the conjugated chain the stronger absorption. Beta-carotene with eleven conjugated carbon–carbon double bonds absorbs strongly with a peak shifted so far that it is not in the UV region but lies in the blue region of the visible spectrum ($\lambda_{max} = 450$ nm) so carotene is bright orange.

◀ *Figure 4.20.11*
The structure of β–carotene

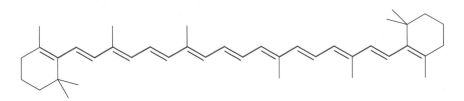

The non-bonding electrons of oxygen and nitrogen atoms in functional groups such as $-OCH_3$, $-NO_2$ and $-CO_2CH_3$ can also extend the conjugated system and so shift the peak of absorption towards longer wavelengths.

UV spectrometers make it possible to extend the techniques of colorimetry to colourless compounds. In the pharmaceutical industry, for example, scientists use UV spectroscopy to check that medicines contain the correct amounts of drugs and that the products do not deteriorate in storage.

Test yourself

5 Estimate λ_{max} for the polyalkene, $C_{10}H_{14}$ with four conjugated double bonds and decide whether or not you would expect the compound to colourless.

6 A UV spectrometer was calibrated with a series of standard solutions of naphthalene. The calibration results were:

Concentration/mol dm^{-3}	Absorbance
3.00×10^{-4}	0.232
5.00×10^{-4}	0.379
7.00×10^{-4}	0.524
9.00×10^{-4}	0.672

The instrument reading for an unknown sample was 0.472 mol dm^{-3}. Estimate the concentration of naphthalene in this sample.

Infra-red (IR) spectroscopy

Infra-red (IR) spectroscopy is an analytical technique used to identify functional groups in organic molecules. Most compounds absorb IR radiation. The wavelengths they absorb correspond to the natural frequencies at which vibrating bonds in the molecules bend and stretch. It is polar bonds which absorb particularly strongly as they vibrate. Examples of polar bonds in organic molecules are O–H, C–O and C=O.

◀ *Figure 4.20.12*
Bond vibrations which give rise to absorptions in the infra-red region

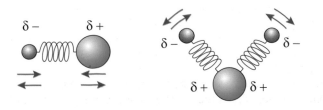

Definitions

Infra-red **wavenumbers** range from 400 cm^{-1} to 4500 cm^{-1}. Spectroscopists find the numbers more convenient than wavelengths. The wavenumber is the number of waves in 1 cm.
Transmittance on the vertical axis measures the percentage of radiation which passes through the sample. The troughs appear at the wavenumbers which the compound absorbs strongly.

Bonds vibrate in particular ways and absorb radiation at specific wavelengths. This means that it is possible to look at an infra-red spectrum and identify particular functional groups.

Spectroscopists have found that it is possible to correlate the absorptions in the region 4000–1500 cm^{-1} with the stretching or bending vibrations of particular bonds. As a result the infra-red spectrum gives valuable clues to the presence of particular functional groups in organic molecules (see Figure 4.20.13 and page 251 in the reference section).

Hydrogen bonding broadens the absorption peaks of –OH groups in alcohols and even more so in carboxylic acids.

Molecules with several atoms can vibrate in many ways because the vibrations of one bond affect others close to it. The complex pattern of vibrations can be used as a 'fingerprint' to be matched against the recorded IR spectrum in a database. Comparing the IR spectrum of a product of synthesis with the spectrum of the pure compound can be used to check that the product is pure.

Figure 4.20.14 ►
Chart to show the regions of the infra-red spectrum and important correlations between bonds and observed absorptions (see also page 251).

Wave number ranges

4000 cm^{-1}	2500 cm^{-1}	1900 cm^{-1}	1500 cm^{-1}	650 cm^{-1}
C – H O – H N – H single bond stretching vibrations	C ≡ C C ≡ N triple bond stretching vibrations	C = C C = O double bond stretching vibrations		fingerprint region

Test yourself   D

7 Figure 4.20.13 shows the infra-red spectra of ethanol, ethanoic acid, propanenitrile and ethyl butanoate.

   a) Which vibrations give rise to the peaks marked with letters?

   b) Which spectrum belongs to which compound?

8 A compound P is a liquid which does not mix with water; its molecular formula is C_7H_6O and it has an infra-red spectrum with strong sharp peaks at 2800 cm^{-1}, 2720 cm^{-1} and 1700 cm^{-1} and a weaker absorption peak between 3000 and 3100 cm^{-1}. Oxidation of P gives a white crystalline solid Q with a strong broad IR absorption band in the region 2500–3300 cm^{-1} and another strong absorption at 1680–1750 cm^{-1}.

   a) Suggest possible structures for P and Q.

   b) What chemical tests could you use to check on your suggestions?

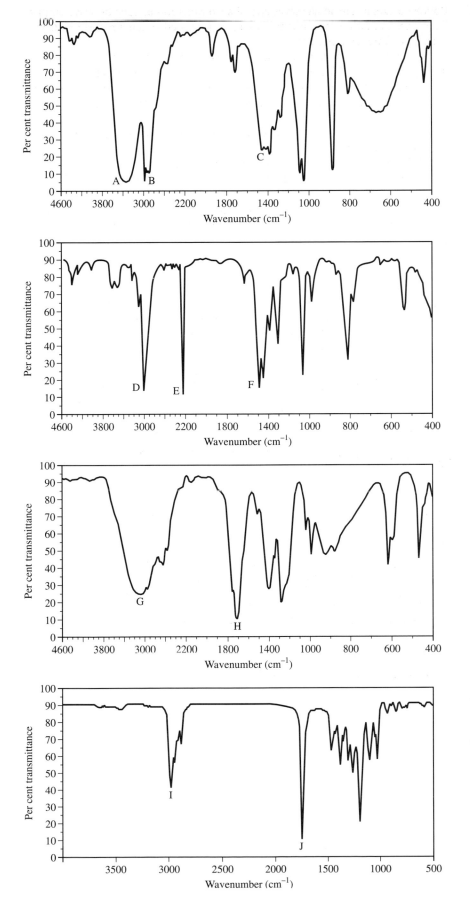

◀ *Figure 4.20.13*
Infra-red spectrum of four organic compounds. The units along the bottom are wavenumbers (in cm⁻¹). Note that the quantity plotted on the vertical axis is transmittance. This means that the line dips at wavenumbers at which the molecules absorb radiation. Chemists often refer to these dips in the line as 'peaks' because they indicate high levels of absorption

Note

In medicine, magnetic resonance imaging uses nmr to detect the hydrogen nuclei in the human body, especially in water and lipids. A computer translates the information from a body scan into 3-D images of the soft tissue and internal organs which are normally transparent to X-rays.

Figure 4.20.15 ▶
Diagram of an nmr spectrometer showing the key features of the technique

Nuclear magnetic resonance spectroscopy (nmr)

Nuclear magnetic resonance spectroscopy (nmr) is a powerful analytical technique for finding the structures of carbon compounds. The technique is used to identify unknown compounds, to check for impurities and to study the shapes of molecules.

Proton nmr

This type of spectroscopy studies the behaviour of the nuclei of atoms in magnetic fields. It is limited to those nuclei which behave like tiny magnets. In common organic compounds the only nucleus to do so is the proton, ^{1}H. The nuclei of carbon-12, oxygen-16 and nitrogen–14 do not show up in nmr spectra.

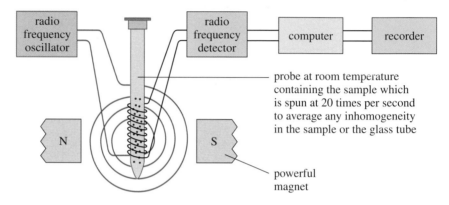

probe at room temperature containing the sample which is spun at 20 times per second to average any inhomogeneity in the sample or the glass tube

powerful magnet

Figure 4.20.16 ▲
Scientists at the controls of an nmr spectrometer being used to study a carboxylic acid

When placed in a very strong magnetic field, the nuclei of hydrogen–1 atoms can line up either in the same direction or in the opposite direction to an external magnetic field. There is a small energy jump when the protons flip from one alignment to the other. The size of the energy jump corresponds to the energies of quanta of radiowaves.

The sample is dissolved in a solvent with no hydrogen-1 atoms. Also in the solution is some tetramethyl silane (TMS) which is a standard reference compound which produces a single, sharp absorption peak well away from the peaks produced by samples for analysis.

The tube with the sample is supported in a strong magnetic field. The operator turns on an oscillator which produces radiation at radio-frequencies. The radio-frequency detector records the intensity of the signal as the oscillator scans across a range of wavelengths. The recorder prints out a spectrum with peaks wherever the sample absorbs radiation strongly. The zero on the scale is fixed by the absorption of hydrogen atoms in the reference chemical TMS. The distances of the sample peaks from this zero are called their 'chemical shifts' (δ).

Each peak corresponds to one or more hydrogen nuclei in a particular chemical environment. Nuclei in different parts of a molecule experience slightly different magnetic fields in an nmr machine because they are shielded to a greater or lesser extent from the field applied by the spectrometer by the tiny magnetic fields associated with the electrons of neighbouring bonds and atoms.

The area under a peak is proportional to the number of nuclei in each chemical environment. The instrument can automatically integrate the curve and plot a line showing the relative numbers of protons in each chemical environment in the molecule.

Chemists have measured chemical shifts for protons in many organic compounds and tabulated values (see page 251).

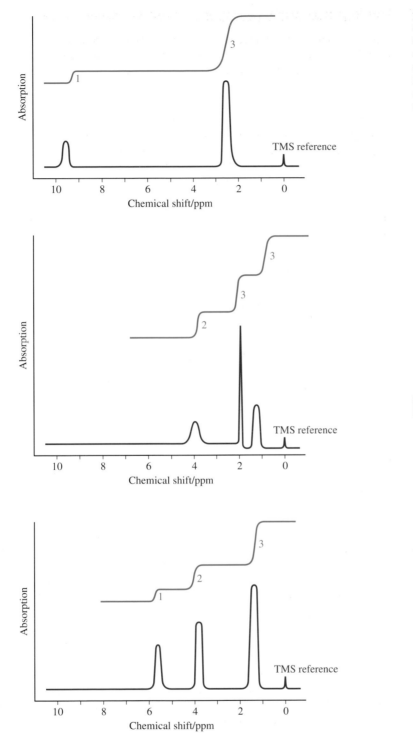

◄ **Figure 4.20.17**
Low resolution nmr spectrum of three organic compounds. Note the green line which is the result of integrating the signal from the instrument to measure the relative area under each peak.

Test yourself **D**

9 Match the nmr spectra in Figure 4.20.17 with these compounds: ethanol, ethanal and ethyl ethanoate. Give your reasons.

10 Why does the reference compound tetramethyl silane $(CH_3)_4Si$ give rise to a single sharp nmr peak?

11 With the help of the table of chemical shifts on page 251, sketch the low-resolution nmr spectrum you would expect for:

a) butanone

b) 2-methylpropan-2-ol.

Note
Protons with the same chemical shift do not couple with each other.

Figure 4.20.18 ▶
High resolution nmr spectrum for an aromatic hydrocarbon. Note the extra peaks compared with the low-resolution spectrum

Coupling

At high resolution it is possible to produce nmr spectra with more detail which provide even more information about molecular structures. Protons connected to neighbouring atoms interact with each other. Chemists call this interaction 'coupling' and they find that the effect is to split the peaks into a number of lines.

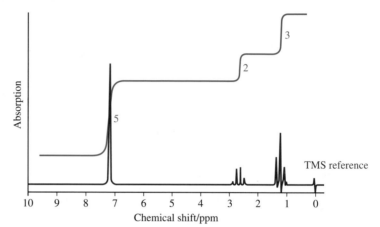

In simple examples the 'n + 1' rule makes it possible to work out the numbers of coupled protons. A peak from protons bonded to an atom which is next to an atom with two protons splits into three lines. A peak from protons bonded to an atom which is next to an atom with three protons splits into four lines.

Figure 4.20.19 ▶
Pascal's triangle predicts the pattern of peaks and the relative peak heights

number of equivalent protons causing splitting	splitting pattern and relative intensity of the peaks
1	1 1
2	1 2 1
3	1 3 3 1
4	1 4 6 4 1

Test yourself — D

12 Use the table of chemical shifts on page 251 to identify the compound in Figure 4.20.18.

13 Sketch the high-resolution nmr spectrum you would expect to observe with:

a) propane

b) ethoxyethane.

Labile protons

Hydrogen bonding affects the properties of compounds with hydrogen atoms attached to highly electronegative atoms such as oxygen or nitrogen. These molecules can rapidly exchange protons as they move from one electronegative atom to another. Chemists describe these protons as labile.

Definition

An atom or compound is **labile** if it quickly and easily moves or reacts.

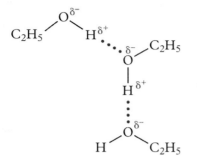

◄ *Figure 4.20.20*
Hydrogen bonding in ethanol

Labile protons do not couple with the protons linked to neighbouring atoms. This means that the nmr peak for a proton in an –OH group appears as a single peak in a high resolution spectrum.

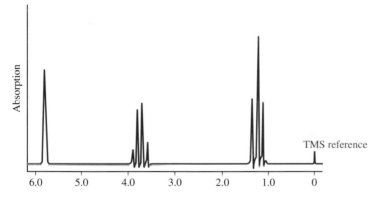

◄ *Figure 4.20.21*
The high resolution nmr spectrum of ethanol. Note that there is no coupling between the proton in the –OH group and the protons in the next door CH₂ group

A useful technique for detecting labile protons is to measure the nmr spectrum in the presence of deuterium oxide (heavy water), D_2O. Deuterium is an isotope of hydrogen, ^2H. Deuterium nuclei can exchange rapidly with labile protons. Deuterium nuclei do not show up in the proton nmr region of the spectrum and so the peaks of any labile protons disappear.

Test yourself **D**

14 Write an equation to show the reversible exchange of deuterium and hydrogen nuclei between ethanol and deuterium oxide.

15 The table shows the main features of the high resolution nmr spectrum of a compound containing carbon, hydrogen and oxygen. The peak with a chemical shift of 11.7 disappears in the presence of ^2H_2O. Deduce the structure of the hydrocarbon.

Chemical shift	Number of lines	Integration
1.2	triplet	3
2.4	quartet	2
11.7	singlet	1

Organic Chemistry

Section four

Section four study guide

Review

This guidance will help you to organise your notes and revision. Check the terms and topics against the specification you are studying. You will find that some topics are not required for your course.

Key terms

Show that you know the meaning of these terms by giving examples. Consider writing the key term on one side of an index card and the meaning of the term with an example on the other side. Then you can easily test yourself when revising. Alternatively use a computer data base with fields for the key term, the definition and the example. Test yourself with the help of reports which just show one field at a time.

- Structural isomerism
- Geometric isomerism
- Optical isomerism
- Polarised light
- Chiral compound
- Asymmetric molecule
- Enantiomer
- Racemic mixture
- Kekulé structure
- Delocalised electrons
- Lewis acid and base
- Homolytic fission
- Heterolytic fission
- Electrophile
- Carbocation
- Nitration
- Friedel–Crafts reaction
- Phenyl group
- Antiseptic
- Disinfectant
- Carbonyl group
- Nucleophile
- Crystalline derivative
- Addition–elimination reactions
- Hydrolysis
- Saponification
- Triglyceride
- Fat
- Vegetable oil
- Fatty acid
- Saturated fatty acid
- Unsaturated fatty acid
- Hofmann degradation
- Quaternary ammonium cation
- Surfactants
- Diazonium salt
- Azo dye
- Amino acid
- Zwitterion
- Polypeptide
- Protein
- Peptide bond
- Hydrogen bonding
- Addition polymerisation
- Free-radical
- Chain reaction
- Condensation reaction
- Condensation polymerisation
- Polyamide
- Polyester
- Cross linking
- Thermoplastic
- Thermoset
- Fibre
- Biodegradable
- Grignard reagent
- Molecular ion
- Mass-to-charge ratio
- Wavenumber
- Conjugated system
- Chemical shift

Symbols and conventions

Make sure that you understand the conventions which chemists use when using symbols and writing equations. Illustrate you notes with examples.

- Names of carbon compounds: alkanes, alkenes, arenes (aromatic hydrocarbons), halogenoalkanes, alcohols, phenols, aldehydes, ketones, carboxylic acids, esters, acyl chlorides, acid anhydrides, nitriles, amines, amino acids.

- Methods for representing the shapes of molecules in two and three dimensions.
- Use of curly arrows to show how bonds break and form when describing reaction mechanisms.
- Methods for representing amino acids and proteins.
- The distinctions between the primary, secondary and tertiary structure of proteins.
- Symbols used to represent monomers and polymers.
- Methods for representing atactic, isotactic and syndiotactic polymer chains.

Facts, patterns and principles

Use charts, annotated equations and spider diagrams to summarise key facts and ideas. Brighten your notes with colour to make them memorable. Make sure that you show clearly the reagents and conditions needed to bring about each reactions. Try writing the equation for a reaction on one side of an index card and the reagents and conditions on the other side. You can carry around a set of cards like this to revise from and use test yourself in odd moments of the day.

- The link between the physical properties of a series of organic compounds and the types of intermolecular forces between the molecules.
- The effects of chiral compounds on polarised light.
- The structure and stability of arenes (aromatic hydrocarbons) and their reactions.
- Reactions of aldehydes and ketones.
- Reactions of carboxylic acids and the interpretation of the acidity of the acidic of carboxylic acid groups in terms of electron delocalisation.
- Reaction of the derivatives of carboxylic acids including acyl chorides, acid anhydrides, esters and nitriles.
- Reactions of alkyl and aryl amines including an interpretation of the relative base strength of amines with different structures.
- Types of reaction mechanism: free radical chain reactions, nucleophilic addition and substitution reactions, electrophilic addition and substitution reactions; the explanation for Markovnikov's rule; addition polymerisation.

Techniques of analysis and synthesis

Use tables to summarise chemical tests and their results. Draw flow diagrams to show the key steps in practical procedures. Find examples to illustrate the significance of instrumental methods.

- Test-tube tests to identify or distinguish functional groups.
- Simple distillation and fractional distillation.
- Recrystallisation to purify a solid product.
- The significance of boiling and melting points.
- Mass spectrometry to determine relative molecular masses and molecular structures.
- Infra-red spectroscopy to identify functional groups and to characterise pure compounds.
- Ultraviolet and visible spectroscopy to measure concentrations and analyse colours.
- Nuclear magnetic resonance spectroscopy to determine molecular structures.
- Planning synthetic routes for organic compounds.

Interpretation of data

Give your own worked examples, with the help of the Test Yourself and examination questions to show that you can work out the following from given data.

- Empirical and molecular formulae.
- Theoretical and percentage yields for organic reactions.
- Molecular structures from mass spectra, IR spectra and nmr spectra.

Chemical applications

Use charts, tables or diagrams to summarise the practical applications of organic chemistry mentioned in the specification for the course you are studying.

- The importance of compounds derived from benzene and other arenes.
- The importance of naturally occurring and synthetic esters: perfumes, flavourings, fats and oils.
- The importance of organic compounds as pharmaceuticals and agrochemicals.
- The uses of polymers related to their properties and structures and the issues arising from the disposal of polymers.

Key skills

Communication

Reading, writing and talking about the applications of organic chemistry allows you to show that you can select and read material that contains information you need. You can demonstrate your ability to pick out the main points and identify lines of reasoning and then synthesise the information in a form relevant to your purpose using specialist vocabulary when appropriate.

Information technology

When studying organic chemistry you can use chemical modelling software from a CD-ROM or the Internet to study the shapes of the molecules. In this way you can explore bond angles, molecular shapes and the effects of isomerism. You can also use a database of spectra to practice interpreting the spectra from mass spectrometers, IR spectroscopes and nmr machines.

Section five

Reference

Contents

Periodic table

Period / **Group**

Key:

Atomic number
Symbol
Name
Relative atomic mass

Transition elements

1	2												3	4	5	6	7	8
1 **H** Hydrogen 1																		2 **He** Helium 4
3 **Li** Lithium 7	4 **Be** Beryllium 9												5 **B** Boron 11	6 **C** Carbon 12	7 **N** Nitrogen 14	8 **O** Oxygen 16	9 **F** Fluorine 19	10 **Ne** Neon 20
11 **Na** Sodium 23	12 **Mg** Magnesium 24												13 **Al** Aluminium 27	14 **Si** Silicon 28	15 **P** Phosphorus 31	16 **S** Sulfur 32	17 **Cl** Chlorine 35.5	18 **Ar** Argon 40
19 **K** Potassium 39	20 **Ca** Calcium 40	21 **Sc** Scandium 45	22 **Ti** Titanium 48	23 **V** Vanadium 51	24 **Cr** Chromium 52	25 **Mn** Manganese 55	26 **Fe** Iron 56	27 **Co** Cobalt 59	28 **Ni** Nickel 59	29 **Cu** Copper 63.5	30 **Zn** Zinc 65.4		31 **Ga** Gallium 70	32 **Ge** Germanium 73	33 **As** Arsenic 75	34 **Se** Selenium 79	35 **Br** Bromine 80	36 **Kr** Krypton 84
37 **Rb** Rubidium 85	38 **Sr** Strontium 88	39 **Y** Yttrium 89	40 **Zr** Zirconium 91	41 **Nb** Niobium 93	42 **Mo** Molybdenum 96	43 **Tc** Technetium	44 **Ru** Ruthenium 101	45 **Rh** Rhodium 103	46 **Pd** Palladium 106	47 **Ag** Silver 108	48 **Cd** Cadmium 112		49 **In** Indium 115	50 **Sn** Tin 119	51 **Sb** Antimony 122	52 **Te** Tellurium 128	53 **I** Iodine 127	54 **Xe** Xenon 131
55 **Cs** Caesium 133	56 **Ba** Barium 137	57 ▶ **La** Lanthanum 139	72 **Hf** Hafnium 178	73 **Ta** Tantalum 181	74 **W** Tungsten 184	75 **Re** Rhenium 186	76 **Os** Osmium 190	77 **Ir** Iridium 192	78 **Pt** Platinum 195	79 **Au** Gold 197	80 **Hg** Mercury 201		81 **Tl** Thallium 204	82 **Pb** Lead 207	83 **Bi** Bismuth	84 **Po** Polonium	85 **At** Astatine	86 **Rn** Radon
87 **Fr** Francium	88 **Ra** Radium 226	89 ▶▶ **Ac** Actinium	104 **Rf** Rutherfor-dium	105 **Db** Dubnium	106 **Sg** Seaborgium	107 **Bh** Bohrium	108 **Hs** Hassium	109 **Mt** Meitnerium	110 **Uun** Ununnilium	111 **Uuu** Unununium	112 **Uub** Ununbium							

▶ Lanthanoid elements

58 **Ce** Cerium 140	59 **Pr** Praseo-dymium 141	60 **Nd** Neo-dymium 144	61 **Pm** Promethium	62 **Sm** Samarium 150	63 **Eu** Europium 152	64 **Gd** Gadolinium 157	65 **Tb** Terbium 159	66 **Dy** Dysprosium 163	67 **Ho** Holmium 165	68 **Er** Erbium 167	69 **Tm** Thulium 169	70 **Yb** Ytterbium 173	71 **Lu** Lutetium 175

▶▶ Actinoid elements

90 **Th** Thorium 232	91 **Pa** Protactinium 231	92 **U** Uranium 238	93 **Np** Neptunium 237	94 **Pu** Plutonium	95 **Am** Americium	96 **Cm** Curium	97 **Bk** Berkelium	98 **Cf** Californium	99 **Es** Einstein-ium	100 **Fm** Fermium	101 **Md** Mendel-evium	102 **No** Nobelium	103 **Lr** Lawren-cium

Note: Relative atomic masses are shown only for elements which have stable isotopes or isotopes with a very long half-life.

Properties of elements and compounds

Atomic and ionic radii are given in picometres (1 pm = 10^{-12} m). The atomic radii are metallic radii for metals and covalent radii for non-metals (except for the noble gases for which the radii are van der Waals radii). Ionic radii are quoted for the common simple ions. Where an element forms more than one simple ion the charge on the ion is shown in brackets.

Properties of selected elements

Element	Molar mass/ g mol^{-1}	Melting point/°C	Boiling point/°C	First ionisation energy/ kJ mol^{-1}	Radii atomic/pm	ionic/pm	Enthalpy of atomisation/ kJ mol^{-1}	Standard molar entropy/ JK^{-1} mol^{-1}
Group 1								
Lithium, Li	7	181	1347	519	155	74	159	29.1
Sodium, Na	23	98	883	496	185	113	107	51.2
Potassium, K	39	64	774	419	230	133	89	64.2
Group 2								
Beryllium, Be	9	1278	2970	900	112	41	324	9.5
Magnesium, Mg	24	649	1090	738	160	72	148	32.7
Calcium, Ca	40	839	1484	590	197	100	178	41.4
Barium, Ba	137	729	1637	503	224	136	180	62.8
Group 3								
Boron, B	11	2300	3658	800	98	23	563	5.9
Aluminium, Al	27	663	2467	578	143	53	326	28.3
Group 4								
Carbon, C(diamond)	12	3550	4827	1086	77	–	717	2.4
Silicon, Si	28	1410	2355	788	118	–	456	18.8
Germanium, Ge	73	938	2830	762	139	–	377	31.1
Tin, Sn	119	232	2270	708	143	93 (2+)	302	51.5
Lead, Pb	207	328	1740	716	175	120 (2+)	195	64.8
Group 5								
Nitrogen, N	14	−210	−196	1402	75	171	473	95.8
Phosphorus, P (white)	31	44	280	1012	110	190	315	41.1
Group 6								
Oxygen, O	16	−218	−183	1314	73	140	249	102.5
Sulfur, S	32	113	445	1000	104	185	279	31.8
Group 7								
Fluorine, F	19	−220	−188	1681	71	133	79	158.6
Chlorine, Cl	35.5	−101	−34	1250	99	180	122	165.2
Bromine, Br	80	−7	59	1140	114	195	112	174.9
Iodine, I	127	114	184	1008	133	215	107	180.7
Group 8								
Helium, He	4	−272	−269	2372	140	–	–	126.0
Neon, Ne	20	−249	−246	2081	160	–	–	146.2
Argon, Ar	40	−189	−186	1521	190	–	–	154.7
d-block elements								
Scandium, Sc	45	1541	2831	631	164	75 (3+)	378	34.6
Titanium, Ti	48	1660	3287	658	145	67 (3+)	470	30.6
Vanadium, V	51	1887	3377	650	135	64 (3+)	514	28.9
Chromium, Cr	52	1857	2672	653	129	62 (3+)	397	23.8
Manganese, Mn	55	1244	1962	717	137	67 (2+)	281	32.0
Iron, Fe	56	1535	2750	759	126	61 (2+)	416	27.3
Cobalt, Co	59	1495	2870	758	125	65 (2+)	425	30.0
Nickel, Ni	59	1453	2732	737	125	70 (2+)	430	29.9
Copper, Cu	63.5	1084	2567	746	128	73 (2+)	338	33.2
Zinc, Zn	65.4	420	907	906	137	75 (2+)	131	41.6

Properties of selected inorganic compounds

Compound	Melting point/°C	Boiling point/°C	$\Delta H_f^{\ominus}/$ kJ mol^{-1}	$\Delta G_f^{\ominus}/$ kJ mol^{-1}	Standard molar entropy/ J mol^{-1} K^{-1}
Aluminium chloride, $AlCl_3$	sublimes	–	–704	–629	111
Aluminium oxide, Al_2O_3	2045	2980	–1676	–1582	51
Ammonia, NH_3	–78	–34	–46.1	–16	192
Barium chloride, $BaCl_2$	963	1560	–859	–810	124
Barium oxide, BaO	1917	2000	–554	–525	70
Calcium carbonate, $CaCO_3$	decomposes	–	–1207	–1129	93
Calcium chloride, $CaCl_2$	782	1600	–796	–748	105
Calcium oxide, CaO	2600	2850	–635	–604	40
Carbon monoxide, CO	–190	–191	–110	–137	198
Carbon dioxide, CO_2	sublimes	–	–394	–394	214
Chlorine oxide, Cl_2O	–20	explosive	+80.3	+98	266
Copper(II) oxide, CuO	1326	–	–157	–130	43
Hydrazine, N_2H_4	2	114	+50.6	+149	121
Hydrogen fluoride, HF	–83	20	–271	–273	174
Hydrogen chloride, HCl	–114	–85	–92.3	–95	187
Hydrogen bromide, HBr	–87	–67	–36.4	–53	199
Hydrogen iodide, HI	–51	–35	+26.5	+2	206
Lithium fluoride, LiF	845	1676	–616	–588	36
Lithium chloride, LiCl	606	1407	–409	–384	59
Lithium iodide, LiI	449	1171	–270	–270	87
Magnesium chloride, $MgCl_2$	714	1418	–641	–592	90
Magnesium oxide, MgO	2800	3600	–602	–569	27
Dinitrogen monoxide, N_2O	–91	–88	+82	+104	220
Nitrogen monoxide, NO	–163	–152	+90.2	+86	211
Nitrogen dioxide, NO_2	–11	21	+33.2	+51	240
Phosphorus(III) oxide, P_4O_6	24	175	–1640	–	–
Phosphorus(V) oxide, P_4O_{10}	sublimes	–	–2984	–2698	229
Phosphorus(III) chloride, PCl_3	–112	76	–320	–272	217
Phosphorus(V) chloride, PCl_5	sublimes	–	–444	–	166
Potassium chloride, KCl	770	1500	–437	–409	83
Potassium iodide, KI	686	1330	–328	–325	106
Silicon(IV) chloride, $SiCl_4$ (silicon tetrachloride)	–70	58	–687	–620	240
Silicon(IV) oxide, SiO_2	1610	2230	–911	–857	42
Sodium fluoride, NaF	993	1695	–574	–544	52
Sodium chloride, NaCl	801	1465	–411	–384	72
Sodium hydroxide, NaOH	318	1390	–426	–379	64
Sodium oxide, Na_2O	sublimes	–	–414	–376	75
Sulfur dioxide, SO_2	–75	–10	–297	–300	248
Sulfur trioxide, SO_3	17	43	–441	–368	96
Sulfur chloride, SCl_2	–78	decomposes	–19.7	–	282
Water, H_2O(l)	273	373	–286	–237	70
Water, H_2O(g)	273	373	–242	–229	189

Properties of selected organic compounds

Compound	Formula	Melting point/°C	Boiling point /°C	$\Delta H_f^{\ominus}$/ kJ mol^{-1}	$\Delta G_f^{\ominus}$/ kJ mol^{-1}	Standard molar entropy/ J mol^{-1}/K^{-1}
Alkanes						
Methane	CH_4	−182	−161	−75	−51	186
Ethane	C_2H_6	−183	−89	−85	−33	230
Propane	C_3H_8	−190	−42	−104	−24	270
Butane	C_4H_{10}	−138	−1	−126	−16	310
Pentane	C_5H_{12}	−130	36	−173	−9	261
Hexane	C_6H_{14}	−95	69	−199	−4	296
Decane	$C_{10}H_{22}$	−30	174	−301	+17	426
2-methylpropane	C_4H_{10}	−160	−12	−135	−18	295
2-methylbutane	C_5H_{12}	−160	28	−179	−14	260
2-methylpentane	C_6H_{14}	−154	60	−205	−8	291
2,2-dimethylpropane	C_5H_{12}	−17	10	−190	−15	306
Alkenes						
Ethene	C_2H_4	−169	−104	+52	+68	220
Propene	C_3H_6	−185	−48	+20	+75	267
But-1-ene	C_4H_8	−185	−6	−0.4	+72	306
cis but-2-ene	C_4H_8	−139	4	−8	+66	301
trans but-2-ene	C_4H_8	−106	1	−12	+63	296
Arenes and related compounds						
Benzene	C_6H_6	6	80	+49	+125	173
Methylbenzene	C_7H_8	−95	111	+12	+110	320
Phenylethene	C_8H_8	31	145	+103	+202	345
Nitrobenzene	$C_6H_5NO_2$	6	211	+12	+142	224
Phenol	C_6H_5OH	43	182	−165	−48	144
Halogenoalkanes						
1-chlorobutane	C_4H_9Cl	−123	78	−188	−	−
2-chloro-2-methylpropane	C_4H_9Cl	−25	51	−191	−	
1-bromobutane	C_4H_9Br	−112	102	−144	−	327
2-bromobutane	C_4H_9Br	−112	91	−155	−	−
1-iodobutane	C_4H_9I	−103	131	−191	−	−
Alcohols						
Methanol	CH_3OH	−94	65	− 239	−166	240
Ethanol	C_2H_5OH	−117	79	−277	−175	161
Propan-1-ol	C_3H_7OH	−127	97	−303	−171	197
Propan-2-ol	C_3H_7OH	−90	82	−318	−180	180
Butan-1-ol	C_4H_9OH	−90	117	−327	−169	228
Aldehydes						
Methanal	HCHO	−92	−21	−109	−113	219
Ethanal	CH_3CHO	−121	20	−192	−128	160
Propanal	C_2H_5CHO	−81	49	−217	−142	213
Butanal	C_3H_7CHO	−99	76	−241	−306	243
Benzenecarbaldehyde	C_6H_5CHO	−26	178	−87	−	221
Ketones						
Propanone	CH_3COCH_3	−95	56	−248	−155	200
Butanone	$C_2H_5COCH_3$	−86	80	−276	−156	240
Pentan-3-one	$C_2H_5COC_2H_5$	−40	102	−297	−	266
Phenylethanone	$C_6H_5COCH_3$	21	203	−142	−	228

Properties of elements and compounds

Compound	Formula	Melting point/°C	Boiling point /°C	$\Delta H_f^\ominus$/ kJ mol^{-1}	$\Delta G_f^\ominus$/ kJ mol^{-1}	Standard molar entropy/ J mol^{-1}/K^{-1}
Amines						
Methylamine	CH_3NH_2	−93	−6	−23.0	+32.1	243
Dimethylamine	$(CH_3)_2NH$	−93	7	−18.5	+59.2	280
Trimethylamine	$(CH_3)_3N$	−117	3	−47.5	–	–
Ethylamine	$C_2H_5NH_2$	−80	17	−47.5	–	–
Butylamine	$C_4H_9NH_2$	−49	78	−128	−81.8	–
Phenylamine	$C_6H_5NH_2$	−6	184	+31.3	–	–
Carboxylic acids						
Methanoic acid	HCO_2H	8	101	−425	−361	129
Ethanoic acid	CH_3CO_2H	17	118	−484	−390	160
Propanoic acid	$C_2H_5CO_2H$	−21	141	−511	−384	–
Esters						
Methyl methanoate	HCO_2CH_3	−99	32	−386	–	–
Methyl ethanoate	$CH_3CO_2CH_3$	−98	57	−446	–	–
Ethyl methanoate	$HCO_2C_2H_5$	−80	55	−371	–	–
Ethyl ethanoate	$C_2H_5CO_2C_2H_5$	−83	77	−479	–	–
Other carboxylic acid derivatives						
Ethanoyl chloride	CH_3CO_2Cl	−112	51	−273	−208	201
Ethanamide	$CH_3CO_2NH_2$	82	221	−317	–	–
Ethanoic anhydride	$(CH_3CO_2)_2O$	−73	140	−637	–	–
Nitriles						
Ethanenitrile	CH_3CN	−46	82	+31.4	–	–
Propanenitrile	C_2H_5CN	−93	97	+15.5	–	–
Aminoacids						
1-aminoethanoic acid (glycine)	$NH_2CH_2CO_2H$	decomposes	–	−529	–	–

2,4-dinitrophenylhydrazine derivatives of carbonyl compounds

Carbonyl compound	Melting point of 2,4-dinitrophenylhydrazine derivative/°C
methanal	166
ethanal	168
propanal	155
butanal	126
benzenecarbaldehyde	237
propanone	126
butanone	115
pentan-3-one	156
phenylethanone	250

Thermodynamic data for energy cycles

Bond lengths and bond energies

Bond	Bond length/pm	Average bond enthalpy/kJ mol^{-1}
H—H	74	435
F—F	142	158
Cl—Cl	199	243
Br—Br	228	193
I—I	267	151
H—F	92	568
H—Cl	127	432
H—Br	141	366
H—I	161	298
O=O	121	498
O—H	96	464
C—H	109	435
C—C	154	347
C=C	134	612
C≡C	120	838
C—Cl	177	346
C—Br	194	290
C—I	214	228
N≡N	110	945
N—N	145	158
N—H	101	391

Electron affinities

Species	Electron affinity/kJ mol^{-1}
O	−141
O$^-$	+798
F	−328
Cl	−349
Br	−325

Lattice enthalpies for ionic compounds

See also the values on page 67

Compound	Lattice enthalpy/kJ mol^{-1}
Halide	
LiF	−1046
LiCl	−861
LiBr	−818
LiI	−759
NaF	−929
KF	−826
MgF$_2$	−2961
MgCl$_2$	−2524
Oxide	
MgO	−3850
CaO	−3461
BaO	−3114
Sulfide	
MgS	−3406
BaS	−2832

Hydration enthalpies

Cation	Hydration enthalpy/ kJ mol^{-1}	Anion	Hydration enthalpy/ kJ mol^{-1}
Li$^+$	−559	F$^-$	−483
Na$^+$	−444	Cl$^-$	−340
K$^+$	−361	Br$^-$	−309
Mg^{2+}	−2003	I$^-$	−296
Ca^{2+}	−1657		
Al^{3+}	−2537		

Chemical equilibria

Dissociation constants for weak acids

Acid	K_a/mol dm^{-3}	pK_a
Ammonium ion	5.6×10^{-10}	6.4
Carbonic acid	4.5×10^{-7}	7.4
Chloric(I) acid	3.7×10^{-8}	7.4
Ethanoic acid	1.7×10^{-5}	4.8
Hydrated aluminium ion	1.0×10^{-5}	5.0
Hydrated iron(III) ion	6.0×10^{-3}	2.2
Hydrocyanic acid	4.9×10^{-10}	9.3
Hydrofluoric acid	5.6×10^{-4}	3.3
Hydrogencarbonate ion	4.8×10^{-11}	10.3
Hydrogensulfate ion	1.0×10^{-2}	2.0
Hydrogen sulfide	8.9×10^{-8}	7.1
Methanoic acid	1.6×10^{-4}	3.8
Nitrous acid	4.7×10^{-4}	3.3
Phenol	1.3×10^{-10}	9.9
Phosphoric(V) acid	7.9×10^{-3}	2.1
Propanoic acid	1.3×10^{-5}	4.9
Sulfurous acid	1.5×10^{-2}	1.8

Standard electrode potentials

Table with the more negative values at the top and the more positive values at the bottom.

Half-cell	Half-reaction	$E^\ominus/V$
$Li^+(aq)\mid Li(s)$	$Li^+(aq) + e^- \longrightarrow Li(s)$	−3.03
$K^+(aq)\mid K(s)$	$K^+(aq) + e^- \longrightarrow K(s)$	−2.92
$Ca^{2+}(aq)\mid Ca(s)$	$Ca^{2+}(aq) + 2e^- \longrightarrow Ca(s)$	−2.87
$Na^+(aq)\mid Na(s)$	$Na^+(aq) + e^- \longrightarrow Na(s)$	−2.71
$Mg^{2+}(aq)\mid Mg(s)$	$Mg^{2+}(aq) + 2e^- \longrightarrow Mg(s)$	−2.37
$Al^{3+}(aq)\mid Al(s)$	$Al^{3+}(aq) + 3e^- \longrightarrow Al(s)$	−1.67
$Zn^{2+}(aq)\mid Zn(s)$	$Zn^{2+}(aq) + 2e^- \longrightarrow Zn(s)$	−0.76
$Cr^{3+}(aq)\mid Cr(s)$	$Cr^{3+}(aq) + 3e^- \longrightarrow Cr(s)$	−0.74
$Fe^{2+}(aq)\mid Fe(s)$	$Fe^{2+}(aq) + 2e^- \longrightarrow Fe(s)$	−0.44
$Cr^{3+}(aq), Cr^{2+}(aq)\mid Pt$	$Cr^{3+}(aq) + e^- \longrightarrow Cr^{2+}(aq)$	−0.41
$V^{3+}(aq), V^{2+}(aq)\mid Pt$	$V^{3+}(aq) + e^- \longrightarrow V^{2+}(aq)$	−0.26
$Sn^{2+}(aq)\mid Sn(s)$	$Sn^{2+}(aq) + 2e^- \longrightarrow Sn(s)$	−0.14
$Pb^{2+}(aq)\mid Pb(s)$	$Pb^{2+}(aq) + 2e^- \longrightarrow Pb(s)$	−0.13
$2H^+(aq), H_2(g)\mid Pt$	$2H^+(aq) + 2e^- \longrightarrow H_2(g)$	−0.00
$Sn^{4+}(aq), Sn^{2+}(aq)\mid Pt$ (in acid)	$Sn^{4+}(aq) + 2e^- \longrightarrow Sn^{2+}(aq)$	−0.15
$Cu^{2+}(aq)\mid Cu^+(aq)$	$Cu^{2+}(aq) + e^- \longrightarrow Cu^+(aq)$	+0.15
$Cu^{2+}(aq)\mid Cu(s)$	$Cu^{2+}(aq) + 2e^- \longrightarrow Cu(s)$	+0.34
$[O_2(aq) + 2H_2O(l)], 4OH^-\mid Pt$	$O_2(aq) + 2H_2O(l) + 4e^- \longrightarrow 4OH^-(aq)$	+0.40
$Cu^+(aq)\mid Cu(s)$	$Cu^+(aq) + e^- \longrightarrow Cu(s)$	+0.52
$I_2(aq), 2I^-(aq)\mid Pt$	$I_2(aq) + 2e^- \longrightarrow 2I^-(aq)$	+0.54
$[O_2(g) + 2H^+(aq)], H_2O_2(l)\mid Pt$	$O_2(g) + 2H^+(aq) + 2e^- \longrightarrow H_2O_2(l)$	+0.68
$Fe^{3+}(aq), Fe^{2+}(aq)\mid Pt$	$Fe^{3+}(aq) + e^- \longrightarrow Fe^{2+}(aq)$	+0.77
$Ag^+(aq)\mid Ag(s)$	$Ag^+(aq) + e^- \longrightarrow Ag(s)$	+0.80
$[ClO^-(aq) + H_2O(l)], [Cl^-(aq) + 2OH^-(aq)]\mid Pt$	$ClO^-(aq) + H_2O(l) + 2e^- \longrightarrow Cl^-(aq) + 2OH^-(aq)$	+0.89
$Br_2(aq), 2Br^-(aq)\mid Pt$	$Br_2(aq) + 2e^- \longrightarrow 2Br^-(aq)$	+1.09
$[O_2(g) + 4H^+(aq)], 2H_2O(l)\mid Pt$	$O_2(g) + 4H^+(aq) + 4e^- \longrightarrow 2H_2O(l)$	+1.23
$Au^{3+}(aq), Au^+(aq)\mid Pt$	$Au^{3+}(aq) + 2e^- \longrightarrow Au^+(aq)$	+1.29
$[Cr_2O_7^{2-}(s) + 14H^+(aq)],$ $[2Cr^{3+}(aq) + 7H_2O(l)]\mid Pt$	$Cr_2O_7^{2-}(s) + 14H^+(aq)] + 6e^- \longrightarrow$ $2Cr^{3+}(aq) + 7H_2O(l)$	+1.33
$Cl_2(aq), 2Cl^-(aq)\mid Pt$	$Cl_2(aq) + 2e^- \longrightarrow 2Cl^-(aq)$	+1.36
$[PbO_2(s) + 4H^+(aq)],$ $[Pb^{2+}(aq) + 2H_2O(l)]\mid Pt$	$PbO_2(s) + 4H^+(aq) + 2e^- \longrightarrow$ $Pb^{2+}(aq) + 2H_2O(l)$	+1.46
$[MnO_4^-(s) + 8H^+(aq)],$ $[Mn^{2+}(aq) + 4H_2O(l)]\mid Pt$	$MnO_4^-(s) + 8H^+(aq) + 5e^- \longrightarrow$ $Mn^{2+}(aq) + 4H_2O(l)$	+1.51
$Au^+(aq)\mid Au(s)$	$Au^+(aq) + e^- \longrightarrow Au(s)$	+1.69
$[H_2O_2(aq) + 2H^+(aq)], [2H_2O(l)]\mid Pt$	$H_2O_2(aq) + 2H^+(aq) + 2e^- \longrightarrow 2H_2O(l)$	+1.77
$S_2O_8^{2-}(aq), 2SO_4^{2-}(aq)\mid Pt$	$S_2O_8^{2-}(aq) + 2e^- \longrightarrow 2SO_4^{2-}(aq)$	+2.01

Standard electrode potentials

Table with the more positive values at the top and the more negative values at the bottom.

Half-cell	Half-reaction	$E^{\ominus}/V$
$S_2O_8^{2-}(aq), 2SO_4^{2-}(aq) \mid Pt$	$S_2O_8^{2-}(aq) + 2e^- \longrightarrow 2SO_4^{2-}(aq)$	+2.01
$[H_2O_2(aq) + 2H^+(aq)], [2H_2O(l)] \mid Pt$	$H_2O_2(aq) + 2H^+(aq) + 2e^- \longrightarrow 2H_2O(l)$	+1.77
$Au^+(aq) \mid Au(s)$	$Au^+(aq) + e^- \longrightarrow Au(s)$	+1.69
$[MnO_4^-(s) + 8H^+(aq)],$ $[Mn^{2+}(aq) + 4H_2O(l)] \mid Pt$	$MnO_4^-(s) + 8H^+(aq) + 5e^- \longrightarrow$ $Mn^{2+}(aq) + 4H_2O(l)$	+1.51
$[PbO_2(s) + 4H^+(aq)],$ $[Pb^{2+}(aq) + 2H_2O(l)] \mid Pt$	$PbO_2(s) + 4H^+(aq) + 2e^- \longrightarrow$ $Pb^{2+}(aq) + 4H_2O(l)$	+1.46
$Cl_2(aq), 2Cl^-(aq) \mid Pt$	$Cl_2(aq) + 2e^- \longrightarrow 2Cl^-(aq)$	+1.36
$[Cr_2O_7^{2-}(s) + 14H^+(aq)],$ $[2Cr^{3+}(aq) + 7H_2O(l)] \mid Pt$	$Cr_2O_7^{2-}(s) + 14H^+(aq)] + 6e^- \longrightarrow$ $2Cr^{3+}(aq) + 7H_2O(l)$	+1.33
$Au^{3+}(aq), Au^+(aq) \mid Pt$	$Au^{3+}(aq) + 2e^- \longrightarrow Au^+(aq)$	+1.29
$[O_2(g) + 4H^+(aq)], 2H_2O(l) \mid Pt$	$O_2(g) + 4H^+(aq) + 4e^- \longrightarrow 2H_2O(l)$	+1.23
$Br_2(aq), 2Br^-(aq) \mid Pt$	$Br_2(aq) + 2e^- \longrightarrow 2Br^-(aq)$	+1.09
$[ClO^-(aq) + H_2O(l)], [Cl^-(aq) + 2OH^-(aq)] \mid Pt$	$ClO^-(aq) + H_2O(l) + 2e^- \longrightarrow Cl^-(aq) + 2OH^-(aq)$	+0.89
$Ag^+(aq) \mid Ag(s)$	$Ag^+(aq) + e^- \longrightarrow Ag(s)$	+0.80
$Fe^{3+}(aq), Fe^{2+}(aq) \mid Pt$	$Fe^{3+}(aq) + e^- \longrightarrow Fe^{2+}(aq)$	+0.77
$[O_2(g) + 2H^+(aq)], H_2O_2(l) \mid Pt$	$O_2(g) + 2H^+(aq) + 2e^- \longrightarrow H_2O_2(l)$	+0.68
$I_2(aq), 2I^-(aq) \mid Pt$	$I_2(aq) + 2e^- \longrightarrow 2I^-(aq)$	+0.54
$Cu^+(aq) \mid Cu(s)$	$Cu^+(aq) + e^- \longrightarrow Cu(s)$	+0.52
$[O_2(aq) + 2H_2O(l)], 4OH^- \mid Pt$	$O_2(aq) + 2H_2O(l) + 4e^- \longrightarrow 4OH^-(aq)$	+0.40
$Cu^{2+}(aq) \mid Cu(s)$	$Cu^{2+}(aq) + 2e^- \longrightarrow Cu(s)$	+0.34
$Cu^{2+}(aq) \mid Cu^+(aq)$	$Cu^{2+}(aq) + e^- \longrightarrow Cu^+(aq)$	+0.15
$Sn^{4+}(aq), Sn^{2+}(aq) \mid Pt$ (in acid)	$Sn^{4+}(aq) + 2e^- \longrightarrow Sn^{2+}(aq)$	−0.15
$H^+(aq), H_2(g) \mid Pt$	$2H^+(aq) + 2e^- \longrightarrow H_2(g)$	−0.00
$Pb^{2+}(aq) \mid Pb(s)$	$Pb^{2+}(aq) + 2e^- \longrightarrow Pb(s)$	−0.13
$Sn^{2+}(aq) \mid Sn(s)$	$Sn^{2+}(aq) + 2e^- \longrightarrow Sn(s)$	−0.14
$V^{3+}(aq), V^{2+}(aq) \mid Pt$	$V^{3+}(aq) + 2e^- \longrightarrow V^{2+}(aq)$	−0.26
$Cr^{3+}(aq), Cr^{2+}(aq) \mid Pt$	$Cr^{3+}(aq) + e^- \longrightarrow Cr^{2+}(aq)$	−0.41
$Fe^{2+}(aq) \mid Fe(s)$	$Fe^{2+}(aq) + 2e^- \longrightarrow Fe(s)$	−0.44
$Cr^{3+}(aq) \mid Cr(s)$	$Cr^{3+}(aq) + 3e^- \longrightarrow Cr(s)$	−0.74
$Zn^{2+}(aq) \mid Zn(s)$	$Zn^{2+}(aq) + 2e^- \longrightarrow Zn(s)$	−0.76
$Al^{3+}(aq) \mid Al(s)$	$Al^{3+}(aq) + 3e^- \longrightarrow Al(s)$	−1.67
$Mg^{2+}(aq) \mid Mg(s)$	$Mg^{2+}(aq) + 2e^- \longrightarrow Mg(s)$	−2.37
$Na^+(aq) \mid Na(s)$	$Na^+(aq) + e^- \longrightarrow Na(s)$	−2.71
$Ca^{2+}(aq) \mid Ca(s)$	$Ca^{2+}(aq) + 2e^- \longrightarrow Ca(s)$	−2.87
$K^+(aq) \mid K(s)$	$K^+(aq) + e^- \longrightarrow K(s)$	−2.92
$Li^+(aq) \mid Li(s)$	$Li^+(aq) + e^- \longrightarrow Li(s)$	−3.03

Spectroscopy

Infra-red spectroscopy correlation data

Bond	Characteristic wavenumber/cm^{-1}
O—H free	3580–3670
N—H amines	3500–3300
O—H hydrogen bonded in alcohols and phenols	3550–3230
C—H aromatic	3150–3000
C—H aliphatic	3000–2850
O—H hydrogen bonded in carboxylic acids	2500–3300
C—N	2280–2200
C—C	2250–2070
C=O aldehydes, ketones, acids and esters	1750–1680
C=C	1680–1610
C—O alcohols, ethers and esters	1300–1000

Chemical shifts for nmr spectroscopy

Type of proton	Chemical shift/ppm
R—**CH$_3$**	0.9
R—**CH$_2$**—R	1.3
R$_3$C—**H**	2.0
CH$_3$—CO$_2$R	2.0
CH$_3$—CO—R	2.1
C$_6$H$_5$—**CH$_3$**	2.3
R—**CH$_2$**—CO—R	2.4
R—**CH$_2$**—Hal	3.2–3.7
R—**CH$_2$**—O—R	3.4
R—O—**CH$_3$**	3.8
R—O—**H**	3.5–5.5
RC**H**=CH$_2$	4.9
RC**H** CH$_2$	5.9
C$_6$H$_5$—O**H**	7.0
C$_6$H$_5$—**H**	7.3
R—C**H**O	9.7
R—CO$_2$**H**	11.0–11.7

Answers to 'Test yourself' questions

In a few instances it is not possible to give a sensible short answer – so there is no answer included. There are also no answers to questions asking for large diagrams, charts or graphs.

Section one Studying chemistry

1.2 Showing what you know and can do

The questions in this section use examples from *AS Chemistry* to illustrate the types of knowledge and understanding required for success in advanced chemistry courses. For the answers check the following references to *AS Chemistry*.

1 a) *AS Chemistry* page 153
   b) page 212   c) page 204
2 a) page 212   b) page 136   c) page 100
3 a) page 59–60   b) page 113–116   c) page 109–110
   d) page 100–101
4 a) page 67–68   b) page 85–86   c) page 185
5 a) page 45–46   b) page 245   c) page 213
6 a) page 116   b) page 222   c) page 167–168
   d) See section 4.19 in **this** book   e) page 204 in this book
7 a) *AS Chemistry* page 38
   b) page 139   c) page 80
8 a) page 119   b) page 130–131   c) page 45–46
   d) page 160–163   e) page 205.

Section two Physical chemistry

2.2 Reaction kinetics – the effect of concentration changes

1 a) Concentrated hydrochloric acid reacts much faster with marble chips than dilute hydrochloric acid.
   b) Any reaction involving gases such as the manufacture of ammonia from nitrogen and hydrogen. Increasing the pressure forces the molecules closer together and increases the rate of collisions which lead to reaction.
   c) Magnesium powder reacts much faster with dilute hydrochloric acid than magnesium ribbon.
   d) Catalytic converters are only effective in speeding up the reactions which remove pollutants from car exhausts once they are hot.
   e) A platinum-rhodium alloy catalyses the oxidation of ammonia to nitrogen oxide.
2 See *AS Chemistry* page 117–119.
3 a) Collect the gas in a graduated syringe.
   b) Follow the disappearance of the colour of bromine with a colorimeter.
   c) Remove samples at intervals, stop the reaction by cooling and then titrate with alkali the acid produced by the reaction.
   d) Measure the conductivity of the solution to follow the increase in the concentration of ions.
4 $0.0087 \ mol \ dm^{-3} \ s^{-1}$
5 a) $2N_2O_5(g) \longrightarrow 4NO_2(g) + O_2(g)$
   b) $7.0 \times 10^{-4} \ mol \ dm^{-3} \ s^{-1}$
6 $3.7 \times 10^{-4} \ s^{-1}$
7 $0.138 \ mol^{-1} \ dm^3 \ s^{-1}$
8 d) $k = 1.8 \times 10^{-5} \ s^{-1}$

9 a) Rate = $k[RBr][OH^-]$ b) Second order
11 a) Graph similar to Figure 2.2.11.
   b) $2NH_3(g) \longrightarrow N_2(g) + 3H_2(g)$
   Rate = k
12 a) $2H_2(g) + 2NO(g) \longrightarrow N_2(g) + 2H_2O(g)$
   b) Rate = $k[H_2(g)] \ [NO(g)]^2$
13 a) Rate = $k[RBr][OH^-]$ b) $k = 3400 \ mol^{-1} \ dm^3 \ s^{-1}$
14 a) Rate = $k[R'Br]$ b) $2020 \ s^{-1}$

2.3 Reaction kinetics – the effect of temperature changes

1 a) E_a/RT becomes smaller in magnitude as T rises, hence $\ln k$ becomes more positive. This means that k gets larger and the rate is faster.
   b) The larger the activation energy the larger the value of E_a/RT and so the smaller the magnitude of $\ln k$. Hence k gets smaller and the rate is less.
2 $E_a = 78 \ kJ \ mol^{-1}$
3 In the presence of a catalyst the reaction pathway has an activation energy which is much lower than when there is no catalyst. Tungsten metal adsorbs hydrogen into the upper layers of the crystal structure as single atoms. So the catalyst breaks the bonds between the atoms in one of the reactants.

2.4 Rate equations and reaction mechanisms

1 a) $NO_2(g) + CO(g) \longrightarrow NO(g) + CO_2(g)$
   b) Rate = $k[NO_2(g)]^2$
   c) Zero order
2 a) $H_2O_2(aq) + 2H^+(aq) + 2I^-(aq) \longrightarrow 2H_2O(l) + I_2(aq)$
   b) Rate = $k[H_2O_2(aq)][I^-(aq)]$
   c) Second order
   d) The first step which includes the two species which feature in the rate equation.

2.5 The equilibrium law

1 The diagram should include: a regular array of particles in the solid, particles moving from the solid into the solution, particles of the solute in solution mixed with many more irregularly arranged solvent particles, solute particles returning to the solid phase as fast as other particles leave the solid phase.
2 Figure 2.5.1: At first there are only H_2 and I_2 molecules. As they collide and react HI molecules form. As the concentrations of H_2 and I_2 fall, the rate of the forward reaction decreases. As the concentration of HI rises there is a growing number of collisions between these molecules so the rate of the back reaction increases. Equilibrium is reached when the rates of the forward and backward reactions are equal. Figure 2.5.2: At first only HI molecules. As they collide and react H_2 and I_2 molecules form. As the concentration of HI molecules falls, the rate of the forward reaction decreases. As the concentrations of H_2 and I_2 molecules rise there is a growing number of collisions between these molecules so the rate of the back reaction increases. Again, equilibrium is reached when the rates of the forward and backward reactions are equal.
3 The three values for K_c (which has no units for this reaction): 46.4, 46.4, 45.1.
4 $K_c = 0.05 \ mol \ dm^{-3}$

5 The system is not at equilibrium. There is a tendency for the equilibrium to shift to the left to increase the concentration of NO_2.

6 **b)** $K_c = 1.26 \times 10^3$ $mol^{0.5}$ $dm^{-1.5}$
  c) $K_c = 6.25 \times 10^{-7}$ mol dm^{-3}

7 **a)** $K_c = \dfrac{[CO(g)][H_2O(g)]}{[CO_2(g)][H_2(g)]}$ no units

  b) $K_c = \dfrac{[NH_3(g)]^2}{[N_2(g)][H_2(g)]^3}$ units: mol^{-2} dm^6

8 **a)** Under these conditions, the forward reaction tends to go essentially to completion.
  b) Under these conditions, the back reaction tends to go essentially to completion.
  c) Under these conditions, the equilibrium mixture contains significant amounts of reactants and products.

9 Correspondingly small. (The value of the equilibrium constant of the reverse reaction is the inverse of the value for the forward reaction.)

10 **a)** $K_c = \dfrac{[H_2(g)]^3}{[H_2O(g)]^3}$

  b) $K_c = \dfrac{[H_2S(g)]}{[H_2(g)]}$

  c) $K_c = \dfrac{[Fe^{3+}(aq)]}{[Ag^+(aq)]}$

11 Pure water contains $(1000\ g \div 18\ g\ mol^{-1}) = 55.6$ mol dm^{-3}. So in dilute solutions even up to 1 mol dm^{-3} of reactants the change in the concentration of water molecules is negligible even if water molecules are involved in the reaction.

12 Adding alkali neutralises $H^+(aq)$ ions. So the concentration of aqueous hydrogen ions in solution falls. As predicted by Le Chatelier's principle the equilibrium shifts in the direction that forms more $H^+(aq)$ ions. Dichromate(VI) ions change to chromate(VI) ions. The solution turns from orange to yellow.

13 **a)** At equilibrium: $K_c = \dfrac{[\text{ester}][\text{water}]}{[\text{alcohol}][\text{acid}]}$

  Adding more ethanol raises [alcohol] in the mixture.

  Temporarily $K_c > \dfrac{[\text{ester}][\text{water}]}{[\text{alcohol}][\text{acid}]}$

  So some of the added ethanol reacts with some acid to form more ester and water until the concentrations of ester and water have risen enough, and the concentrations of alcohol and water fallen enough until once again: $\dfrac{[\text{ester}][\text{water}]}{[\text{alcohol}][\text{acid}]} = K_c$

  b) The answer to part **a)** shows that the equilibrium shifts to the right producing more product. As predicted by Le Chatelier's principle this tends to counteract the effect of adding extra ethanol.

2.6 Gaseous equilibria

1 **a)** $K_p = \dfrac{p_{SO_3}{}^2}{p_{SO_2}{}^2\, p_{O_2}}$

  b) $K_p = \dfrac{p_{N_2}{}^2\, p_{H_2O}{}^6}{p_{NH_3}{}^4\, p_{O_2}{}^3}$

  c) $K_p = p_{CO_2}$

2 $K_p = 18$ kPa
3 $K_p = 27$ kPa

4 **a)** The new equilibrium mixture contains more methanol and less carbon monoxide.
  b) The new equilibrium mixture contains more methanol and less hydrogen and carbon monoxide.
  c) The new equilibrium mixture will contain less methanol and more hydrogen and carbon monoxide.

2.7 The effects of temperature changes and catalysts on equilibria

1 Raising the temperature causes the equilibrium to shift to the left (lowering the partial pressure of HI at equilibrium). This shows that the reaction forming HI is exothermic.

2 Raising the temperature causes the equilibrium to shift to the right (raising the partial pressure of CO_2 at equilibrium). This shows that the decomposition of calcium carbonate is endothermic.

2.8 Acids, bases and the pH scale

1 **a)** Dative covalent bond.

  b)
  $$\overset{\displaystyle H}{H\,{}^x_\bullet\overset{\bullet\,x}{\underset{\bullet\bullet}{O}}{}^\bullet_\bullet H}\ ^+$$

  c)

2
$$\underset{CO_2H}{\overset{CO_2H}{|}}(aq) + 2OH^-(aq) \longrightarrow \underset{CO_2^-}{\overset{CO_2^-}{|}}(aq) + 2H_2O(l)$$

3 **a)** Hydrogen chloride and sodium hydrogensulfate.
  b) Protons transfer from H_2SO_4 molecules to chloride ions.
  c) HCl is a gas so it escapes from the reaction mixture as it forms. This means that the back reaction does not start so the forward reaction continues essentially to completion.

4 **a)** The oxide ion is a base, accepting protons from oxonium ions in dilute hydrochloric acid forming water molecules.
  b) Oxonium ions in dilute sulfuric acid donate protons to ammonia molecules forming ammonium ions.
  c) Ammonium ions are proton donors giving protons to hydroxide ions thus forming ammonia molecules and water molecules.
  d) Oxonium ions in dilute hydrochloric acid give protons to carbonate ions. The carbonic acid formed decomposes to carbon dioxide and water.

5 **a)** Dative covalent bond.

  b)
  $$\overset{\displaystyle H}{\underset{\displaystyle H}{H\,{}^\bullet_x\overset{x}{N}{}^x_\bullet H}}\ ^+$$

  c) Tetrahedral. The H—N—H bond angle is 109.5°

6 Nitrate ion, ethanoate ion, hydrogensulfate ion, carbonate ion.

7 Hydroxide ion, water molecule, ammonium ion, hydrogencarbonate ion, carbonic acid, hydrogensulfate ion.

Answers

8 a) pH = 1 **b)** pH = 2 **c)** pH = 3
9 pH = 1.1
10 a) Concentration = 5.0×10^{-4} mol dm^{-3}
  b) Concentration = 4.0×10^{-6} mol dm^{-3}
  c) Concentration = 2.0×10^{-7} mol dm^{-3}
  d) Concentration = 3.2×10^{-11} mol dm^{-3}
11 a) Ionisation is endothermic.
  b) Extent of ionisation increases as the temperature rises. The oxonium concentration rises as the temperature rises. The number value of the pH falls.
12 a) pH = 14 **b)** pH = 12.3 **c)** pH = 11.3

2.9 Weak acids and bases

1 A solution with pH = 3 could, for example, be a dilute solution of a strong acid or a more concentrated solution of a weak acid. To determine the strength of an acid it is necessary to know the concentration of the acid as well as the pH.
2 pH = 5.7
3 pH = 3.0
4 $K_a = 1.59 \times 10^{-4}$ mol dm^{-3}
5 $K_a = 1.4 \times 10^{-4}$ mol dm^{-3}
6 $pK_a = 3.8$
7 $K_a = 6.3 \times 10^{-5}$ mol dm^{-3}
8 trichlorethanoic acid > dichlorethanoic acid > chlorethanoic acid > ethanoic acid

2.10 Acid–base titrations

2 At the equivalence point the solution just contains a neutral salt with an anion which is a very weak base and a cation which is not an acid. With hydrochloric acid and sodium hydroxide, for example, the solution at the equivalence point is a neutral solution of sodium chloride.
3 a) pH = 3.7 **b)** pH = 10.3
4 pH = 12.2
6 pH = 2.9
7 pH = 4.6
8 The ethanoate ion is the conjugate base of a weak acid so it is a relatively strong base. Ethanoate ions in solution take protons from water to form ethanoic acid molecules and hydroxide ions.
9 In a titration the equivalence point is determined by the concentrations of the acid and the alkali and not by their strengths.
10 The ammonium ion is the conjugate acid of a weak base so it is a relatively strong acid. Ammonium ions in solution give protons to water to form oxonium ions and ammonia molecules.
11 $CH_3CO_2H(aq) + NH_3(aq) \longrightarrow CH_3CO_2^-NH_4^+(aq)$

2.11 Indicators

1 In a more acid solution the oxonium ion concentration is higher. This tends to shift the equilibrium in Figure 2.11.5 to the right giving more of the protonated red form. In a less acid solution the oxonium ion concentration is lower. This tends to shift the equilibrium in Figure 2.11.5 to the left giving more of the unprotonated yellow form.
2 a) The indicator would change colour before the titration reached the equivalence point.
  b) The indicator would not change colour until the titration has passed the equivalence point.
  c) Strong acid/strong base titration: methyl orange, methyl red or bromothymol blue.
  Weak acid/strong base titration: phenolphthalein.

Strong acid/weak base titration: methyl orange or methyl red.
  d) There is only a slight, and rather gradually change of pH at the end-point of the titration. There needs to be a steeply rising part of the graph spanning at least two pH units to give a sharp colour change with an indicator.
3 The theory showing that an indicator typically changes colour over two pH units assumes that the two colours are equally intense so that the eye is equally sensitive to both of them.
4 a) $CO_3^{2-}(aq) + H^+(aq) \longrightarrow HCO_3^-(aq)$
  $HCO_3^-(aq) + H^+(aq) \longrightarrow H_2CO_3(aq)$
  b) The initial concentrations of sodium carbonate and hydrochloric acid are the same. 25 cm^3 of the acid provides one mole of hydrogen ions per mole of carbonate ions. 50 cm^3 of the acid provides a total of two moles of hydrogen ions per mole of carbonate ions.
5 a) $HO_2C—CO_2H(aq) + NaOH(aq) \longrightarrow$
      $HO_2C—CO_2Na(aq) + H_2O(l)$
  $HO_2C—CO_2Na(aq) + NaOH(aq) \longrightarrow$
      $NaO_2C—CO_2Na(aq) + H_2O(l)$
  b) The initial concentrations of ethanedioic acid and sodium hydroxide are the same. 25 cm^3 of the alkali removes one mole of hydrogen ions per mole of acid. 50 cm^3 of the acid removes a total of two moles of hydrogen ions per mole of acid.
6 a) pH = pK_a for ethanedioic acid = 1.2
  b) pH = pK_a for $HO_2C—CO_2Na$ = 4.2

2.12 Buffer solutions

1 a) pH = 7.2 **b)** pH = 3.9 **c)** pH = 3.9
2 Ratio salt:acid = 3.98

2.13 Neutralisation reactions

1 a) pH = 7 **b)** pH above 7 **c)** pH below 7
  d) pH above 7
2 In the apparatus shown some of the energy from the reaction heats up the container and the surroundings and this is not allowed for in the calculation.
3 HCl is a strong acid and fully ionised before mixing with the alkali. Ethanoic acid is only very slightly ionised in solution. The base has to break the O–H bond in weak acid.
4 $HBr(aq) + NaOH(aq) \longrightarrow NaBr(aq) + H_2O(l)$
      $\Delta H^\ominus = -57.6$ kJ mol^{-1}
  $HCl(aq) + NH_3(aq) \longrightarrow NH_4Cl(aq) + H_2O(l)$
      $\Delta H^\ominus = -53.4$ kJ mol^{-1}
  $CH_3CO_2H(aq) + NH_3(aq) \longrightarrow CH_3CO_2NH_4(aq)$
      $\Delta H^\ominus = -50.4$ kJ mol^{-1}

2.14 Electrode potentials

1 a) Mg atoms oxidised: $Mg(s) \longrightarrow Mg^{2+}(aq) + 2e^-$
  Copper(II) ions reduced: $Cu^{2+}(aq) + 2e^- \longrightarrow Cu(s)$
  b) Bromide ions oxidised: $2Br^-(aq) \longrightarrow Br_2(aq) + 2e^-$
  Chlorine molecules reduced: $Cl_2(aq) + 2e^- \longrightarrow 2Cl^-(aq)$
  c) Cu atoms oxidised: $Cu(s) \longrightarrow Cu^{2+}(aq) + 2e^-$
  Silver(I) ions reduced: $2Ag^+(aq) + 2e^- \longrightarrow 2Ag(s)$
2 a) Cu atoms oxidised: $Cu(s) \longrightarrow Cu^{2+}(aq) + 2e^-$
  Silver(I) ions reduced: $2Ag^+(aq) + 2e^- \longrightarrow 2Ag(s)$
  b) $Cu(s)|Cu^{2+}(aq) \, \| \, Ag^+(aq)|Ag(s) \; E_{cell}^\ominus = +0.46$ V

3 a) $Sn^{2+}(aq) + 2e^- \longrightarrow Sn(s)$ $E^{\ominus} = -0.14\,V$
 b) $Br_2(aq) + 2e^- \longrightarrow 2Br^-(aq)$ $E^{\ominus} = +1.07\,V$
 c) $Cr^{3+}(aq) + 3e^- \longrightarrow Cr(s)$
                 $E^{\ominus} = (-1.01\,V + 0.27\,V) = -0.74\,V$

4 $E^{\ominus} = -0.13\,V$

5 a) $Zn(s)\,|\,Zn^{2+}(aq)\,\|\,V^{3+}(aq), V^{2+}(aq)\,|\,Pt$
                             $E_{cell}^{\ominus} = +0.50\,V$
    $Zn(s) + 2V^{3+}(aq) \longrightarrow Zn^{2+}(aq) + 2V^{2+}(aq)$
 b) $Pt\,|\,2Br^-(aq), Br_2(aq)\,\|\,I_2(aq), 2I^-(aq)\,|\,Pt$
                             $E_{cell}^{\ominus} = +0.53\,V$
    $Br_2(aq) + 2I^-(aq) \longrightarrow 2Br^-(aq) + I_2(aq)$
 c) $Pt\,|\,2Cl^-(aq), Cl_2(aq)\,\|\,[PbO_2(s) + 4H^+(aq)],$
         $[Pb^{2+}(aq) + 2H_2O(l)]\,|\,Pt\;E_{cell}^{\ominus} = +0.10\,V$
    $PbO_2(s) + 4H^+(aq) + 2Cl^-(aq) \longrightarrow$
                     $Pb^{2+}(aq) + 2H_2O(l) + Cl_2(aq)$

6 a) Li > K > Ca > Na > Mg
 b) Zn > Fe > Sn > Pb > Cu

7 a) $H_2O_2 > MnO_4^- > Cr_2O_7^{2-} > Fe^{3+}$
 b) $H_2O_2 > Cl_2 > O_2 > Br_2 > ClO^-$

8 Sodium reacts violently with water and cannot be used as an electrode with aqueous solutions.

9 a) $Zn(s) + 2Ag^+(aq) \longrightarrow Zn^{2+}(aq) + 2Ag(s)$
 b) No reaction
 c) $2Cr(s) + 6H^+(aq) \longrightarrow 2Cr^{3+}(aq) + 3H_2(g)$.
    Very, very slow because of the protective oxide film on the surface of the metal.
 d) $Ca(s) + 2H^+(aq) \longrightarrow Ca^{2+}(aq) + H_2(g)$
 e) No reaction
 f) No reaction
 g) $Cl_2(aq) + 2I^-(aq) \longrightarrow 2Cl^-(aq) + I_2(aq)$

10 Hydrogen peroxide can oxidise itself. It does tend to disproportionate but the reaction is only rapid in the presence of a catalyst.

11 Gold(I) ions tend to disproportionate. Gold(I) ions changing to gold metal can oxidise gold(I) ions to gold(III).

2.15 Cells and batteries

1 a) (i) lead metal   **(ii)** lead and lead(IV) oxide
   (iii) sulfuric acid
 b) $Pb(s) + PbO_2(s) + 2SO_4^{2-}(aq) + 4H^+(aq) \longrightarrow$
                        $2PbSO_4(s) + 2H_2O(l)$
 c) 2.05 V
 d) Connecting six cells in series.
 e) $PbSO_4(s) + 2e^- \longrightarrow Pb(s) + 2SO_4^{2-}(aq)$
    $PbSO_4(s) + 2H_2O(l) \longrightarrow$
            $PbO_2(s) + SO_4^{2-}(aq) + 4H^+(aq) + 2e^-$

2.16 Corrosion

1 More negative: $Fe^{2+}(aq) + 2e^- \longrightarrow Fe(s)$
                         $E^{\ominus} = -0.44\,V$
   More positive: $O_2(aq) + 2H_2O(l) + 4e^- \longrightarrow 4OH^-$
                         $E^{\ominus} = +0.40\,V$
   So the stronger oxidising agent (oxygen) oxidises the stronger reducing agent (iron).

2 Ions in the solution raise the conductivity of the water. This means that corrosion currents can flow more quickly and the metal rusts faster.

3 The standard electrode potentials show that magnesium and zinc are stronger reducing agents than iron. They reduce oxygen in preference to iron. Also, when connected to the iron, they make the whole iron surface cathodic thus keeping the metal in its reduced state.

2.17 Enthalpy changes, bonding and stability

1 a) $\Delta H_1^{\ominus}$: Standard enthalpy change of formation of magnesium chloride
    $\Delta H_2^{\ominus}$: Standard enthalpy change of atomisation of magnesium
    $\Delta H_3^{\ominus}$: First ionisation enthalpy of magnesium
    $\Delta H_4^{\ominus}$: Second ionisation enthalpy of magnesium
    $\Delta H_5^{\ominus}$: Twice the standard enthalpy change of atomisation of chlorine (also the bond dissociation enthalpy for a chlorine molecule)
    $\Delta H_6^{\ominus}$: Twice the electron affinity (electron addition enthalpy) of chlorine
    $\Delta H_7^{\ominus}$: Lattice enthalpy of magnesium chloride
 b) Lattice enthalpy $= -2552\,kJ\,mol^{-1}$

2 Calculations based on a Born–Haber cycle assume that the overall enthalpy change for the formation of an ionic compound from its elements is the same whether this happens in one step, or in a series of stages via gaseous atoms and gaseous ions.

3 The two main enthalpy terms which determine the value of the standard enthalpy change of formation of an ionic compound are the total enthalpy change needed to ionise the metal atoms and the lattice enthalpy.

4 The standard enthalpy of atomisation is the energy needed to produce 1 mole of gaseous atoms from the element in its standard state under standard conditions. Bromine is a liquid consisting of diatomic molecules. The bond dissociation enthalpy of bromine is the energy needed to break up 1 mol of gaseous molecules into 2 mol gaseous atoms.

5 a) Lattice enthalpy of MgCl
    Standard enthalpy of atomisation of chlorine
    Electron affinity (electron addition enthalpy) of chlorine
    Standard enthalpy of atomisation of magnesium
    First ionisation enthalpy of magnesium
   All these quantities can be determined by experiment except for the lattice enthalpy for MgCl which does not exist.
 b) Standard enthalpy change of formation of MgCl $= -92\,kJ\,mol^{-1}$
 c) $\Delta H_f^{\ominus} = -458\,kJ\,mol^{-1}$
 d) The enthalpy change shows that MgCl is unstable relative to decomposition to Mg plus $MgCl_2$.

6 The formation of $MgCl_3$ is energetically very unfavourable for two reasons. Firstly, the third ionisation enthalpy of magnesium is very much greater than the first and second ionisation enthalpies because an electron in the second shell has eight fewer shielding electrons and has to be removed from an effective nuclear charge of 10+, whereas the two electrons in the outer shell only experience an effective nuclear charge of 2+. Secondly, removing the two outer electrons takes away the whole outer shell leaving an ion much smaller than the atom. Removing a third electron, one of the eight electrons from the second shell, does not bring about a significant reduction in size. So despite the 3+ charge the increase in lattice enthalpy is nowhere near enough to compensate for the extra energy needed to form a 3+ ion.

7 To form the stable compound $MgCl_2$ the magnesium atoms (2.8.2) lose the two electrons in their outer shell giving the ions (2.8) eight electron in the outer full shell. Mg^{2+} is significantly smaller than Mg^+ which still has one electron in the outer shell. The smaller size and larger charge of Mg^{2+} means that the lattice enthalpy of $MgCl_2$ is very much greater in magnitude than the lattice enthalpy of MgCl. This more than compensates for the extra energy needed to remove two electrons from an Mg atom.

The formation of $MgCl_3$ is energetically very unfavourable as explained in the answer to question 6. Thus the formula predicted by the octet rule is the one which is energetically stable.

8   Order of magnitude of lattice enthalpies:
$MgO > MgS > BaO > BaS$

9   **a)** The iodide ion is larger than the fluoride ion. So in LiI the 1+ and 1– charges are further apart and the attraction between them is weaker. So less energy is given out per mole when 1 mol LiI forms than when 1 mol of LiF forms.
b) The larger iodide ion is more polarisable than the small fluoride ion. So the bonding in LiI is likely to have a higher degree of electron sharing. So the discrepancy between the two values is likely to be greater for LiI.

10   $\Delta H_f^\ominus = + 143 \text{ kJ mol}^{-1}$

11   **a)** $\Delta H_f^\ominus = - 48 \text{ kJ mol}^{-1}$ **b)** $\Delta H_f^\ominus = - 244 \text{ kJ mol}^{-1}$
c) $\Delta H_f^\ominus = - 271 \text{ kJ mol}^{-1}$
The trend in values reflects two other trends:
- strength of triple bond in nitrogen molecules, the weaker double bond in oxygen molecules and the very weak single bond in fluorine molecules,
- the increasing strength of the X—H bond from N—H to O—H to F—H.

12   **a & b)** The value from enthalpies of formation is more accurate because the data refers to the specific elements and compounds involved. Average bond energies are averages covering a range of molecules and are only suitable for making estimates.

13   **a)** N_2O is unstable relative to its elements. A glowing splint is hot enough to speed up the reaction. The gas mixture produced consists of one-third oxygen by volume and this is a high enough proportion for the splint to burst into flame.
b) CCl_4 is more stable than the free elements however there is no mechanism which allows the reaction to proceed at a significant rate. This may be to do with the giant structure of graphite.
c) Hydrogen iodide is unstable relative to its elements but the activation energy is too high for the reaction to proceed at a measurable rate at room temperature. A red hot wire heats up the gas and speeds up the reaction.

14   **a)** The two sets of values correlate. The more positive the value of enthalpy change the higher the temperature needed to decompose the compound. See sections 2.19 and 2.10 for explanations.
b) No. See sections 2.19 and 2.20 for theories to account for the changes in the stability of compounds with temperature.

15   A magnesium ion is smaller than a barium ion. So in MgO the 2+ and 2– ions are closer together and the forces of attraction between them are stronger.

16   The oxide and carbonate ions both carry a 2– charge but the oxide ion is smaller. With a smaller anion the ions in the oxide pack closer together so the charges are closer and the attraction between them also greater. So more energy is given out per mole when the ions of

the oxide come together to form a crystal lattice.

17   Carbonate decomposes to form an oxide and carbon dioxide. Energy is needed to break up the carbonate ion. This energy comes from the release of energy as 2+ ions and 2– ions get closer together when the larger carbonate ions turn into the smaller oxide ions.

2.18 The solubility of ionic crystals

1   **a)** $\Delta H_{solution}^\ominus[\text{LiF}] = + 4 \text{ kJ mol}^{-1}$
$\Delta H_{solution}^\ominus[\text{LiI}] = - 96 \text{ kJ mol}^{-1}$
b) The fluoride ions and iodide ions are both singly charged. The iodide ion is the larger. So the lattice enthalpy for lithium iodide is smaller in magnitude as is the hydration enthalpy of the iodide ion.
c) LiI is more soluble than LiF. Chemists often use the value of ΔH to predict the direction of change. The more negative ΔH the more likely it is that the change will take place. For these two, very similar compounds the values of $\Delta H_{solution}^\ominus$ are consistent with the difference in solubility.

2   The iodates(V), chromates(V)

3   The fluorides.

2.19 Entropy changes

1   **a)** Spontaneous, fast   **b)** Not spontaneous
c) Spontaneous, slow   **d)** Not spontaneous
e) Spontaneous, fast

2   **a)** Bromine gas   **b)** Liquid water
c) Ammonia gas which can rotate and vibrate in more complex ways than HF.
d) Ethane gas which can rotate and vibrate in more complex ways than methane.
e) Aqueous sodium chloride.

3   **a)** Increase   **b)**   Increase   **c)** Decrease
d) Increase   **e)** Increase

4   $\Delta S^\ominus = +693.3 \text{ J K}^{-1} \text{ mol}^{-1}$. The reactions leads to an increase in the total number of moles of gas.

5   **a)** A solid reacts with a gas to form a single, solid product. Solids have lower standard molar entropies than gases.
b) $-T\Delta S^\ominus \approx +65 \text{ K}$. $\Delta H^\ominus$ is substantially more negative than this so overall $\Delta G^\ominus$ is negative.

6   **a)** A solid changes to a solid and a gas. Gases have higher standard molar entropies than solids.
b) At 298 K, $\Delta G^\ominus$ is positive ($\approx +67 \text{ kJ mol}^{-1}$)
c) At about 696 K when $\Delta H^\ominus \approx T\Delta S$.

2.20 Free energy

1   No. If the entropy change is negative the free energy change becomes more positive as the temperature rises.

2   **a)** $\Delta G^\ominus = +218 \text{ kJ mol}^{-1}$
b) No. $\Delta H^\ominus$ is positive at 298 K.
c) At about 1600 K when the magnitude of $T\Delta S$ is more positive than ΔH, so that ΔG is negative.
d) The decomposition temperature of $BaCO_3$ is much higher than the decomposition temperature for $MgCO_3$. This is consistent with the observation that magnesium carbonate decomposes on heating with a Bunsen flame while barium carbonate does not. The group 2 carbonates become more stable down the group.

3   **a)** $\Delta H^\ominus = -332 \text{ kJ mol}^{-1}$, $\Delta S^\ominus = -114.5 \text{ J mol}^{-1} \text{ K}^{-1}$
b) At about 2900 K (remember to work in consistent units bearing in mind that the units for entropy changes are $\text{J mol}^{-1} \text{ K}^{-1}$ whereas the units for ΔH are kJ mol^{-1})

4 Reactant: steam, products: hydrogen and oxygen.
Reactants: hydrogen and chlorine at 298 K, products: hydrogen chloride.
Reactants: PCl_5 at 298 K, products: PCl_3 and chlorine.
Reactants: ammonia and hydrogen chloride, product: ammonium chloride.

5 See *AS Chemistry* page 120 Figure 3.19.2.

Section three Inorganic chemistry

3.2 Metal ions in water

1 See page 89. The changes shown on that page are reversed. Adding protons to $[Al(H_2O)_2(OH)_4]^-(aq)$ first produces a precipitate of $[Al(H_2O)_3(OH)_3](s)$ and then redissolves the precipitate to give a solution of hydrated aluminium(III) ions.

2 a) $Mg^{2+}(aq) + CO_3^{2-}(aq) \longrightarrow MgCO_3(s)$
b) The hydrated aluminium(III) ion is an acid. It is sufficiently acidic to react with carbonate ions to give carbon dioxide gas.

3 Sodium and potassium salts are generally soluble in water. Most aluminium(III) and iron(III) salts are insoluble.

4 Polarity: the polar water molecules can hydrate the ions of salts such as sodium chloride. The energy released when ions are hydrated (the hydration energy) is often large enough to compensate for the energy needed to separate the ions from the crystal lattice (–lattice energy) (see page 74).
Lone pairs: the lone pairs allow water molecules to act as ligands forming co-ordinate bonds with transition metal ions. Salts with aquo complexes are often soluble.
Hydrogen bonding: this helps to dissolve solutes with groups which can take part in hydrogen bonding. Alcohols, organic acids, sugars and amines are all soluble in water because they can hydrogen bond with water molecules.

5 Aluminium hydroxide is a neutral complex, effectively an uncharged molecule. In general, molecules are less soluble in water than ions.

6 a) Polarising power increases as the charge on the ions increase while the size of the ions decreases.
b) Polarising power decreases as the ions, with the same charge, get larger.

3.3 Complex formation

1 Oxidation numbers of the metal ions in the order printed: +2, +1, +1, +3, +2.

2 $Ag^+(aq) + Cl^-(aq) \longrightarrow AgCl(s)$ – white precipitate
$AgCl(s) + 2NH_3(aq) \longrightarrow [Ag(NH_3)_2]^{2+}(aq) + Cl^-(aq)$
– precipitate redissolves

3 $AgBr(s) + 2S_2O_3^{2-}(aq) \longrightarrow$
$\qquad\qquad [Ag(S_2O_3)_2]^{3-}(aq) + Br^-(aq)$

4 The reagent is used to distinguish aldehydes from ketones. Aldehydes reduce the silver(I) ions to metallic silver. If not present as a complex the silver(I) ions would precipitate as silver oxide under alkaline conditions.

5 In order that the complexes appear in the questions: linear, octahedral, tetrahedral

6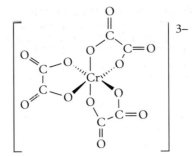

7 a) Stability is in reverse order to the sequence of complexes in the question i.e. most stable = $[Ni\text{-}edta]^{2-}$
b) Equilibrium constants indicate the position of equilibrium for a reversible reaction (see page 24). The more positive the equilibrium constant for the formation of a complex, the more stable the complex.

8 a) $[Co(H_2O)_6]^{2+}(aq) + 6NH_3(aq) \longrightarrow$
$\qquad\qquad [Co(NH_3)_6]^{2+}(aq) + 6H_2O(l)$
b) $[Co(NH_3)_6]^{2+}(aq) + 4Cl^-(aq) \longrightarrow$
$\qquad\qquad [Co(Cl)_4]^{2-}(aq) + 6NH_3(aq)$
c) $[Fe(H_2O)_6]^{2+}(aq) + 6CN^-(aq) \longrightarrow$
$\qquad\qquad [Fe(CN)_6]^{4-}(aq) + 6H_2O(l)$

9 $[Cu(H_2O)_6]^{2+}(aq) + 2OH^-(aq) \longrightarrow$
$\qquad\qquad [Cu(OH)_2(H_2O)_4]^{2+}(s) + 2H_2O(l)$
$[Cu(OH)_2(H_2O)_4]^{2+}(s) + 4NH_3(aq) \longrightarrow$
$\qquad [Cu(NH_3)_4(H_2O)_2]^{2+}(aq) + 2H_2O(l) + 2OH^-(aq)$

10 a) The hard water contains 228 mg dm^{-3} hardness as $CaCO_3$.
b) 1 mg dm^{-3} is 10^{-3} g in 10^3 g water. This is 1 part by mass of calcium carbonate in 10^6 g water.

11 Amount of edta added from the burette = 2.290×10^{-4} mol. The capsule contains the same amount of zinc in moles. Mass of zinc = 0.1498 g

3.4 Acid–base reactions

1 $Zn(s) + 2H^+(aq) \longrightarrow Zn^{2+}(aq) + H_2(g)$
– a redox reaction
$CuO(s) + 2H^+(aq) \longrightarrow Cu^{2+}(aq) + H_2O(l)$
– the oxide ion accepts two protons to form water
$CaCO_3(s) + 2H^+(aq) \longrightarrow Ca^{2+}(aq) + CO_2(g) + H_2O(l)$
– the carbonate ion accepts two protons to form carbonic acid which decomposes to carbon dioxide and water.

2 These are all strong acids which ionise to give $H^+(aq)$ in solution. Dilute solutions of these acids are all similar because they all show the reactions of the aqueous hydrogen ion.

3 See the answer to question 1.

4 The equations on page 91 are all proton transfer reactions with protons being removed progressively from ligand water molecules turning H_2O into OH^-.

5 The proton donor in the aqueous hydrochloric acid is $H_3O^+(aq)$. The proton acceptor in the limestone is the carbonate ion.

6 a)
$\qquad Na^+ \left[\ddot{\underset{..}{\overset{..}{Cl}}}\!: \right]^- \qquad \ddot{\underset{..}{\overset{..}{Cl}}}\!\!\overset{\overset{\displaystyle \ddot{\underset{..}{\overset{..}{Cl}}}}{}}{\underset{\underset{\displaystyle \ddot{\underset{..}{\overset{..}{Cl}}}}{}}{\times\!C\!\times}}\!\!\ddot{\underset{..}{\overset{..}{Cl}}}$

b)

c) The outer electron configuration of a noble gas is ns^2np^6, so for periods 2 and 3 there are 8 electrons in the outer shell. In simple examples atoms gain, lose or share electrons to achieve the ns^2np^6 electron configuration of the nearest noble gas in the periodic table.

7

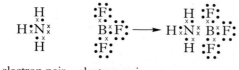

electron pair donor   electron pair acceptor

8 The reactions of aqueous acids with metal oxides and metal carbonates are Lewis acid-base reactions. In each case H^+ is a Lewis acid. The Lewis bases are the oxide ion and the carbonate ion. At the same time the metal ions also act as Lewis acids accepting protons from water molecules (Lewis bases) to form aquo complexes.

3.5 Redox reactions

1 Oxidation: gain of oxygen by carbon to form carbon monoxide. This gains further oxygen from the iron ore and is oxidised to carbon dioxide.
Reduction: loss of oxygen from the iron oxide in the metal ore to obtain the free metal.

2 a) Oxidation at the anode: $2Cl^-(aq) \longrightarrow Cl_2(g) + 2e^-$
Reduction at the cathode: $2H^+(aq) + 2e^- \longrightarrow H_2(g)$
b) Oxidation at the anode: $Cu(s) \longrightarrow Cu^{2+}(aq) + 2e^-$
Reduction at the cathode: $Cu^{2+}(aq) + 2e^- \longrightarrow Cu(s)$

3 Some of the bromide ions (oxidation state −1) are oxidised to bromine (oxidation state 0).
The oxidising agent is sulfuric acid (sulfur in oxidation state +6) which is reduced to sulfur dioxide (sulfur in oxidation state +4).
(See *AS Chemistry* pages 151 – 2).

4 a) +5 **b)** +5 **c)** +2 **d)** −1 **e)** +1

5 a) sulfurous acid (+4 state)
b) sulfuric acid (+6 state)
c) hydrogen sulfide (−2 state)

6 a) Nitrogen:
+5	HNO_3, NO_3^-
+4	NO_2, N_2O_4
+3	HNO_2, NO_2^-
+2	NO
+1	N_2O
0	N_2
−1	
−2	N_2H_4
−3	NH_3

b) Chlorine:
+7	$KClO_4$, ClO_4^-
+6	
+5	$KClO_3$, ClO_3^-
+4	
+3	$KClO_2$, ClO_2^-
+2	
+1	$HOCl$, OCl^-
0	Cl_2
−1	HCl, $NaCl$, Cl^-

7 a) Oxidised **b)** Reduced
c) Reduced **d)** Oxidised

8 a) $Fe^{3+}(aq) + e^- \longrightarrow Fe^{2+}(aq)$
b) $Br_2(aq) + 2e^- \longrightarrow 2Br^-(aq)$

c) $H_2O_2(aq) + 2H^+(aq) + 2e^- \longrightarrow 2H_2O(l)$

9 Chlorine oxidises iodide ions to iodine. Iodine then gives a blue–black colour with starch. Any oxidising agent which can oxidise iodide ions gives a positive result.

10 a) $Zn(s) \longrightarrow Zn^{2+}(aq) + 2e^-$
b) $2I^-(aq) \longrightarrow I_2(s) + 2e^-$
c) $SO_2(aq) + 2H_2O(l) \rightarrow SO_4^{2-}(aq) + 4H^+(aq) + 2e^-$

11 a) $2IO_3^-(aq) + 12H^+(aq) + 10I^-(aq) \longrightarrow$
$$6I_2(s) + 6H_2O(l)$$
b) $H_2O_2(aq) + 2H^+(aq) + 2Fe^{2+}(aq) \longrightarrow$
$$2H_2O(l) + 2Fe^{3+}(aq)$$
c) $Cr_2O_7^{2-}(aq) + 2H^+(aq) + 3SO_2(aq) \longrightarrow$
$$2Cr^{3+}(aq) + H_2O(l) + 3SO_4^{2-}(aq)$$
d) $MnO_2(s) + 4H^+(aq) + 2Cl^-(aq) \longrightarrow$
$$Mn^{2+}(aq) + Cl_2(g) + 2H_2O(l)$$
e) $Cu(s) + 2NO_3^-(aq) + 4H^+(aq) \longrightarrow$
$$Cu^{2+}(aq) + 2NO_2(g) + 2H_2O(l)$$

12 a) 2 cm^3
b) 32 cm^3
c) 80 cm^3

13 1.54 mol dm^{-3}
14 11.2%
15 72.7%

3.6 The periodic table

1 Oxygen: $1s^2 2s^2 2p^4$
Fluorine: $1s^2 2s^2 2p^5$
Magnesium: $1s^2 2s^2 2p^6 3s^2$
Titanium: $1s^2 2s^2 2p^6 3s^2 3p^6 3d^2 4s^2$
Cobalt: $1s^2 2s^2 2p^6 3s^2 3p^6 3d^7 4s^2$

2 a) They all have the electron configuration of neon: $1s^2 2s^2 2p^6$
b) Along the sequence the number of protons in the nuclei increases by one from one entity to the next. So the charge on the nucleus increases along the sequence while the electron configuration stays the same. So from N^{3-} to Mg^{2+} the radius decreases from 171 pm to 72 pm.

3 Atomic radii decrease across a period as electrons are fed into the same main shell as the charge on the nucleus increases. There is a jump up in atomic radius at the beginning of each period as a new shell starts to fill.
Metals to the left of the period lose electrons to form positive ions. Non-metals to the right form negative ions by gaining electrons. Elements in the same group generally form simple ions with the same charge.

4 Across a period the added electrons go into the same shell as the charge on the nucleus increases. Shielding by electrons in the same shell is limited and so the electrons in the outer shell are held increasingly strongly across a period. So the general trend is for ionisation enthalpies to increase across a period.
Down a group elements have the same outer electron configuration but increasing numbers of full inner shells. Inner shells shield the outer electrons from the full attraction of the nuclear charge. As a result down a group the outer electrons get further and further away from the same effective nuclear charge. So down a group ionisation enthalpies decrease.

5 a) The oxide layer on the surface of beryllium prevents attack by air, water and acids. Beryllium oxide is amphoteric while the oxides of the other group 2 metals are basic. The small size of the beryllium ion means that bonding in the chloride is polar covalent

and the solid has a layer lattice.
Fluorine rapidly oxidises water and forms ionic compounds with elements such as aluminium which form covalently bonded compounds with other halogens. Hydrogen bonding means that hydrogen fluoride has a much higher boiling point than expected for its molecular size. Hydrogen fluoride is a weak acid while the other hydrogen halides are strong acids.

b) Elements in period 2 have only 2s and 2p orbitals available for bonding and no d orbitals. This means that they can form a maximum of four covalent bonds.

3.7 Period 3: the elements

1 a) (i) $2Na(s) + 2H_2O(l) \longrightarrow 2NaOH(aq) + H_2(g)$
  (ii) $Mg(s) + 2H_2O(l) \longrightarrow Mg(OH)_2(aq) + H_2(g)$
  (iii) $Cl_2(aq) + 2H_2O(l) \longrightarrow$
  $ HOCl(aq) + H^+(aq) + Cl^-(aq)$
  b) (i) Sodium is oxidised from the 0 to the +1 state. Hydrogen is reduced from +1 to 0.
  (ii) Magnesium is oxidised from 0 to +2. Hydrogen is reduced from +1 to 0.
  (iii) Chlorine disproportionates from 0 to the +1 and −1 states.

2 Melting points and boiling points rise from sodium to silicon. They then fall sharply to phosphorus and remain low for the other non-metals in the period.

3

← Giant structures →	Molecular → Single atoms
Na Mg Al | Si | P$_{white}$ S Cl Ar	

← Metallic bonding → \| ← Covalent bonding →

4 Metals with giant structures have high boiling points and high enthalpies of vaporisation. Vaporising a giant structure requires much energy to break up the network of strong bonds. Molecular substances are much easier to vaporise because the molecules are held together by weak intermolecular forces.

5 Both chlorine molecules and argon atoms are non-polar. Diatomic chlorine molecules are larger than argon atoms so there is a larger surface area over which intermolecular forces can act between molecules in contact.

6
$$\overset{x\,x}{\underset{x\,x}{\times}}\!Cl\!\times\!\overset{\bullet\bullet}{\underset{\bullet\bullet}{Cl}}\!\bullet$$

7 The bonding electrons in metals are delocalised and free to move when subject to an applied electrical potential difference. The electron pairs in covalent bonds are normally localised between two atoms. There are no free electrons to carry a current.

8 a) $4Na(s) + O_2(g) \longrightarrow 2Na_2O(s)$
  $2Na(s) + O_2(g) \longrightarrow Na_2O_2(s)$
  b) $4P(s) + 10O_2(g) \longrightarrow P_4O_{10}(s)$

9 See *AS Chemistry* Figure 4.3.6 on page 135.

10
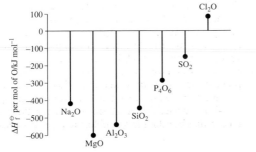

11
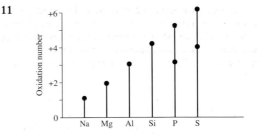

12 a) $2Al(s) + 3Cl_2(g) \longrightarrow 2AlCl_3(s)$
  b) $Si(s) + 2Cl_2(g) \longrightarrow SiCl_4(s)$
  c) $2P(s) + 3Cl_2(g) \longrightarrow 2PCl_3(s)$
  d) $2P(s) + 5Cl_2(g) \longrightarrow 2PCl_5(s)$

13 a) Calcium chloride is a drying agent. The tube lets excess gas escape from the apparatus but prevents moisture entering. Water hydrolyses hot aluminium chloride.
  b) Silicon tetrachloride is a liquid. The liquid can be condensed by replacing the sample tube with a delivery tube (with bungs) to carry the vapours into a vertical side arm test tube cooled in ice cold water. The tube with the drying agent is then attached to the side arm.

3.8 Period 3: the oxides

1 a) $Na_2O(s) + 2HCl(aq) \longrightarrow 2NaCl(aq) + H_2O(l)$
  b) $MgO(s) + 2HNO_3(aq) \longrightarrow Mg(NO_3)_2(aq) + H_2O(l)$
  c) $MgO(s) + H_2O(l) \longrightarrow Mg(OH)_2(s)$

2 See page 91.

3 a) The larger the charge and the smaller the radius the larger the polarising power. So the number ratio of charge to atomic radius (here in nm) can be used as a measure of polarising power. Na^+: 8.8, Mg^{2+}: 27.8, Al^{3+}: 56.6, Si^{4+}: 100.
  b) Note that polarising power increases along this series. The polarising power of the silicon ion is so large that bonds between silicon atoms with other elements are covalent.

4 P_2O_5

5 a) $P_2O_5(s) + 3H_2O(l) \longrightarrow 2H_3PO_4(aq)$
  b) $SO_2(g) + H_2O(l) \longrightarrow H_2SO_3(aq)$
  c) $SO_3(g) + H_2O(l) \longrightarrow H_2SO_4(aq)$

6
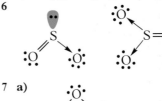

7 a)
$$\underset{x\,x\,x}{\overset{x\,x}{\times}}\!Cl\diagdown\overset{\bullet\bullet}{\underset{\bullet\bullet}{O}}\diagup Cl\!\underset{x\,x}{\overset{x\,x}{{}_{x}}}$$

b) $\Delta H_f^{\ominus}[Cl_2O(g)] = +80.3$ kJ mol^{-1}. This shows that the decomposition of this compound into its elements is exothermic. The compound is unstable relative to decomposition into its elements (but see sections 2.19 and 2.20). $\Delta H_f^{\ominus}$, or more reliably $\Delta G_f^{\ominus}$, cannot be used to predict the rates of reaction and so cannot suggest whether or not a compound is likely to decompose explosively.

8 In period 3: the ionic oxides of the metals of groups 1, 2 and 3 have high melting points; silicon oxide, with its giant covalent structure also has a high melting point; the molecular oxides of the other non-metals have low melting points. The oxides of groups 6 and 7 are gases.

9 Sodium oxide reacts with water to form sodium hydroxide which is a strong alkali. Magnesium oxide also combines with water to form a basic oxide which is sparingly soluble but makes the water alkaline. Aluminium and silicon oxides are insoluble and do not affect the pH of water. Phosphorus(V) oxide reacts with water to form a soluble oxoacid.

3.9 Period 3: the chlorides

1

$$[:\overset{\bullet\bullet}{\underset{\bullet\bullet}{Cl}}^{x}]^{-} \quad Mg^{2+} \quad [:\overset{\bullet\bullet}{\underset{\bullet\bullet}{Cl}}:]^{-}$$

2 The charge on the magnesium is 2+. The attraction between this smaller metal ion and chloride ions is stronger than the attraction between singly charged, larger sodium ions and chloride ions.

3 $Na^+ + e^- \longrightarrow Na$
$Cl^- \longrightarrow \frac{1}{2}Cl_2 + e^-$

4 The aluminium ion has a high polarising power (see page 68 and answer to question 3 in section 3.8). The aluminium ion distorts the electron cloud round the chloride ion giving rise to a large degree of electron sharing.

5 $Al_2Cl_6(g) + 3H_2O(g) \longrightarrow Al_2O_3(s) + 6HCl(g)$

6 Aluminium(III) ions are hydrated in aqueous solution. The small, highly charged aluminium ion draws electrons towards itself from the ligand water molecules. This means that the ligand water molecules ionise more readily than free water molecules. The hydrated ion is an acid (see page 91).

7 a)

b) $SiCl_4$: tetrahedral. PCl_3: pyramidal. PCl_5: trigonal bipyramid. S_2Cl_2: non-linear with the two Cl—S—S angles less than 109°.

c) Polar: PCl_3 and S_2Cl_2.

8 $PCl_3(l) + 3H_2O(l) \longrightarrow H_3PO_3(aq) + 3HCl(aq)$

9 PCl_4^+: tetrahedral.
PCl_6^-: octahedral.

10 PCl_5 is a reagent which replaces an –OH group with a chlorine atom and this is what happens both when it reacts with water and with ethanol. In both reactions one of the products is HCl.

11 Hydrogen chloride: fumes in moist air, turns blue litmus red and forms a white smoke with ammonia gas.

12 The ionic chlorides of sodium and magnesium have high melting points. Polar covalent aluminium chloride is a solid which does not melt but sublimes. The molecular covalent chlorides $SiCl_4$, PCl_3 and S_2Cl_2 are liquids at room temperature.

3.10 Group 4

1 Tin and lead are clearly metals. They are shiny when polished, bendable and they conduct electricity. Silicon and germanium also look shiny but they are brittle and, when pure, they are semi-conductors. Diamond is a clearly non-metallic form of carbon. It does not have the silvery shine of a metal. It is brittle and does not conduct electricity. Graphite though a brittle solid, does have some metallic features; it is a shiny-grey colour (looking a bit like lead) and it conducts electricity.

2 a) The trend is for melting points to fall down the group.
   b) Carbon (as diamond) with its giant lattice held by rigid, directional and strong covalent bonds between small atoms has the highest melting point. Lead has a giant structure but the atoms are larger and the bonding is non-directional metallic bonding.

3 Si: $1s^2 2s^2 2p^6 3s^2 3p^2$
   Sn: $1s^2 2s^2 2p^6 3s^2 3p^6 3d^{10} 4s^2 4p^6 4d^{10} 5s^2 5p^2$

4 a) The general trend is for ionisation enthalpies to decrease down the group.
   b) The charge on the nucleus increases down the group, but so does the number of inner full shells of electrons. The shielding effect of the inner full shells means that the outer electrons feel more or less the same effective nuclear charge of 4+. Down the group the outer electrons get further and further from the same effective nuclear charge and so are easier to remove.
   c) The trend means that it becomes progressively easier for the elements to form positive ions and show typically metallic properties.

5 a) $CO_2(g) + OH^-(aq) \longrightarrow HCO_3^-(aq)$
   b) $SnO(s) + 2HCl(aq) \longrightarrow Sn^{2+}(aq) + 2Cl^-(aq) + H_2O(l)$
   c) If cold:
      $PbO_2(s) + 4HCl(aq) \longrightarrow PbCl_4(l) + 2H_2O(l)$
      On heating lead(IV) oxide oxidises chloride ions to chlorine.
   d) $PbO(s) + 2OH^-(aq) + H_2O(l) \longrightarrow Pb(OH)_4^{-}(aq)$

6 a) Under acid conditions the $Fe^{3+}(aq)$, $Fe^{2+}(aq) | Pt$ electrode system is more positive than the $Sn^{4+}(aq)$, $Sn^{2+}(aq) | Pt$ electrode system. Iron(III) can oxidise tin(II) to tin(IV). See pages 57 and 58.
   b) Under acid conditions the $[PbO_2(s) + 4H^+(aq)]$, $[Pb^{2+}(aq) + 4H_2O(l)] | Pt$ electrode system is more positive than the $Cl_2(aq), 2Cl^-(aq) | Pt$ electrode system. Lead(IV) can oxidise chloride ions to chlorine.

7 CCl_4 is a non-polar molecular compound. The attractions between the molecules are the result of temporary dipoles. The intermolecular forces between these relatively large molecules are sufficient to keep the compound in the liquid state at room temperature. CCl_4 molecules cannot take part in hydrogen bonding so they cannot interact strongly enough with water molecules to break into the relatively strongly hydrogen-bonded system in liquid water.

8 Melt the solid in a crucible and test to see if the melt conducts electricity. Electrolysis should give tin at the cathode and chlorine at the anode.

9 See pages 82–83. This means that the reaction tends to go but very slowly. The standard enthalpy change (or better the standard free energy change) for the reaction is negative but the activation energy is so high that the big energy barrier prevents the reaction going at a measurable rate.

10 They all have the outer electron configuration $ns^2 np^2$ and form compounds in the +2 and +4 states.

11 Non-metal oxides, chlorides and hydrides are characterised by covalent bonding. The equivalent compounds of reactive metals are ionic. A 4+ ion would be highly polarising. So much so that, if it could exist, it would polarise a neighbouring anion to such an extent that the bonding would be covalent (see page 68). There are no simple ionic compounds with 4+ ions. The polarising power of the 2+ ions of relatively large atoms such as the atoms of tin and lead means

that ionic compounds of these elements can exist with electronegative atoms such as fluorine, oxygen and chlorine.

12 Carbon is exceptional in its ability to form long chains and rings of atoms (organic chemistry). The small carbon atoms can also form double and triple bonds with each other (alkenes and alkynes) and with oxygen (carbon dioxide and carbonyl compounds). Carbon forms molecular oxides while the other elements in the group form oxides with giant structures. The tetrachloride of carbon does not hydrolyse in water, unlike the equivalent compounds of other elements in the group.

13 Broadly speaking the trend down all three groups is for the metallic characteristics to become more prominent and the non-metal characteristics to become less prominent. Down group 2 the elements become more reactive as metals. Down group 7 the elements become less reactive as non-metals. Down group 4 the trend is from non-metals to metals.

3.11 Period 4: d-block elements

1 **a)** $1s^2 2s^2 2p^6 3s^2 3p^6 3d^6 4s^2$
   b) $1s^2 2s^2 2p^6 3s^2 3p^6 3d^6$
   c) $1s^2 2s^2 2p^6 3s^2 3p^6 3d^5$

2 First ionisation energies rise sharply from Na to Ar. Across the period the electrons are fed into the same outer shell as the nuclear charge increases. Shielding by the added electrons is very limited and so the electrons in the outer shell feel the attraction of the rising nuclear charge.
   First ionisation energies rise much less markedly from Sc to Zn. Across this part of period 4 the electrons are fed into an inner d-shell as the nuclear charge increases. Shielding by the added electrons means that the outer 4s electrons in the outer 4th shell do not feel the full attraction of the rising nuclear charge.

3 Most d-block elements can exist in more than one oxidation state. The +2 state becomes more stable relative to the +3 state across the series. From Sc to Mn the highest oxidation state corresponds to using all the 3d and 4s electrons in bonding.

4 **a)** +3: Cr_2O_3 +6: K_2CrO_4
   b) +2: $MnSO_4$ +4: MnO_2 +7: $KMnO_4$
   c) +2: $FeSO_4$ +3: $FeCl_3$
   d) +1: Cu_2O +2: $CuSO_4$

5 $Cr^{3+}(aq), Cr^{2+}(aq) | Pt \ E^{\ominus} = -0.41$ V.
   $Fe^{3+}(aq), Fe^{2+}(aq) | Pt \ E^{\ominus} = +0.77$ V.
   Chromium(II) more strongly reducing tending to turn to cobalt(III). Iron(III) more strongly oxidising tending to turn to iron(II).

6 **a)** Tetrahedral (see page 95) **b)** Octahedral.

7 Zn^{2+}: $1s^2 2s^2 2p^6 3s^2 3p^6 3d^{10}$
   Cu^+: $1s^2 2s^2 2p^6 3s^2 3p^6 3d^{10}$
   Sc^{3+}: $1s^2 2s^2 2p^6 3s^2 3p^6$
   The d-orbitals are either full or empty in all three ions, so no transitions between d-energy levels with slightly different energies are possible.

8 **a)** Change of oxidation state.
   b) Change of ligand.
   c) Change of ligand.
   d) Change of both oxidation state and ligand.

9 Prepare a series of accurately diluted solutions from an accurate standard solution with concentrations in the required range. Take samples at each dilution and take colorimeter readings. Plot a graph of meter reading against concentration.

10 Top left: $Cr_2O_7^{2-}(aq)$ – orange
   Top right: $Ni^{2+}(aq)$ – green
   Bottom left: $Co^{2+}(aq)$ – pink
   Bottom right: $Cu^{2+}(aq)$ – blue

11 **a)** Heat the solution and/or add some Mn^{2+} ions.
   b) (i) At first the solution will turn the purple colour of unreacted but diluted $KMnO_4$.
   (ii) So long as the ethanedioate is in excess, once the reaction begins the purple colour will disappear as manganate(VII) reacts to form the almost colourless Mn^{2+} ions. Bubbles of carbon dioxide will form as the solution is swirled.

12 For a continuous flow process, the reaction mixture (gas or liquid) can flow through a solid heterogeneous catalyst in a reactor so that there is no problem in separating the products from the catalyst. Alternatively for a batch process, the solid catalyst can be suspended in the reaction mixture and then recovered by filtration.

13 Vanadium(V) oxide.

3.12 Transition metals

1 V: $1s^2 2s^2 2p^6 3s^2 3p^6 3d^3 4s^2$
   V^{3+}: $1s^2 2s^2 2p^6 3s^2 3p^6 3d^2$

2 +4

3 $2VO_2^+(aq) + 8H^+(aq) + 3Zn(s) \longrightarrow$
   $2V^{2+}(aq) + 4H_2O(l) + 3Zn^{2+}(aq)$

4 $Cr_2O_7^{2-}(aq) + 14H^+(aq) + 6Fe^{2+}(s) \longrightarrow$
   $2Cr^{3+}(aq) + 7H_2O(l) + 6Fe^{3+}(aq)$

5 See pages 171–172. Oxidation of alcohols to carbonyl compounds and aldehydes to acids.

6 Sulfur dioxide is a reducing agent. It reduces orange dichromate(VI) ions to green chromium(III) ions. (Any reducing gas gives a positive result with this test but sulfur dioxide is the most likely example.)

7 **a)** $2MnO_4^-(aq) + 6H^+(aq) + 5H_2O_2(aq) \longrightarrow$
   $2Mn^{2+}(aq) + 8H_2O(l) + 5O_2(g)$
   b) During the titration the purple manganate(VII) ions are reduced to such a dilute solution of manganese(II) ions that it appears colourless. At the end point the slightest excess of $MnO_4^-(aq)$ gives rise to a permanent pale purple colour.
   c) 0.0438 mol dm^{-3}

8 $3MnO_4^{2-}(aq) + 4H^+(aq) \longrightarrow$
   $2MnO_4^-(aq) + MnO_2(s) + 2H_2O(l)$

9 The initial precipitate is $Mn(OH)_2(s)$ which is then oxidised to brown $MnO_2(s)$ by oxygen in the air.

10 $[Fe(H_2O)_6]^{3+}(aq)$ loses three protons to become the neutral complex $[Fe(H_2O)_3(OH)_3](s)$ which is insoluble and precipitates as an orange-brown solid.

11 The more highly charged iron(III) ion has a greater polarising power than the iron(II) ion. The greater the polarising power of the metal ion, the more electrons are pulled from ligand water molecules and the greater the tendency of these ligand water molecules to donate protons to free water molecules thus increasing the $H_3O^+(aq)$ concentration and making the solution more acidic.

12 The hydrated iron(III) ion is more acidic (see answer to question 11). It is so acidic that it can react with carbonate ions to form carbon dioxide.

13 $[Fe(H_2O)_6]^{3+}(aq) + SCN^-(aq) \longrightarrow$
   $[Fe(H_2O)_5(SCN)]^{2+}(aq) + H_2O(l)$

14 The oxidising agent of a more positive electrode system tends to oxidise the reduced form of a less positive (or more negative) electrode system. Here $S_2O_8{}^{2-}$(aq) ions can oxidise iron(II) to iron(III) which can then oxidise iodide ions to iodine. The second reaction regenerates iron(II) ions and so the iron ions are only needed in small amounts to act as the catalyst because the are not used up.

$$I_2(aq) + 2e \longrightarrow 2I^-(aq) \qquad E^\ominus = +0.54\,V$$
$$Fe^{3+}(aq) + 2e^- \longrightarrow Fe^{2+}(aq) \qquad E^\ominus = +0.77\,V$$
$$S_2O_8{}^{2-}(aq) + 2e^- \longrightarrow 2SO_4{}^{2-}(aq) \quad E^\ominus = +2.01\,V$$

15 Iron is used as a catalyst in the Haber process for the manufacture of ammonia from nitrogen and hydrogen gases.

16 See the list of characteristic properties on page 131 and pick examples from text on page 134 and the answers to questions 10 to 15.

17 Cobalt(II) associated with chloride ions is blue. Adding water to the chloride produces hexaaquacobalt(II) ions which are pink.

18 This is an application of Le Chatelier's principle. See the equation for the reversible ligand exchange reaction involving cobalt(II) water molecules and chloride ions on page 135. Adding concentrated hydrochloric acid increases the concentration of chloride ions so that the equilibrium shifts to the right producing more of the blue $[CoCl_4]^{2-}$ ions.

19 See the electrode potentials quoted on page 135. Converting the aquo complexes to ammine complexes makes the standard electrode potential for the chromium(III), chromium(II) | Pt electrode less positive. In other words chromium is less powerfully oxidising. So chromium(III) then has a small tendency to turn to chromium(II). In the presence of ammonia it is therefore easier to oxidise the +2 state to the +3 state.

20 The hydrogenation of unsaturated oils (see page 186).

21 Octahedral, tetrahedral (see page 95).

22 The electrode potential for the Cu^+(aq), Cu(s) | Pt electrode is more positive than the electrode potential for the Cu^{2+}(aq), Cu^+(aq) | Pt. So copper(I) turning to copper can oxidise itself to copper(II) so copper(I) can oxidise itself

$$Cu^{2+}(aq) + 2e^- \longrightarrow Cu^+(aq) \qquad E^\ominus = +0.15\,V$$
$$Cu^+(aq) + e^- \longrightarrow Cu(s) \qquad E^\ominus = +0.52\,V$$

23 Adding a few drops of copper(II) sulfate to a mixture of zinc and sulfuric acid greatly speeds up the formation of hydrogen gas. A little of the zinc displaces copper metal which deposits on the surface of the zinc. Hydrogen gas forms much more readily on the surface of copper than on the surface of zinc.

24 See page 97.

25 Fehling's solution is strongly alkaline. Unless present as a stable complex, the copper(II) ions would precipitate as copper(II) hydroxide.

Section four Organic chemistry

4.2 Structures of organic molecules

1 $C_2H_6O_2$

2 Empirical and molecular formulae both: $C_2H_5O_2N$.

3 Empirical formula, CHO_2; molecular formula, $C_2H_2O_4$; ethanedioic acid.

4 Empirical formula, C_2H_5; molecular formula, C_4H_{10}; 2-methylpropane.

5 **a)** 2-bromobutane **b)** propyl ethanoate
c) 3-methylbutan-2-ol **d)** pent-2-ene

6 **a)** primary alcohol, secondary alcohol, aldehyde
b) carbonyl group as in a ketone, carboxylic acid group.

7 carbonyl group as in a ketone, carbon-carbon double bond, carboxylic acid group.

8 Isomers: hexane; 2-methylpentane; 3-methylpentane; 2,3-dimethylbutane.

9 Isomers: 2-bromopentane, 3-bromopentane.

10 Isomers include: the alcohol butan-1-ol and the ether ethoxyethane.

11 In general, the more branched the hydrocarbon the lower its boiling point: 2,2-dimethylpropane (10 °C); 2-methylbutane (28 °C); pentane (36 °C).

4.3 Stereochemistry

1 The compounds with geometric isomers are: but-2-ene and 1,2-dichloroethene.

2 Chiral objects: glove, shoe, screw, spiral binding of the notebook.

3 Chiral molecules: $CH_3CHClBr$, $CH_3CHOHCO_2H$.

4 Isoleucine – the carbon atom at the centre of the tetrahedron with four different groups attached to it; carvone – the carbon atom at the bottom of the hexagon as drawn in Figure 4.3.10 (do not forget the hydrogen atom not shown in a skeletal formula); dextropropoxyphene – note the two chiral centres at the two carbon atoms each with four different groups arranged tetrahedrally around them as in Figure 4.3.11.

5 Glycine has a plane of symmetry; it has two hydrogen atoms attached to the central carbon atom.

6 Only butan-2-ol is chiral.

4.4 Organic reaction mechanisms

1 **a)** The nucleus of an oxygen-18 atom has two more neutrons than an oxygen-16 atom.
b) They have the same number of protons in the nucleus so they have the same number of electrons. As a result the atoms of the isotopes have the same electron configuration and hence the same chemical properties.

2 **a)** The $CH_3CO-OC_2H_5$ bond breaks
b) The oxygen-18 atoms would have ended up in ethanol molecules instead of in ethanoic acid molecules.

4 The expected product is 1,2-dibromoethane.

5 **a)** The reaction produces a mixture of liquids with different boiling points. The liquids could be separated by fractional distillation and the compounds identified by recording their boiling points. Mass spectrometry and infra-red spectroscopy could be used to confirm the identity of the compounds.
b) The results show that chloride ions can take part in the reaction leading to a product in which one bromine atom is replaced by chlorine. Chloride ions do not react with ethene. This suggests that bromine molecules first react with ethene leading to an intermediate which can react with negative ions: normally bromide ions but chloride ions too if they are present.
c) A positively charged intermediate can be expected to combine with negative ions such as chloride ions.

6 In a tertiary halogenoalkane the halogen atom is attached to a carbon atom which is itself attached to three other carbon atoms.

7 **a)** $(CH_3)_3C—Br(l) + OH^-(aq) \longrightarrow$
$(CH_3)_3C—OH(aq) + Br^-(aq)$
b) Acidify with nitric acid, then add silver nitrate solution. A cream-coloured precipitate indicates that bromide ions are present. The precipitate redissolves in concentrated ammonia solution (unlike silver iodide which is yellow).
c) One of the reactants, $OH^-(aq)$, does not appear in the rate equation. This suggests that the step involving hydroxide ions is fast. So there must be at least one other step: the rate-determining step.
8 Energy is needed to break covalent bonds. This energy comes either from highly energetic collisions between fast moving molecules or from high-energy photons.
9

$CH_3—CH_3 \longrightarrow 2CH_3^{\bullet}$

10 The reaction produces halogenoalkanes which are useful for synthesis of other organic chemicals. Halogenoalkanes are more polar, and so more reactive than alkanes.
11 **a)** Chloromethane and hydrogen chloride
b) A termination step in which two methyl radicals combine produces ethane.
c) Each of the hydrogen atoms in methane can be replaced in turn to substitute one, two, three or four chlorine atoms for hydrogen.
12 The second propagation step regenerates free chlorine atoms.
13 When a C—Cl bond breaks the chlorine atom takes both the bonding electrons. One electron came originally from the carbon atom and the other from the chlorine atom. So after bond breaking, the chlorine atom has gained an electron and is negatively charged, while the carbon atom has lost an electron and is positively charged.
14 Free radicals: Cl, Br, CH_3
Nucleophiles: H_2O, Br^-, CN^-, OH^-, NH_3
Electrophiles: Br_2, H^+, NO_2^+
15 The polar bonds are: C–Cl, C–O, O–H, C=O, C–N and N–H. The $\delta-$ is on the more electronegative atom. Cl, O or N.
16 **a)** If chloride ions are present they can compete with bromide ions in the second step.
b) Only bromine molecules are involved in the first step. Chloride ions alone do not react with ethene. The product has to contain at least one bromine atom.
17 It is the electrons in the π-bond which take part in the reaction.
18 **a)** 2-bromobutane **b)** 2-bromobutane
c) 2-bromopentane
19 Order of stability: $(CH_3)_3C^+ > CH_3CH_2CH^+CH_3 > CH_3CH_2CH_2CH_2^+$
20 Redraw Figure 4.4.11 with CN^- in place of OH^-.
21 A tertiary halogenoalkane produces a tertiary carbocation which is stabilised by the inductive effect of three alkyl groups (see question 19). Also in the S_N2 mechanism the nucleophile attacks the $\delta+$ carbon atom from behind and if there are three bulky alkyl groups around this carbon atom the incoming reagent may not be able to get at the carbon atom.
22 R—Cl is the most polar and the strongest bond. R—I is the weakest and least polar. The trend correlates better with bond strength.

4.5 Aromatic hydrocarbons

1 The molar mass. Chemists use mass spectrometry to determine relative molar masses (see section 4.20).
2 One possibility is $CH_2{=}CH—C{\equiv}C—CH{=}CH_2$. In this structure the six hydrogen atoms are not in equivalent positions. Also the double and triple bonds would be expected to be highly reactive.
3 **a)** C_7H_8 **b)** C_7H_8 **c)** $C_6H_5—CH_3$
4 **b)** Benzene is about 150 kJ mol^{-1} more stable than a Kekulé molecule with three alkene-like double bonds.
c) -240 kJ mol^{-1}
5 **a)** All the bonds are equivalent and have the same length (shorter than a normal C—C bond but longer than a normal C=C bond).
b) Since all six carbon—carbon bonds are equivalent the molecules with two chlorine atoms on any pair of adjacent carbon atoms are identical.
c) The delocalised electrons stabilise the molecule. Forming an addition product breaks the ring of delocalised electrons leading to an energetically less favoured product.
6 When benzene molecules collide with each other the distortion of the electron clouds sets up temporary dipoles (as with any other non-polar molecules). The attractions between these temporary dipoles give rise to weak intermolecular forces.
7 The attractions between water molecules are relatively strong because of hydrogen bonding. The hydrogen bonding keeps out the benzene molecules which are non-polar and cannot interact with polar water molecules.
8 Benzene mixes freely with non-polar, hydrocarbon solvents such as hexane.
9 **b)** $C_6H_6(l) + Cl_2(g) \longrightarrow C_6H_5Cl(l) + HCl(g)$
c) The electrophile is Cl^+ formed in the presence of iron(III) chloride (as in Figure 4.5.12 but with Cl in place of Br).
10 Sulfuric acid is the acid. Nitric acid is the base.
11 Water hydrolyses and inactivates the catalyst.
12 **a)** See Figure 3.4.5. The aluminium chloride acts as an electron pair acceptor. So it is a Lewis acid.
b) The proton is a stronger Lewis acid and takes the extra chloride ion from $AlCl_4^-$:
$AlCl_4^- + H^+ \longrightarrow AlCl_3 + HCl$
13 **a)** $C_6H_5COCH_2CH_3$
b) $C_6H_5C(CH_3)_3$
14 Iodoalkanes are more reactive because the C—I bond is weaker. The intermediate carbocation forms more readily.

4.6 Phenols

1 **a)** Hydrogen bonding between the –OH groups of phenol molecules gives rise to much stronger intermolecular forces than the weak attractions between temporary dipoles in benzene.
b) Hydrogen bonding between phenol molecules and water molecules makes phenol soluble to some extent. The relatively large, non-polar benzene ring makes phenol less soluble than ethanol.
2 About 7.8 tonnes of benzene needed and the yield of propanone is about 5 tonnes assuming that it too is produced with a yield of 86%.
3 The phenol first melts and then burns with a very smoky flame.
4 Similarities: reacts with sodium forming hydrogen, reacts with acyl chlorides forming esters.

Differences: unlike ethanol, phenol is acidic enough to form salts with aqueous alkalis, also phenol cannot be esterified directly with a carboxylic acid in the presence of an acid catalyst.

5 A milky white precipitate of phenol separates from the aqueous solution.

6 CH_3CO- is an acyl group. An acylation reaction replaces the H of an $-OH$ or $-NH_2$ group with an acyl group.

7 Delocalised electrons between the benzene ring and the lone pairs on the oxygen atom of the $-OH$ group in phenol make it more difficult to break the C—O bond.

8 Phenol nitrates much more readily than benzene. The nitrating agent for benzene is a mixture of concentrated nitric and sulfuric acids.

9 Paracetamol has a phenolic $-OH$ group and gives a violet colouration. Aspirin does not.

4.7 Carbonyl compounds

1

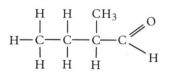

2 **a)** Propanone and propanal are liquids at room temperature, boiling at 56 °C and 49 °C respectively. Butane, which has the same molar mass, is a gas boiling at –0.5 °C. Propan-1-ol with a very similar molar mass boils at 97 °C.

b) The C=O bond is polar so that aldehyde molecules have a permanent dipole. Molecules with permanent dipoles have stronger intermolecular forces than alkanes which are non-polar. The hydrogen bonding such as between the $-OH$ groups of alcohols is considerably stronger than other intermolecular forces.

3 In any aldehyde solubility is a balance between the attraction for water molecules of the electronegative oxygen atom and the hydrophobic properties of the hydrocarbon part of the molecule. The benzene ring means that the hydrocarbon part of benzaldehyde is the more significant.

4 **a)** $CH_3CH_2CH_2OH(l) + 2[O] \longrightarrow$
$CH_3CH_2CHO(l) + H_2O(l)$

b) The oxidising agent can also oxidise propanal to propanoic acid (see page 172).

c) See *AS Chemistry* page 216.

5

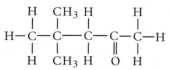

6 Both have the molecular formula C_3H_8O but one is a ketone and the other an aldehyde

7 $CH_3CH_2CHOHCH_3(l) + 2[O] \longrightarrow$
$CH_3CH_2 COCH_3(l) + H_2O(l)$

8 This is a Friedel–Crafts reaction. The reagent is ethanoyl chloride. The catalyst is aluminium chloride. The reaction mixture needs heating.

9 **a)** $C_8H_8O_3$ **b)** $C_{10}H_{16}O$

10 Citral and retinol: aldehyde groups; menthone: ketone group.

11 Butanal give butan-1-ol (primary). Butanone give butan-2-ol (secondary).

12

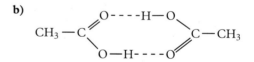

13 The cyanide ion gains a proton: $CN^- + H^+ \rightarrow HCN$. So it acts as a base.

14 **a)** The product (see Figure 4.7.10) has four different groups round the central carbon atom so it is asymmetric and chiral.

b) There is an equal chance that the cyanide ion will attack the carbon atom of the carbonyl group from either side of the planar molecule. This leads, on average, to equal amounts of the two mirror image molecules (see Figure 4.19.3).

15 Same as Figure 4.7.8 but with propanone in place of ethanal.

16 Oxidation: see Figure 4.7.7; reduction: see Figure 4.7.6; addition: see Figure 4.7.8.

17 Same as Figures 4.7.9 and 4.7.10 but with H:⁻ in place of NC:⁻ in the first step and H_2O in place of HCN in the second.

18 Butanone (see data section pages 245–246)

19 2-chlorobutane hydrolyses to the secondary alcohol, butan-2-ol, which oxidises to the ketone, butanone.

20 **a)** The reagent first oxidises the alcohol to a carbonyl compound which can then take part in the tri-iodomethane reaction.

b) Sodium chlorate(I) is a solution of chlorine in sodium hydroxide. The chlorate(I) ions oxidise iodide ions to iodine.

21 Pentan-2-one and 3-methylbutan-2-one

22 Ethanal

4.8 Carboxylic acids

1 Hexanoic acid, octanoic acid and decanoic acid.

2 $C_3H_6O_2$, $CH_3CH_2CO_2H$, propanoic acid.

3 **a)**

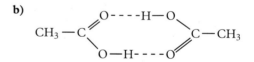

b)

4 $C_6H_5CO_2H(s) + OH^-(aq) \longrightarrow$
$C_6H_5CO_2^-(aq) + H_2O(l)$.
The uncharged, molecular acid is relatively insoluble. Converting it to negative ions produces the sodium salt which is more soluble.

5 $2CH_3CH_2CO_2H(aq) + CO_3^{2-}(aq) \longrightarrow$
$2CH_3CH_2CO_2^-(aq) + CO_2(g) + H_2O(l)$

6 **a)** Through the inductive effects, alkyl groups tend to push electrons towards neighbouring groups. The methyl group in ethanoic acid thus increases the electron density of the carboxylic acid group so that it holds onto its proton more strongly than the carboxylic acid group in methanoic acid which has a hydrogen atom in place of an alkyl group.

b) A chlorine atom is an electron withdrawing group. It tends to reduce the the electron density of the carboxylic acid group in chloroethanoic acid. So that the acid gives up its proton more readily than ethanoic acid.

7 When PCl₅ reacts with a compound with an –OH group it replaces the –OH group with a chlorine atom. Another product is HCl which can be identified as a colourless gas which is acidic, fumes in moist air and forms a white smoke with ammonia gas.

8 Using an excess of one reactant: adding more of one reactant forces the equilibrium to the right increasing the proportion of the other (more valuable) reactant converted to the ester.

Using more concentrated sulfuric acid than needed for its catalytic effect: sulfuric acid is a dehydrating agent so it removes water (one of the products) from the reaction mixture; this shifts the equilibrium to the right.

Distilling off the product at formed: this removes the ester (the other product) from the reaction mixture also shifting the equilibrium to the right.

4.9 Acylation

1 a)

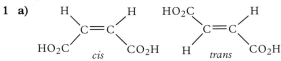

b) The *cis* isomer with both carboxylic acid groups close together.

2 CH₃COCl(l) + H₂O(l) → CH₃CO₂H(aq) + HCl(g)
The water (hydro) splits (lysis) apart the ethanoyl chloride.

3 The reaction goes at room temperature in the presence of alkali.

4 CH₃COCl(l) + 2NH₃(aq) →
CH₃CO₂NH₂(aq) + NH₄Cl(aq)

5

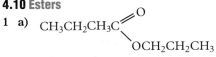

6 Hydrogen chloride is a strongly acidic and corrosive gas.

7 About 62%.

8 Because it is bonded to two highly electronegative elements: oxygen and chlorine.

9 As for Figure 4.9.8 but with H₃N: in place of the ethanol. During the elimination step the HCl combines with ammonia to form ammonium chloride.

4.10 Esters

1 a)

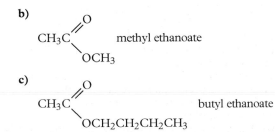

propyl butanoate

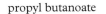

b)

CH₃C O OCH₃ methyl ethanoate

c)

CH₃C O OCH₂CH₂CH₂CH₃ butyl ethanoate

2 Ethyl ethanoate is a larger molecule than either ethanol or ethanoic acid and it has a permanent dipole but no

–OH groups so molecules of the ester cannot form hydrogen bonds with each other. Hydrogen bonding is stronger than the attraction between permanent dipoles. Both ethanol and ethanoic acid molecules are held together by hydrogen bonding. So ethanol has a higher boiling point than might be expected from its molecular size. An ethanoic acid molecule can form more hydrogen bonds than an ethanol molecule.

3 a) A Heating the mixture of reagents under reflux.
B Distilling off the product from the reaction mixture.
C Removing acidic impurities and then unchanged ethanol.
D Drying with a dehydrating agent.
E Final distillation and measurement of boiling point.

b) It is the catalyst.

c) Bubbles of gas (carbon dioxide) form when shaking with sodium carbonate. It is important to release the pressure ion the tap funnel.

d) Formation of a complex in.

e) This is an ether C₂H₅—O—C₂H₅ which has a lower boiling point than the ester.

f) About 65%.

4 a) Propan-1-ol and butanoic acid.
b) Ethanol and sodium methanoate.

5 a) Excess of dilute acid so that there is plenty of water.
b) To make the ester no water is added and the acid is concentrated.

6

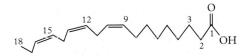

7 CH₃(CH₂)₇CH=CH(CH₂)₇CO₂H + H₂(g) →
CH₃(CH₂)₁₆CO₂H
The product is octadecanoic (stearic) acid.

8 Alkaline hydrolysis is not reversible and goes to completion. Also using alkali produces the salt of the fatty acid which is the soap.

4.11 Amides

1 a) Ethanamide (see Figure 4.11.1)
b) N-methyl ethanamide

2 Under acid conditions the products are an ammonium salt and the organic acid. Under alkaline conditions the products are ammonia and the salt of the organic acid.

3 a) Ammonia and butanoic acid
b) Methylamine and potassium propanoate

4 a) Butanenitrile
b) Propylamine

5 a) Heat with bromine in aqueous sodium hydroxide.
b) Heat with phosphorus(V) oxide

6 P is propanoyl chloride. Q is propanamide. For the third step: heat with bromine in aqueous sodium hydroxide

7 See Figure 4.11.6. Equation goes from right to left.

8 a) To make proteins and nucleic acids.
b) Slow hydrolysis releases the ammonia for plant growth slowly.
c) Ammonium nitrate: 28 parts N in 80 parts of compound (35%). Urea: 28 parts N in 60 (47%). So urea has the higher proportion of nitrogen by mass.

4.12 Nitriles

1
$$H-\underset{\underset{H}{|}}{\overset{\overset{H}{|}}{C}}-\underset{\underset{H}{|}}{\overset{\overset{H}{|}}{C}}-\underset{\underset{H}{|}}{\overset{\overset{H}{|}}{C}}-C\equiv N$$

2 Ethanenitrile

3 **a)**

$$\underset{H}{\overset{H}{\diagdown}}C=\underset{\underset{H}{|}}{\overset{\overset{H}{|}}{C}}-C\equiv N$$

b)

$$-\underset{\underset{H}{|}}{\overset{\overset{H}{|}}{C}}-\underset{\underset{CN}{|}}{\overset{\overset{H}{|}}{C}}-\underset{\underset{H}{|}}{\overset{\overset{H}{|}}{C}}-\underset{\underset{CN}{|}}{\overset{\overset{H}{|}}{C}}-\underset{\underset{H}{|}}{\overset{\overset{H}{|}}{C}}-\underset{\underset{CN}{|}}{\overset{\overset{H}{|}}{C}}-$$

4 **a)** Nucleophilic substitution **b)** Dehydration
5 **a)** Sodium butanoate **b)** Butylamine.

4.13 Amines

1 **a)**

$$C_2H_5-\underset{\underset{\cdot\cdot}{|}}{\overset{\overset{C_2H_5}{|}}{N}}-H$$

b)

$$H_2N-CH_2-CH_2-CH_2-CH_2-CH_2-CH_2-NH_2$$

c)

$$CH_3-CH_2-CH_2-\underset{\underset{NH_2}{|}}{\overset{\overset{H}{|}}{C}}-CH_2-NH_2$$

d)

$$\langle\text{benzene ring}\rangle-CH_2-\underset{\underset{|}{}}{\overset{\overset{NH_2}{|}}{CH}}-CH_3$$

2 Alkanes are non-polar. Hydrogen bonding links the
–NH$_2$ groups in amines. Hydrogen bonding is stronger
than the weak intermolecular forces between non-polar
molecules.
3 **a)** See for example Figure 4.4.10 with H$_3$N: as the
nucleophile and 1-bromopropane as the
halogenoalkane.
b) A solution of ammonia in water is alkaline. Alkali
catalyses the hydrolysis of halogenoalkanes to alcohols.
4 Order of base strength: butylamine > ammonia >
4-methylphenylamine.
5 $C_2H_5NH_2(g) + HCl(g) \rightarrow C_2H_5NH_3Cl(s)$
The reactants are molecular. The product is ionic.
6 $CH_3CH_2CH_2CH_2NH_2 + CH_3CH_2CH_2Br \longrightarrow$

$$CH_3CH_2CH_2CH_2-\underset{\underset{}{\overset{\overset{H}{|}}{N}}}-CH_2CH_2CH_3 + HBr$$

7 Excess ammonia favours the formation of the primary
amine and limits the extent to which the
halogenoalkane reacts with the products to form
secondary and tertiary amines.
8 **a)** In both there are four covalent bonds from the
central nitrogen atom. The fourth bond is a dative
covalent bond. In both the nitrogen atom carries a
positive charge.

b) It is an ionic solid with a positively charged
quaternary ammonium ion and a negatively charged
chloride ion. Ionic bonding is strong bonding.
9 See figure 4.9.8. The same mechanism but with
CH$_3$CH$_2$NH$_2$ as the nucleophile. There is a lone pair
on the nitrogen atom.
10 $NO_2^-(aq) + H^+(aq) \longrightarrow HNO_2(aq)$
11 **a)** $3HNO_2(aq) \longrightarrow HNO_3(aq) + 2NO(g) + H_2O(l)$
b) $2NO(g) + O_2(g) \longrightarrow 2NO_2(g)$. NO is colourless.
NO$_2$ is brown.
12

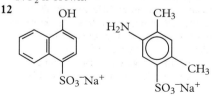

4.14 Amino acids and proteins

1 **a)** Aminoethanoic acid
b) 2-aminopropanoic acid
2 There are two hydrogen atoms attached to the central
carbon atom so the molecule has a plane of symmetry.
3 The two forms rotate the plane of polarised light in
opposite directions.
4 Alanine forms a zwitterion so the attraction between
the molecules involves strong ionic bonding rather than
weak intermolecular forces.
5 **a) (i)**

$$H_2N-\underset{\underset{}{\overset{\overset{CH_3}{|}}{CH}}}-CO_2H\ (aq) + H_3O^+\ (aq) \longrightarrow$$

$$H_3N^+-\underset{\underset{}{\overset{\overset{CH_3}{|}}{CH}}}-CO_2H\ (aq) + H_2O(l)$$

(ii)

$$H_2N-\underset{\underset{}{\overset{\overset{CH_3}{|}}{CH}}}-CO_2H\ (aq) + OH^-\ (aq) \longrightarrow$$

$$H_2N-\underset{\underset{}{\overset{\overset{CH_3}{|}}{CH}}}-CO_2^-\ (aq) + H_2O(l)$$

b) In acid, alanine molecules gain protons and become
positively charged. In alkaline solution, alanine
molecules lose protons and become negatively charged.
In neutral solution, alanine molecules exist as
zwitterions which are overall uncharged with both
–NH$_3^+$ and –CO$_2^-$.

6

$$H_2N-\underset{\underset{H}{|}}{\overset{\overset{CH_3}{|}}{C}}-\overset{\overset{O}{||}}{C}-\underset{\underset{H}{|}}{\overset{\overset{|}{|}}{N}}-\underset{\underset{H}{|}}{\overset{\overset{CH_2}{|}}{C}}-CO_2H \qquad H_2N-\underset{\underset{H}{|}}{\overset{\overset{CH_2}{|}}{C}}-\overset{\overset{O}{||}}{C}-\underset{\underset{H}{|}}{\overset{\overset{|}{|}}{N}}-\underset{\underset{H}{|}}{\overset{\overset{CH_3}{|}}{C}}-CO_2H$$

7

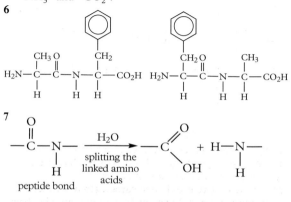

peptide bond

8 **a)** As well as a carboxylic acid group, an amine in a
group, a benzene ring and a peptide bond the molecule
has an ester group. It is the ester group which means
that this sweetener is not simply a dipeptide.

b) Hot water hydrolyses the molecule splitting the peptide bond and the ester bond.

c) Hydrolysis during digestion releases the amino acid phenylalanine.

9 **a)** serine, phenylalanine, alanine, glutamic acid.

b) (i) phenylalanine, and alanine (ii) serine and glutamic acid, (iii) glutamic acid.

10 A change of conditions which alters the shape of the enzyme molecule and its active site can cause an enzyme to lose its activity. A rise in temperature can break the interchain forces which help to keep the shape of the polypeptide chains. Extremes of pH mean that acidic or basic side chains in amino acids may be protonated or deprotonated thus altering the distribution of ionic charges which attract or repel each other and help to control the shape of the molecule.

4.15 Polymers

1 **a)**
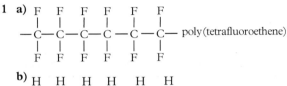
poly(tetrafluoroethene)

b)
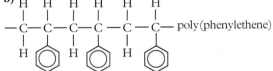
poly(phenylethene)

2 **a)** Polymer: poly(propene); monomer: propene.

b) Polymer: poly(chloroethene); monomer: chloroethene.

3 Homolytic fission: splitting a covalent bond such that each atom is left with one electron. See Figure 4.15.4. Free-radical: a reactive species with unpaired electrons.

4 **a)** The hardener initiates polymerisation. This must not start until a few minutes before using the filler.

b) Benzoyl peroxide

c) Factors affecting the rate of setting include the concentration of the hardener and the temperature.

5

6 **a)** Carboxylic acid and amine groups

b) Water

c)
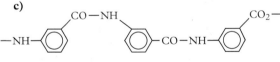

7 In both the monomers are linked together by this group:

In nylon there are more carbon atoms between each linking group (often six). Nylon does not have side groups attached to the carbon atoms.

In a protein there is only one carbon atom between each peptide group. Each amino acid in a protein chain has a side group.

8 **b)** The 1 and 4 positions are opposite each other and so the chains are more or less straight. A polymer

based on functional groups at the 1 and 2, or the 1 and 3, positions would have a very irregular shape.

9 **a)** Weak intermolecular attractions between temporary induced dipoles in non-polar molecules.

b) Hydrogen bonding between N—H and C=O groups.

10 The backbone of a poly(phenylethene) molecule is a chain of carbon atoms linked by non-polar C—C bonds which are inert to aqueous reagents. Polyester molecules have ester links which can be hydrolysed by alkali breaking up the polymer chain.

11 Hydrolysis can split ester and amide bonds thus breaking up the polymer chains. The C—C bonds in polyalkenes are not affected by aqueous reagents.

12 Addition polymers such as polythene, polystyrene and pvc are probably more common in household waste. These polymers are not biodegradable and are often hard to recycle. Incineration is an option worth considering but is not popular because of fears of toxic emissions. Soft drinks bottles consist of the condensation polymer PET which can be recycled if separated from other waste.

13 Plastics have to be sorted by type before they can be recycled in to high value products. Plastics often contain pigments, fillers and plasticisers which make them hard to identify. Separating plastics is hard to mechanise and automate.

4.16 Synthesis

1 **a)** propene, elimination, base.

b) bromoethane, addition, electrophile.

c) propan-1-ol, substitution, nucleophile.

d) nitrobenzene, substitution, electrophile.

3 **a)** First convert to ethanol, use concentrated sulfuric acid followed by water to hydrolyse the addition product (See *AS Chemistry* page 204); then oxidise by refluxing with excess acidified potassium dichromate(VI) (see *AS Chemistry* page 216).

b) First dehydrate to but-1-ene by heating with concentrated phosphoric acid (see *AS Chemistry* page 214); then treat with concentrated sulfuric acid followed by water to hydrolyse the addition product (according to Markovnikov's rules this will give mainly butan-2-ol).

c) Divide the ethanol into two parts. Oxidise one part to ethanoic acid by refluxing with excess acidified potassium dichromate(VI). Then reflux the ethanoic acid with the remaining ethanol in the presence of concentrated sulfuric acid to act as the catalysts (see page 184).

4 **a)** Step 1: convert ethanol to bromoethane by heating under reflux with sodium bromide and concentrated sulfuric acid (see *AS Chemistry* page 213). Step 2: convert to propanenitrile by heating under reflux with potassium cyanide in ethanol. Step 3: reduce to propylamine with LiAlH$_4$ in dry ether.

b) Step 1: convert propan-1-ol to 1-bromopropane by heating under reflux with sodium bromide and concentrated sulfuric acid. Step 2: convert to butanenitrile by heating under reflux with potassium cyanide in ethanol. Step 3: hydrolyse to butanoic acid by heating under reflux with dilute hydrochloric acid.

5 X could be methyl benzene (or other arene with a hydrocarbon side chain). A Friedel–Crafts reaction with chloromethane in the presence of aluminium chloride converts benzene to X.

6 a) These are the conditions used by Levine and Taylor. Step 1: heat under reflux with ethanol in the presence of concentrated sulfuric acid. Step 2: Sodium in the presence of ethanol as the reducing agent for this step. Step 3: treat with a mixture of iodine and red phosphorus (equivalent to PI_3). Step 4: Heat under reflux with KCN in ethanol. Step 5: Heat under reflux with aqueous sodium hydroxide, then acidify with dilute hydrochloric acid.

b) The melting point of a pure sample can be used to identify a compound. In a series of closely related compounds the melting points may not differ greatly from one compound to the next and so they need to be determined with great accuracy.

c) Only a very small sample is required to measure a melting point. Impurities lower the melting point of a sample. Purity was essential for accurate values.

7 Treat with PCl_5 at room temperature to make butanoyl chloride; then add ammonia solution to make butanamide. Finally warm with bromine and concentrated sodium hydroxide (see page 188) to make propylamine.

4.17 Grignard reagents

1 The normal C—C—C bond angle in an alkane is 109.5°. In butane the angle is 90°. So cyclobutane is a 'strained' structure.

2 a) Butan-1-ol **b)** Primary

3 a) 3-methylbutan-2-ol. **b)** Secondary

4 Treat butanone with propyl magnesium bromide, then hydrolyse the product with dilute acid. See Figure 4.17.5: $R = CH_3$, $R' = CH_3CH_2$, $R'' = CH_3CH_2CH_2$.

5 The cyanide ion is a strong base because hydrogen cyanide is a weak acid. So cyanide ions can bring about elimination as well as substitution reactions. Elimination takes place more readily with secondary and tertiary halogenoalkanes.

6 In each case start by making a Grignard reagent from the 1-bromopropane. Then react the Grignard reagent with the compound indicated below. Finally (except in e) hydrolyse the intermediate product with dilute acid.

a) methanal **b)** carbon dioxide
c) butanone **d)** propanone
e) hydrolyse the Grignard reagent with dilute acid.

4.18 Synthetic techniques

1 a) Treat butanoic acid with PCl_5 at room temperature; then react the acid chloride with methanol at room temperature.

b) The required yield is about 0.1 mol. After two steps the yield is:
$(0.55)^2 \times$ starting amount in moles. So the initial mass of butanoic acid should be about 30 g.

c) Butanoic acid is corrosive. PCl_5 is corrosive. Acid chlorides are flammable and corrosive. The hydrogen chloride produced during the synthesis is corrosive. This is a synthesis to carry out in a fume cupboard.

2 Possible reasons:
- the reaction may not go to completion but reach an equilibrium state
- there may be side reactions producing by-products
- there are losses when separating the product from the reaction mixture – complete separation may not be possible
- there are further losses during the purification

stages as the product is transferred from one vessel to another.

3 a) Step 1: warm with a mixture of concentrated nitric and sulfuric acids. Step 2: heat under reflux with tin and concentrated hydrochloric acid (see Figure 4.13.3).

b) Step 1: theoretical – 28 g, percentage – 78%. Step 2: theoretical from the nitrobenzene – 17 g, percentage – 71%.

c) Overall: theoretical from the benzene – 21 g, percentage – 57%.

4 b) Nucleophilic substitution.
$(CH_3)_3COH(l) + HCl(aq) \longrightarrow (CH_3)_3CCl(l) + H_2O(l)$

c) (i) Sodium hydrogencarbonate: neutralises the hydrochloric acid and to convert it to sodium chloride in the aqueous layer.

(ii) Anhydrous sodium sulfate: takes up any water to dry the product

(iii) Filtering: removes the drying agent.

d) (i) Warm with aqueous sodium hydroxide to release halide ions; add dilute nitric acid to acidify the aqueous layer; add silver nitrate and look for a white precipitate (of AgCl) soluble in dilute ammonia solution.

(ii) Test a sample with a very small piece of sodium (look for bubbles of gas – hydrogen) or with a small measure of solid PCl_5 (look for the formation of a fuming, acidic gas – HCl).

4.19 Drugs for medicines

1 Salbutamol: $-C\star HOH-$
Levodopa: $-C\star HNH_2-$
Chloramphenicol: $-C\star HOH-C\star H(CH_2OH)-$

2 In solution the attacking reagent can approach the target molecule from the starting material from either side. Approaching from one side produces one optical isomer. Attacking from the other side produces the other. On average, with many molecules reacting, the overall result is an equal mixture of the two optical isomers.

3 a) A weak acid.

b) In the presence of hydrochloric acid (a strong acid) in the stomach, the $-CO_2^-$ group in the sodium salt gains a proton and changes to $-CO_2H$.

c) The non-ionic, acid form which is more fat soluble.

d) The ionic, salt form of the drug which is more water soluble.

4 a) Amide: $-CO-NH-$. Carboxylic acid: $-CO_2H$.

b) Ester: $-CH_2O-CO-$. Amide: $-CO-NH-$. Amine: $-NH_2$.

c) Ampicillin: to the right of $-CO-NH-$, the same as Penicillin-G. To the left of $-CO-NH-$, the same as Pivampicillin.

4.20 Instrumental analysis

1 a) P: CH_3^+, Q: $C_2H_5^+$, R: $C_3H_7^+$.

b) CH_3^+ is as relatively unstable primary carbocation. Generally, fragmentation favours secondary and tertiary carbocations.

c) For example:
$CH_3CH_2CH_2CH_3^+ \longrightarrow CH_3CH_2CH_3^+ + CH_3^\bullet$

2 a) Chlorine consists of two isotopes: chlorine-35 and chlorine-37 with abundances in the ratio 3:1. In a molecular ion of $C_2H_2Cl_2$ with two chlorine atoms the possibilities are: most likely two atoms of chlorine-35 (96), next one of chlorine-35 and one of chlorine-37 (98) and least likely two atoms of chlorine-37 (100).

The adjacent peaks differ by two mass units and the expected ratio is 9:6:1.

b) These two peaks are fragments formed by the loss of one chlorine atom leaving a single chlorine atom in the fragment so the ratio of abundances corresponds to the ratio of the abundance of the isotopes.

3 Top spectrum: chloroethane. Molecular ion peaks at 64 and 66 are in the ratio 3:1 suggesting that a single chlorine atom is present with chlorine-35 and chlorine-37 present in this ratio. The same ratio of peak heights is seen for CH_2Cl^+ at 49 and 51. The peak at 29 corresponds to $C_2H_5^+$. Loss of further hydrogen atoms gives the peaks at 28, 27 and 26.

Bottom spectrum: methyl bromide. Peaks at 94 and 96 are the molecular ion peaks with two isotopes of bromine which occur which roughly equal abundance: bromine-79 and bromine-81. The peak at 15 is CH_3^+. The peaks at 79 and 81 correspond to the isotopic free bromine atoms.

4 **a)** Top spectrum: benzoic acid – see answer to **b)**. Bottom spectrum: 3-hydroxybenzaldehyde – see answer to **c)**.

b) Peak at 105: $C_6H_5CO^+$. Peak at 77 is $C_6H_5^+$

c) Peak at 93: $HO\text{—}C_6H_4^+$. Peak at 39: $C_3H_3^+$

5 About 300 nm. Colourless

6 6.28×10^{-4} mol dm^{-3}

7 **a)** A O—H stretch, hydrogen bonded
    B C—H stretch
    C C—H and O—H bend
    D C—H stretch
    E C—N stretch
    F C—H bend
    G O—H stretch hydrogen bonded
    H C=O stretch
    I C—H stretch
    J C=O stretch

b) From top to bottom: ethanol, propanenitrile, ethanoic acid, ethyl butanoate.

8 **a)** P: C_6H_5CHO. Q: $C_6H_5CO_2H$

b) Benzene carbaldehyde gives an orange precipitate with 2,4-dinitrophenylhydrazine. Benzene carboxylic acid is soluble in hot water to give an acidic solution which releases carbon dioxide when mixed with sodium carbonate.

9 From top to bottom: ethanal, CH_3CHO with hydrogens in two chemical environments in the ratio 3:1; ethyl ethanoate, $CH_3CO_2CH_2CH_3$, with hydrogen atoms in three chemical environments in the ratio 2:3:3; ethanol, CH_3CH_2OH, with hydrogen atoms in three chemical environments in the ratios 1:2:3. See also page 251 for chemical shift values.

10 TMS has 12 hydrogen atoms all in the same chemical environment.

11 a)

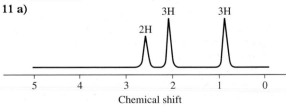

b)

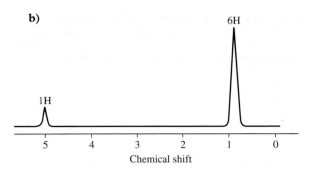

12 $C_6H_5CH_2CH_3$. Five hydrogen atoms at a shift of just over 7 suggests the 5 hydrogen atoms of a benzene ring with one substituent. There are two hydrogen atoms with a chemical shift corresponding to hydrogen atoms bonded to a carbon atom which is attached to a benzene ring. There are three hydrogen atoms which might be part of a methyl group.

13 a)

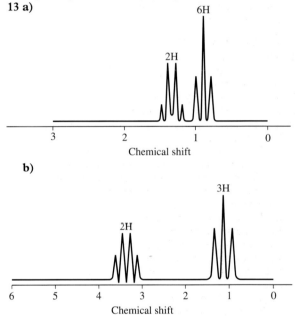

b)

14 $CH_3CH_2OH + D_2O \rightleftharpoons CH_3CH_2OD + HDO$

15 $CH_3CH_2CO_2H$

Index

Index

Index

Photo Acknowledgements

The publishers would like to thank the following individuals, institutions and companies for permission to reproduce photographs in this book. Every effort has been made to trace and acknowledge ownership of copyright. The publishers will be glad to make suitable arrangements with any copyright holders whom it has not been possible to contact.

Advertising Archives (168); Andrew Lambert (25, 41, 45 top, 45 middle, 45 bottom, 60, 63, 76 bottom left, 94 top left, 97, 105 top, 105 middle, 119 top, 119 bottom, 120middle left, 120 bottom right, 121, 122, 124 bottom, 131, 132, 133, 134, 135, 136, 148, 162, 166, 174, 175 all four photos, 177, 196, 204 bottom left, 112); Baker Refractories (168); Corbis/ NASA/Roger Ressmeyer (48 right)/ Roger Tidman (73)/ Matthew Mckee; Eye Ubiquitous (89 middle)/ Ecoscene (92)/ London Aerial Photo Library (120 bottom left)/ Farrell Grehan (124 top)/ Charles O'Rear (152)/ Patrick Bennett (163); Geoscience Photo Library (99); Grant Sutherland (62); H & S (204 bottom right); Herbie Knott (204 top left); Holt Studios/ Bob Gibbons (36 left)/ Peter Wilson (36 right)/ Rosemary Mayer (176 right); Life file/Emma Lee (183); The Natural History Museum London (75 middle); Photodisc (1, 6, 87, 139, 241); Science Photo Library (27, 204 top right)/ Alfred Pasieka (75 top, 76 top left, 120 top left)/ Arnold Fisher (118)/ Biosym Technologies, Inc. (129)/ Charles D. Winters (89 top)/ Clive Freeman (89 bottom)/ Colin Cuthbert (30, 217, 229)/ David Campione (24)/ David Scharf (176 left)/ Dr Jeremy Burgess (203, 207)/ Dr Jurgen Scriba (3)/ Eye of Science (208)/ G. Brad Lewis (59, 76 centre)/ Geoff Tompkinson (5, 7 centre)/ James Bell (140) / James Holmes/Celltech (234)/ James Holmes/Zedcor (205)/ James King-Holmes (209 middle)/ Ken Eward (141, 201, 202)/ Leonard Lessin (222)/ Martin Bond (8 top, 59 left, 120 top right)/ Martyn F. Chillmaid (59 left)/ Maximilian Stock Ltd. (8 bottom)/ Pascal Goetgheluck (147)/ Peter Menzel (210)/ Peter Ryan (209 top)/ Philippe Plailly (7 top)/ Prof. P. Motta/ Dept. of Anatomy/ University of 'La Sapienza', Rome (88)/ Prof. Stewart Lowther (102)/ Profs P.M. Motta & S. Correr (93)/ Roberto de Guglirmo (116)/ Russ Lappa (111)/ Sidney Moulds (37)/ Simon Fraser (32)/ Tek Image (47)/ Tony Hertz (186)/ Victor de Schwanberg (61); Still Pictures/ Thomas Raupach (90); Tate Picture Library (94 top right); Victoria & Albert Museum Picture Library (95, 159); Wellcome Photo Library (48 left).